COMPUTABILITY
THEORY

CHAPMAN & HALL/CRC MATHEMATICS

OTHER CHAPMAN & HALL/CRC MATHEMATICS TEXTS:

**Functions of Two Variables,
Second edition**
S. Dineen

Network Optimization
V. K. Balakrishnan

**Sets, Functions, and Logic:
A foundation course in mathematics,
Second edition**
K. Devlin

**Algebraic Numbers and Algebraic
Functions**
P. M. Cohn

Computing with Maple
Francis Wright

**Dynamical Systems:
Differential equations, maps, and
chaotic behaviour**
D. K. Arrowsmith and C. M. Place

Control and Optimization
B. D. Craven

Elements of Linear Algebra
P. M. Cohn

Error-Correcting Codes
D. J. Bayliss

**Introduction to Calculus of
Variations**
U-Brechtken-Mandershneid

Integration Theory
W. Filter and K. Weber

Algebraic Combinatorics
C. D. Godsil

**An Introduction to Abstract
Analysis-PB**
W. A. Light

The Dynamic Cosmos
M. Madsen

Algorithms for Approximation II
J. C. Mason and M. G. Cox

Introduction to Combinatorics
A. Slomson

**Elements of Algebraic Coding
Theory**
L. R. Vermani

**Linear Algebra:
A geometric approach**
E. Sernesi

**A Concise Introduction to
Pure Mathematics**
M. W. Liebeck

Geometry of Curves
J. W. Rutter

**Experimental Mathematics with
Maple**
Franco Vivaldi

**Solution Techniques for Elementary
Partial Differential Equations**
Christian Constanda

**Basic Matrix Algebra with
Algorithms and Applications**
Robert A. Liebler

Computability Theory
S. Barry Cooper

*Full information on the complete range of Chapman & Hall/CRC Mathematics books is
available from the publishers.*

COMPUTABILITY THEORY

S. BARRY COOPER

CHAPMAN & HALL/CRC

A CRC Press Company

Boca Raton   London   New York   Washington, D.C.

Library of Congress Cataloging-in-Publication Data

Cooper, S. B. (S. Barry)
  Computability theory / by S. Barry Cooper.
    p. cm. — (Chapman & Hall/CRC mathematics)
  Includes bibliographical references and index.
  ISBN 1-58488-237-9 (alk. paper)
  1. Computable functions. I. Title. II. Series.

  QA9.59.C68 2003
  511.3—dc22   2003055823

Visit the CRC Press Web site at www.crcpress.com

© 2004 by Chapman & Hall/CRC

No claim to original U.S. Government works
International Standard Book Number 1-58488-237-9
Printed in the United States of America 1 2 3 4 5 6 7 8 9 0
Printed on acid-free paper

Preface

This book may be your first contact with computability theory, although certainly not with computability. On the other hand, you may already have a good working knowledge of the subject. Either way, there should be plenty here to interest and inform everyone, from the beginner to the expert. The treatment is unashamedly contemporary, and many topics were previously available only through articles in academic journals or weighty specialist reference works.

As you will probably know already, the history of the computer is unusual, in that the theory — the Universal Turing Machine — preceded its physical embodiment by around ten years. I have tried to write *Computability Theory* very much in the spirit of Alan Turing, with a keen curiosity about patterns and underlying theory, but theory firmly tied to a real-world context. New developments are strongly featured, without sacrificing readability — although you the reader will be the best judge of that! Anyway, I do believe that there is no other book like this currently. There are admirable specialist volumes you will be able to take in hand later.

But this is primarily a book to be *read*, and not just to be referred to. I have tried to give some historical and intuitive background for all the more technical parts of the book. The exercises, both routine and more challenging, have been carefully chosen to complement the main text, and have been positioned so that you can see some *point* in your doing them. If you get half as much mental satisfaction and stimulation to further explorations as I got in selecting the topics, you will be more than adequately rewarded for your hard work. And of course, what cannot be hidden from you — computability theory sometimes has a reputation for being a difficult subject understood by a few people. I hope you will soon be joining those for whom it is also exciting and relevant, like no other subject, to the world we live in.

I should say that what you have here is a very personal view of an increasingly large subject. Although I have tried to choose a wide range of topics relevant to current research, there are inevitably many gaps — I will not advertise them, but the experts will notice! Also, this is not an academic article, so not all results have their discoverers' names attached. I have chosen to credit people with the reader's interests in mind, not those of the expert looking for his or her name in the index. Hopefully I have mentioned just enough names to make the reader fairly well informed about the history of the subject and to give an impression of a living breathing subject full of real people.

If you get through to the *Further Reading* selection at the end of the book, you will find lots of ways of filling out the picture given here. Please send me any comments or corrections. You never know, I may one day be preparing a second edition. Many thanks to all at CRC, especially Jasmin Naim and Mimi Williams, for their ever-present help with what you now have in your hands.

Finally, many thanks to everyone who has helped me and encouraged me with the writing of this book, without whom — as is often said, and never more truly — this book would not have been written. Too many to mention, but you know who you are. Yes, and that includes other writers of books, friends and colleagues, and — more than anyone! — my family who have bravely put up with me over the last three years.

Barry Cooper
Leeds

Contents

Part I

Computability and
Unsolvable Problems

Chapter 1

Hilbert and the Origins of Computability Theory

It is only in the last century that computability became both a driving force in our daily lives and a concept one could talk about with any sort of precision. Computability as a *theory* is a specifically twentieth-century development. And so of course is the computer, and this is no coincidence. But this contemporary awareness and understanding of the algorithmic content of everyday life has its roots in a rich history.

1.1   Algorithms and Algorithmic Content

We can see now that the world changed in 1936, in a way quite unrelated to the newspaper headlines of that year concerned with such things as the civil war in Spain, economic recession, and the Berlin Olympics. The end of that year saw the publication of a thirty-six page paper by a young mathematician, Alan Turing, claiming to solve a long-standing problem of the distinguished German mathematician David Hilbert. A by-product of that solution was the first machine-based model of what it means for a number-theoretic function to be computable, and the description of what we now call a *Universal Turing Machine*. At a practical level, as Martin Davis describes in his 2001 book *Engines of Logic: Mathematicians and the Origin of the Computer*, the logic underlying such work became closely connected with the later development of real-life computers. The stored-program computer on one's desk is a descendant of that first universal machine. What is less often remembered is Turing's theoretical contribution to the understanding of the *limitations* on what computers can do. There are quite easily described arithmetical functions which are not computable by *any* computer, however powerful. And even the advent of quantum computers will not change this.

Before computers, computer programs used to be called *algorithms*. Algorithms were just a finite set of rules, expressed in everyday language, for performing some general task. What is special about an algorithm is that its rules can be applied in potentially unlimited instances of a particular situation. We talk about the *algorithmic content* of Nature when we recognise patterns in natural phenomena which appear to follow general rules. Ideally algorithms and algorithmic content need to be captured precisely in the language of mathematics, but this is not always easy. There are areas (such as sociology or the biological sciences) where we must often resort to language dealing with concepts not easily reducible to numbers and sets. One of the main tasks of science, at least since the time of Isaac Newton, is to make mathematically explicit the algorithmic content of the world about us. A more recent task is to come to terms with, and analyse, the theoretical obstacles to the scientific approach. This is where the discovery of *incomputability*, and the theory which flows from that discovery, play such an important role.

Of course, algorithms have played an explicit role in human affairs since very early times. In mathematics, the first recorded algorithm is that of Euclid, for finding the greatest common factor of two integers, from around 300 BC. The word *algorithm* is derived from the name of the mathematician al-Khwarizmi, who worked at the court of Mamun in Baghdad around the early part of the 9th century AD. The sort of algorithms he was interested in had a very practical basis. This we can see from his description of the content of his most famous book *Hisab al-jabr w'al-muqabala* ("The calculation of reduction and restoration"), said to be the first book on algebra, from which the subject gets its name:

> ... what is easiest and most useful in arithmetic, such as men constantly require in cases of inheritance, legacies, partition, lawsuits, and trade, and in all their dealings with one another, or where the measuring of lands, the digging of canals, geometrical computations, and other objects of various sorts and kinds are concerned.

Much of the mathematical knowledge of medieval Europe was derived from Latin translations of the works of al-Khwarizmi.

We must mention two modern age contributors to the history of algorithms before Hilbert's decisive intervention. The famous German philosopher and mathematician Gottfried Leibniz (1646–1716) believed that the principles of reasoning could be reduced to a formal symbolic system, an algebra or calculus of thought, in which controversy would be settled by calculations. His lifelong interest in and pursuit of this idea led him to take a practical interest in computing machines. His main achievement here was to produce a machine that did not do just additions and subtractions (as did Pascal's calculating machine), but one which was also able to perform multiplications and divisions.

But ambitious as was Leibniz's dream of reducing all reasoning to calculation, it was Charles Babbage (born 26 December 1791, in Teignmouth, Devon) who has become deservedly known as the "Father of Computing". His Difference Engine No.1, completed in 1832, was invented to compile mathematical tables and remains one of the finest examples of precision engineering of the time. But it was Babbage's later design of the Analytical Engine, the attempted construction of which consumed most of his time and large fortune after 1856, that really anticipated today's computers. What was so new was that the Analytical Engine was intended to perform not just one mathematical task but any kind of calculation. An example of this versatility was provided by his devoted follower Ada Byron (daughter of the poet), Lady Lovelace, who, in describing how the engine might calculate Bernoulli numbers, is credited with the first "computer program". But it was not until much later that the anticipatory nature of all this work was recognised, and Babbage, denied government funding, died a disappointed and embittered man in London on October 18, 1871.

1.2 Hilbert's Programme

Before one gets the right answer, one must ask the right question. And the right question being asked for the wrong reasons may be an essential part of the process of discovery. Over one hundred years after Hilbert's famous address to the 1900 International Congress of Mathematicians in Paris, the mathematical agenda he set out there is still of fundamental importance to ongoing research. Many of the twenty-three problems he posed have been solved. But two emerging themes running through them still preoccupy us, although in ways very different from what Hilbert would have expected.

The Computability Theme. The first of these themes concerned the *scope* of algorithms (that is, computer programs) in mathematics. Are there algorithms for solving certain general classes of problems in mathematics? Hilbert's tenth problem, asking for an algorithm for locating solutions to Diophantine equations, is one example. Another is the question (Hilbert's "Entscheidungsproblem") of whether there is an algorithm for deciding if a given sentence is logically valid or not. More generally, do there exist *unsolvable* problems in mathematics? Do there exist computational tasks for which there is no valid program?

The Provability Theme. The second theme concerned the role of *formalism* (typically embodied in axiomatic theories) in mathematics. The period roughly 1904–1928 saw the development of what came to be called *Hilbert's Programme*. The aim of this, simply put, was to capture mathematics in complete, consistent theories. We here understand a "theory" to have a given set of basic statements, or axioms, which we can recognise, and from which, using a small number of logical rules of deduction, we can get all the theorems of the theory. By *capture* we mean derive precisely all mathematical knowledge via theorems of such theories. By a theory being *complete*, we mean that it

is capable of proving or disproving any statement in its language via a finite
deduction from the axioms of the theory. And by *consistent*, we mean that
the theory cannot prove a contradiction (or, equivalently for most purposes,
that the theory has a model). For various reasons, particularly that of having
models close to everyday mathematical practice, it is usual to focus on theories which are *first order*; that is, only allow us to quantify over basic objects
of the intended model, such as numbers (in the case of arithmetic) rather than
over composite entities (such as sets of numbers, or functions). A particular
aim of Hilbert's Programme was that of proving the consistency of everyday
arithmetic (for which there is a standard first order theory) *within* arithmetic.
There were proofs of the consistency of arithmetic, but these used arguments
not formalisable within the first order theory.

It is clear that Hilbert himself saw his formalisation as midwife to a new
era in which all mathematics would be reducible to mechanical manipulations
within suitable theories, and become subject to more and more general algorithms. People have pointed to parallel social trends of the time, such as the
colonialism and European expansion across continents, and view this scientific outlook as part of that larger picture. However valid that might be, it is
very striking that the disintegration of Hilbert's Programme was also within a
social and wider scientific context of growing confusion, complexity and doubt
about traditional certainties. We are thinking about many things — changing
attitudes to imperialism, capitalism, racism and war; accelerating modernism
in music, painting, theatre and the arts generally; the supplanting of Newtonian certainty with quantum ambiguity and relativistic complexity; and the
growing challenge to traditional moral and religious frameworks presented
by psychological and humanistic outlooks. Anyway, never has there been a
mathematical theorem so appropriate to its age and so widely used for extra-
mathematical purposes as Gödel's Incompleteness Theorem. And no theorem
could have been more unwelcome to Hilbert. The day after Gödel commu-
nicated his now famous result to a philosophical meeting in Königsberg, in
September 1930, Hilbert could be found in another part of the same city deliv-
ering the opening address to the Society of German Scientists and Physicians,
famously declaring:

> For the mathematician there is no Ignoramibus, and, in my opin-
> ion, not at all for natural science either. ... The true reason why
> [no one] has succeeded in finding an unsolvable problem is, in my
> opinion, that there *is no* unsolvable problem. In contrast to the
> foolish Ignoramibus, our credo avers:
> We must know,
> We shall know.

At this point we give an informal statement of Gödel's result. We will look
at it more closely in Chapter 8. Taking a "theory" to be the standard kind
described above:

> **THEOREM 1.2.1** *(Gödel's Incompleteness Theorem)*
> *Any consistent theory containing enough of the basic theorems of arithmetic is incapable of even proving its own consistency.*

Not only does Theorem 1.2.1 reduce the provability theme (Hilbert's Programme) to a modest rear-guard action pursued by specialist logicians. But there were also technical features of the proof which were to lead to some very negative consequences for the computability theme, and for how we regard the above quote of Hilbert.

1.3 Gödel and the Discovery of Incomputability

The most important of these features was the first formalisation of the notion of a computable function. This enables us not just to work algorithmically, but to possess a higher level of consciousness about how we do that. We can now *talk about* computability from the outside. It was not long before there were a number of formalisations, or models, of computability. They often came out of different concerns and contexts, and looked quite different from each other. Here are some of the more important ones:

I. The Recursive Functions. This was Gödel's formalisation and, based in logic, is the closest one gets to the way one would describe an algorithm in everyday language. It was Stephen Kleene who gave this model of computability its final form, and who, in a bizarre historical twist, developed the theory of computability under the name of *recursive function theory* (or just *recursion theory*). It was not until the late 1990s that the nineteenth century term "recursion" lost its dominance over the terminology of computability theory and resumed something like its original technical meaning.

II. The λ-Computable Functions. The lambda calculus, first developed by Alonzo Church (founder of the Association for Symbolic Logic, the main organisation of mathematical logicians) and his student Kleene, is not needed until Chapter 11. It has become a very important way of presenting computations so as to minimise distinctions between notations, and is basic to certain programming languages.

III. The Turing Computable Functions. Turing machines provided the first such machine model of computability. They are not very useful for actually doing computations, but that is not the point. What makes them still basic to theoretical computer science is their very prolixity, every computational act carried out in atomic detail. If one is interested in measuring complexity of computations, then Turing machines give a true measure of the work done.

IV. URM Computable Functions. Unlimited Register Machines were invented a little later, by Shepherdson and Sturgis in 1950. The memory of a URM functions more obviously like that of a modern-day computer, and some people find them more useful than Turing machines for formalising computations in a theoretical context.

All of these frameworks enable one to effectively list *all possible* algorithms of that kind, and to use the list to devise a problem which cannot be solved by such an algorithm. This is essentially what Turing did in constructing a *universal* Turing machine and, hence, finding a problem unsolvable by such a machine. By arguing convincingly that *any* algorithm could be performed by a suitable Turing machine, he was able to conclude that there existed problems that were *unsolvable* by any algorithm. Church, also in 1936, did something similar using λ-computability instead.

What is so remarkable about these discoveries of 1936 is the divergence of consequences. On the one hand we have an analysis of computability which makes explicit the underlying logic of Babbage's Analytic Machine and ushers in the age of the stored-program computer. On the other, the analysis leads even more directly to an understanding of how limited the computable universe really is. As we shall see later, there turns out to be a wide range of easily describable unsolvable problems. You should read the article by Robin Gandy, one of the few students supervised by Turing, on *The confluence of ideas in 1936* for detailed and fascinating discussion of how the existence of unsolvable problems was first discovered.

It is clear that the two themes mentioned above relate very closely. It is obvious that theme one has a lot of implications for theme two. We will see this in Chapter 8. But formalism is not just inadequate, it is a bad model of how we do mathematics, and this brings into sharper focus the inadequacies of present-day computers. Formalism is a good model on which to base proofs in mathematics that we want to communicate to others. It does not model how we go about formulating concepts, discovering relationships between them, and proving unexpected new results. So theme two makes us confront the fact that complex physical systems (such as the mathematician's brain) may be mechanical in some sense we do not yet understand. Which brings us back to the computability theme.

1.4   Computability and Unsolvability in the Real World

Basic to much of the mathematics used by scientists are the *real numbers*. One way in which Turing brought home to us the relevance of incomputability was by talking about computable reals and showing that almost all reals are not computable. We will not pretend that there is a simple transition from theorems about incomputable reals to conclusions about "computability" in the material universe. People even disagree about what role incomputable reals play in the *mathematics* used by scientists. What is certain is that there is no longer any satisfactory conceptual model of many aspects of the Universe in terms of purely mechanical interactions. This new situation has only arisen in the last hundred years, since the inadequacies of the Newtonian world-view became clear. The resulting theoretical vacuum is basic to active research of the most fundamental importance. This work is based on two related approaches:

(a) Looking for naturally occurring incomputability. This can be regarded as the "bottom up" approach to constructing a new model. It involves bridging the gap between the mathematically simple way in which incomputability arises in computability theory, and the more basic descriptions of the seemingly simple surface of physical reality. In Chapter 6 we will describe how in 1970 the negative solution to Hilbert's Tenth Problem showed that incomputability is encountered in elementary arithmetic, probably the most basic component of such descriptions. On the other hand, particularly since the 1950s, we have become increasingly aware of how the underlying algorithmic content of nature, based on the iteration of very basic rules, leads to quite deep — nonlinear — mathematical descriptions. Chaos theory, and such mathematically intriguing objects as the Mandelbrot and Julia sets, have provided promising hunting grounds for the still elusive natural examples we believe to exist.

(b) Analysing the relationship between the computable and the incomputable. In 1939 Turing wrote a deep, and still not fully understood, paper attempting to quantify the distance between the aims of Hilbert's Programme and the reality of Gödel's Theorem. One can trace back to this seminal piece of work an important strand of current research in proof theory. More relevant for us, he also introduced the notion of an *oracle Turing machine*. This was a computing machine which operates in the real world, in that it allows inputs to its computations from outside the machine. This became the basis from the 1940s onwards of an increasingly rich model of computationally complex environments — that is, those containing algorithmically connected, but possibly incomputable, information content. This model has tremendous explanatory power, which potentially allows us to leap over the unproved (and perhaps unprovable) assumptions of incomputability on which it is based. Science provides a whole range of apparent anomalies in search of a suitably general theoretical context. In basic science these can arise quite concretely, such as in the incompleteness of the quantum description of the Universe.

More generally the scientific outlook itself is under attack. Only a new model, can make *sense* of the apparent inability of scientific observation to fully capture physical reality. This is the only way to regain the former respect and trust with which the very real achievements of science were formerly regarded. See Chapter 10 for an introduction to the basic properties of the Turing model.

The highly technical side of more recent research is beyond the scope of this book, but anyone persevering to the later chapters will be equipped to grapple with it. You should go to the 2003 paper by Cooper and Odifreddi on "Incomputability in Nature" for an approachable introduction to the search for a new mathematical model.

Most books about computability theory belong in outlook to the recursion theoretic era, the period 1936 to 1997, roughly. We hope this book is different. We have tried to write it in the spirit of Turing, combining an eye for mathematical beauty, an organic feel for the underlying intuitions of the subject, and a compelling sense of the subject's real-world context. Many topics appear for the first time in such a text. Others we have tried to present in a new light with a more modern approach. But we will start in traditional fashion in the following chapter by introducing some of the more important basic models of computability.

Chapter 2

Models of Computability and the Church–Turing Thesis

For convenience, we need to decide on one type of mathematical object — functions, relations or sets — to discuss the computability of. We choose computable functions to describe first, and then get notions of computability for sets and relations as an immediate consequence.

We also need to decide on a suitable *domain* for our functions to operate over. The domain we choose is the set $\mathbb{N} = \{0, 1, 2, \dots\}$ of all *natural numbers*. This is the most basic set we can choose capable of expressing a full range of mathematical practice — and what is computability about if it is not about what we do in practice? For most practical purposes approximations in the form of rational numbers, which can be easily encoded as natural numbers, suffice. In Chapter 10 we will show how to extend our model to deal with situations in which the approximation process fails to reflect reality.

At first sight any description of a function over $\mathbb{N}$ tells us how to go about computing it. So the only question regarding its computability is how we go about formulating the details of the computer program. This impression may be based on years of mathematical practice, but is a very misleading one. Later we will see examples of functions we can easily describe, but whose descriptions are no help at all in computing the functions. We do need to be more precise about what we mean by a computable function to be able to analyse such examples.

Why do we need more than one model of computability? A particular formulation may help us in various ways. It may enable us to formulate an algorithm for computing a given function. Or by expressing an algorithm in a particular way, we may learn something about the complexity of the function. Most importantly — different models reinforce our intuitive grasp of what it means for a function to be computable. It will turn out that the notion of computability is in fact independent of the language used to describe it. But it only becomes real to us when brought into focus from the differing perspectives of our various models.

The particular models chosen below are both historically important and particularly suited to at least one of the above roles.

2.1   The Recursive Functions

The language of logic is abstracted from what we use in everyday life. Gödel's *recursive functions* emerge from the logic, and so are very useful for formalising algorithms of which we have intuitively natural descriptions.

The definition of the recursive functions is what we call an *inductive definition*. We start by defining a small class of very simple functions, called *initial functions*, to be recursive (this is the base of the induction). And then we give a small number of rules for deriving new recursive functions from those already obtained via the inductive process. Of course, the induction is over functions rather than numbers. But the idea of inductively passing on a property step-by-step to the members of an infinite set is the same. Just to complicate matters, we start with a smaller more easily handled class of functions called *primitive recursive* functions.

DEFINITION 2.1.1 (*The Primitive Recursive Functions*)

1) The **initial functions** *(a) – (c) are primitive recursive:*

(a) The **zero function** *defined by*

$$\mathbf{0}(n) = 0, \quad \forall n \in \mathbb{N},$$

(b) The **successor function** *defined by*

$$n' = n + 1, \quad \forall n \in \mathbb{N},$$

(c) The **projection functions** U_i^k *defined by*

$$U_i^k(\overrightarrow{m}) = m_i, \quad \text{each } k \geq 1, \text{ and } i = 1, \ldots, k,$$

(where we write $\overrightarrow{m} = m_1, \ldots, m_k$*).*

2) If $g, h, h_0, \ldots, h_l$ *are primitive recursive, then so is* f *obtained from* $g, h, h_0, \ldots, h_l$ *by one of the rules:*

(d) **Substitution,** *given by:*

$$f(\overrightarrow{m}) = g(h_0(\overrightarrow{m}), \ldots, h_l(\overrightarrow{m})),$$

(e) **Primitive recursion,** *given by:*

$$
\begin{aligned}
f(\overrightarrow{m}, 0) &= g(\overrightarrow{m}), \\
f(\overrightarrow{m}, n + 1) &= h(\overrightarrow{m}, n, f(\overrightarrow{m}, n)).
\end{aligned}
\tag{2.1}
$$

The initial functions (a) and (b) are familiar basic functions of arithmetic. The projection functions in (c) are notationally off-putting, but have a boring technical role which we shall suppress when we get used to it. Rule (d) is just usual substitution.

For the computer scientist *recursion* is a familiar programming technique, which has a computational role going back to the nineteenth century. We call the pair of Equations 2.1 in rule (e) a *primitive recursive scheme*. The scheme describes how we inductively define the values of f, getting $f(\overrightarrow{m}, n+1)$ via known primitive recursive functions in terms of the given *parameters* $\overrightarrow{m}$, the argument n, and the previously computed value $f(\overrightarrow{m}, n)$. If we wanted, we could easily write out an inductive proof that this scheme does define a function f.

We write PRIM for the set of all primitive recursive functions.

Examples of Primitive Recursive Functions

EXAMPLE 2.1.2 *Let $k \in \mathbb{N}$, and let $\mathbf{k}$ denote the constant function defined by $\mathbf{k}(x) = k$ for all $x \in \mathbb{N}$. Show that $\mathbf{k}$ is in PRIM.*

SOLUTION   We show it by induction on k.

Since $\mathbf{0}$ is an initial function, we have $\mathbf{0} \in$ PRIM.

Say $\mathbf{k} \in$ PRIM, some given k. Then $(\mathbf{k} + \mathbf{1})(x) = (\mathbf{k}(x))'$ for each $x \in \mathbb{N}$. So $\mathbf{k} + \mathbf{1} \in$ PRIM (by substitution of $\mathbf{k}$ into $'$).   ▯

EXAMPLE 2.1.3 *Show that the addition function $+ \in$ PRIM.*

SOLUTION   The result follows immediately from the primitive recursive scheme:

$$m + 0 = m,$$
$$m + (n + 1) = (m + n) + 1 = (m + n)'. \tag{2.2}$$

[To see that this is a perfectly good scheme, we could tie it up with Equations 2.1 by writing more formally: $f(m, n) = m + n$, and translating Equations 2.2 into

$$f(m, 0) = U_1^1(n),$$
$$f(m, n + 1) = (f(m, n))' = (U_3^3(m, n, f(m, n)))'.$$

Note how the projection functions can be used to expand the range of variables appearing to that explicitly appearing in the formal definition. We will suppress this level of formality from now on.]   ⬜

EXAMPLE 2.1.4   *Show that the multiplication function* $\times \in$ PRIM.

SOLUTION   The result follows immediately from the primitive recursive scheme:

$$m \times 0 = 0,$$
$$m \times (n+1) = (m \times n) + m,$$

and the result of Example 2.1.3.   ⬜

EXERCISE 2.1.5   *Show that the exponentiation function* m^n *is primitive recursive.*

Before racing on to get all the usual functions of arithmetic primitive recursively, we have a small obstacle — arithmetic on the natural numbers is a little different from what we are used to. For instance, how do we deal with subtraction? Some functions need translating, and we need some very simple functions with technical roles (going back to Gödel's 1931 paper).

EXAMPLE 2.1.6   *If we define a* **predecessor** *function by*

$$\delta(m) = \begin{cases} m - 1 & \text{if } m > 0, \\ 0 & \text{if } m = 0, \end{cases}$$

show that $\delta \in$ PRIM.

SOLUTION   Primitive recursive schemes do not come much simpler than the following:

$$\delta(0) = 0,$$
$$\delta(m+1) = m \quad (= U_1^2(m, \delta(m))).$$

This is what we call a 0-parameter recursion, and there is not even any real reliance on previous values of δ in the second line of the scheme.   ⬜

But we can now get our modified version of subtraction.

EXAMPLE 2.1.7 *Show that* **recursive difference**, *defined by*

$$m \dot{-} n = \begin{cases} m - n & \text{if } m \geq n, \\ 0 & \text{if } m < n, \end{cases}$$

is primitive recursive.

SOLUTION The primitive recursive scheme

$$m \dot{-} 0 = m,$$
$$m \dot{-} (n + 1) = \delta(m \dot{-} n),$$

together with the previous example, gives the result. ▯

EXAMPLE 2.1.8 *Show that* **absolute difference**, *defined by*

$$|m - n| = \begin{cases} m - n & \text{if } m \geq n, \\ n - m & \text{if } m < n, \end{cases}$$

is primitive recursive.

SOLUTION In this case we get the function via substitution using already proven primitive recursive functions:

$$|m - n| = (m \dot{-} n) + (n \dot{-} m).$$

Not every example needs a primitive recursive scheme! ▯

The next exercise introduces two very useful little functions.

EXERCISE 2.1.9 *Show that the functions defined by*

$$\text{sg}(n) = \begin{cases} 0 & \text{if } n = 0, \\ 1 & \text{if } n \neq 0, \end{cases}$$

$$\overline{\text{sg}}(n) = \begin{cases} 1 & \text{if } n = 0, \\ 0 & \text{if } n \neq 0, \end{cases}$$

are primitive recursive.

We can now get some quite complicated primitive recursive functions.

EXAMPLE 2.1.10   *Show that the **remainder function**, defined by*

$$\mathrm{rm}(m, n) = \begin{cases} \text{the remainder upon division of } n \text{ by } m, & \text{if } m \neq 0, \\ n & \text{otherwise}, \end{cases}$$

is primitive recursive.

SOLUTION   The primitive recursive scheme

$$\mathrm{rm}(m, 0) = 0,$$
$$\mathrm{rm}(m, n+1) = \mathrm{rm}(m, n)' \times \mathrm{sg}(|n - \mathrm{rm}(m, n)'|)$$

uses functions already known to be primitive recursive.   ▯

EXERCISE 2.1.11   *Show that $m \,|\, n$ — "m divides n " — is in* PRIM.

EXAMPLE 2.1.12 (Bounded Sums)   *Show that if $f(\overrightarrow{m}, n)$ is known to be in* PRIM, *then $h(\overrightarrow{m}, p) = \displaystyle\sum_{n \leq p} f(\overrightarrow{m}, n)$ is primitive recursive.*

SOLUTION   The primitive recursive scheme

$$h(\overrightarrow{m}, 0) = f(\overrightarrow{m}, 0),$$
$$h(\overrightarrow{m}, p+1) = \sum_{n \leq p} f(\overrightarrow{m}, n) + f(\overrightarrow{m}, p+1)$$
$$= h(\overrightarrow{m}, p) + f(\overrightarrow{m}, p+1)$$

uses functions already known to be primitive recursive.   ▯

EXERCISE 2.1.13 (Bounded Product)   *Show that if $f(\overrightarrow{m}, n) \in$ PRIM, then so is $h(\overrightarrow{m}, p) = \displaystyle\prod_{n \leq p} f(\overrightarrow{m}, n) \ (= f(\overrightarrow{m}, 0) \times \cdots \times f(\overrightarrow{m}, p))$.*

EXERCISE 2.1.14   *Show that if $D(m) =$ the number of divisors of m, then $D \in$ PRIM.*

We might expect this process of building up more and more complicated functions to continue until we have *all* computable functions. However, as far back as 1928 Wilhelm Ackermann defined a computable function which is not primitive recursive. To define the *Ackermann function A*, he used a *nested recursion*. Here is a simplified version due to the Hungarian mathematician Rósza Péter, a largely forgotten co-founder of computability theory:

$$A(m,0) = m + 1$$
$$A(0, n+1) = A(1, n)$$
$$A(m+1, n+1) = A(A(m, n+1), n).$$

The nesting in the last line leads to $A(m, m)$ being much faster growing than any primitive recursive function $f(m)$ could possibly be. One can get an impression of just *how* fast by computing just a few values. To do this, use the fact that the above nested recursion gives the equivalent equations:

$$A(m,0) = m + 1$$
$$A(m,1) = 2 + (m+3) - 3$$
$$A(m,2) = 2 \times (m+3) - 3$$
$$A(m,3) = 2^{(m+3)} - 3$$
$$A(4,n) = 2^{2^{\cdot^{\cdot^2}}} - 3 \quad (m+3 \text{ terms})$$
$$\vdots$$

from which we get the values:

$$A(0,0) = 1, \quad A(1,1) = 3, \quad A(2,2) = 7, \quad A(3,3) = 61, \quad A(4,4) = 2^{2^{2^{65536}}} \ (!),$$

using $2^{2^{2^2}} = 2^{\left(2^{\left(2^{2}\right)}\right)} = 65536$.

We can remedy this inadequacy of PRIM by adding just one more rule for getting new functions from old.

The μ-operator, partial recursive and recursive functions

At this point you might quite reasonably ask why we did not give all the rules when we first stated Definition 2.1.1. The reason is that the class of primitive recursive functions, while including all the functions we commonly meet, has mathematically nice properties that our new rule spoils. For instance, we cannot even be sure our new rule gives us a fully defined function!

DEFINITION 2.1.15 *We say that a function $f : A \to B$ is **total** if $f(x)$ is defined (written $f(x) \downarrow$) for every $x \in A$.*

*Otherwise — that is, if $f(x)$ is undefined ($f(x) \uparrow$) for some $x \in A$ — we say that f is **partial**.*

What we do now is define a larger class of (possibly partial) functions, called the *partial recursive functions*. This is done by adding the following rule to the ones in Definition 2.1.1:

DEFINITION 2.1.16 (μ-**Operator or Minimalisation**)

(f) If $g(\overrightarrow{n}, m)$ is partial recursive, then so is f given by:

$$f(\overrightarrow{n}) = \mu m\, [g(\overrightarrow{n}, m) = 0],$$

where

$$\mu m\, [g(\overrightarrow{n}, m) = 0] = m_0 \iff \text{defn. } g(\overrightarrow{n}, m_0) = 0 \text{ and}$$
$$(\forall m < m_0)[g(\overrightarrow{n}, m) \downarrow \neq 0]. \tag{2.3}$$

REMARK 2.1.17 Intuitively, the μ-operator is a *search* operation — it says compute $g(\overrightarrow{n}, 0), \ldots$ etc. until we find $g(\overrightarrow{n}, m_0) = 0$ — then m_0 is the value wanted. Of course, this search may go on for ever if no such m_0 exists, in which case $f(\overrightarrow{n})$ does not get defined. This is why our new class of functions contains possibly partial functions. But it is the total functions we are interested in, and we will extract these in part (2) of the following definition. ▯

DEFINITION 2.1.18 *Let f be a (possibly partial) function.*

*(1) f is **partial recursive** (p.r.) if it can be defined from the initial functions (a), (b), (c) of Definition 2.1.1, using a finite number of applications of rules (d), (e) from Definition 2.1.1, and of rule (f) (μ-operator).*

*(2) A total p.r. function f is said to be **recursive**.*

So we get the partial recursive functions as the smallest class of functions which contains the initial functions, and which is *closed* under the operations of substitution, primitive recursion and minimalisation.

Having completed our description of the recursive model of the computable functions, we are ready to talk about Church's Thesis. It is important to do so early on, so that we can get plenty of experience of the practical way in which we can combine formal and informal approaches to describing algorithms. The aim is to maximise both clarity and precision in such descriptions.

2.2 Church's Thesis and the Computability of Sets and Relations

It is a remarkable fact that computability exists independently of any language used to describe it. Any sufficiently general model of the computable functions gives the *same* class of functions, it seems. This is the key to the durability of a bold conjecture of Alonzo Church back in the early 1930s.

> **CHURCH's THESIS:**
> (1) f is recursive $\iff$ f is total and effectively computable.
> (2) f is p.r. $\iff$ f is effectively computable.

What we mean by f being "effectively" computable is that there exists *some* description of an algorithm, in *some* language, which can be used to compute any value $f(x)$ for which $f(x) \downarrow$. In a global sense Church's Thesis says that all sufficiently general models of computability are equivalent. So in the absence of any counterexample, we feel justified in assuming it to be true. In a practical sense — which is what is important to us now — it says that if one can give an intuitively convincing description of an algorithm for computing f, then we can find a description of f as a p.r. function. We will find this version invaluable later on.

Let us now, armed with the practical version of Church's Thesis, set out to extend our notions of recursiveness to the other main mathematical objects — sets and relations. We want to know how we can assert, for instance, that the set of all non-negative even numbers is primitive recursive. Or discuss the recursiveness of the usual ordering relation $<$ on the natural numbers.

Recursive Relations and Sets

We have already got definitions of *primitive recursive* and *recursive* for functions. So why not represent sets and relations as functions, so that we can re-use the same definitions?

DEFINITION 2.2.1 *Let S be any set, and R be any relation.*

(1) The **characteristic functions** *of S and R are given by*

$$\chi_S(x) = \begin{cases} 1 & \text{if } x \in S, \\ 0 & \text{if } x \notin S, \end{cases}$$

$$\chi_R(\overrightarrow{x}) = \begin{cases} 1 & \text{if } R(\overrightarrow{x}) \text{ holds}, \\ 0 & \text{if } R(\overrightarrow{x}) \text{ is false}. \end{cases}$$

(2) We say that S or R is **recursive** *(or* **primitive recursive***) if the respective characteristic function χ_S or χ_R is.*

One can think of χ_S, χ_R as truth-functions for the relations $x \in S$, $R(\overrightarrow{x})$, with 0, 1 having their usual interpretations of false and true, respectively. We could of course have used the fact that $x \in S$ is a one-place relation to get the definition of χ_S from that of χ_R.

EXAMPLE 2.2.2 *Show that the relation $n < m$ is recursive.*

SOLUTION We have

$$\chi_<(n, m) = \text{sg}(m \dot- n).$$

So $<$ is primitive recursive, and so recursive. ⬚

EXERCISE 2.2.3 *Show that the identity relation $=$ on $\mathbb{N}$ is primitive recursive.*

EXAMPLE 2.2.4 *Let $\text{Pr}(n)$ be the relation which holds if and only if n is a prime number.*
Show that Pr is primitive recursive.

SOLUTION We notice that $\text{Pr}(n)$ holds $\Longleftrightarrow$ n has ≤ 2 divisors, and $n \neq 0$ or 1.
So $\chi_{\text{Pr}} = \overline{\text{sg}}(\text{D}(n) \dot- 2) \times \text{sg}(n \dot- 1)$. ⬚

EXERCISE 2.2.5 *Show that the relation* $m \mid n$, *which holds if and only if m is a divisor of n, is primitive recursive.*

Before looking at more complicated relations, it is useful to extend Church's Thesis to relations — and hence sets — via Definition 2.2.1.

THEOREM 2.2.6 (Church's Thesis for Relations)
Relation R is recursive $\iff$ *we have an effective procedure for deciding whether $R(\overrightarrow{n})$, each $\overrightarrow{n}$.*

REMARK 2.2.7 (Logical connectives) Just as we can combine functions, using operations like substitution and primitive recursion to get more complicated functions, so we can get more and more complicated relations by using everyday language. For instance, say we are given k-ary relations P, Q, say, where $P(\overrightarrow{m})$ or $Q(\overrightarrow{m})$ might be true or false for a given k-tuple $\overrightarrow{m}$. Then we can get the relation "P and Q" as the one which holds for $\overrightarrow{m}$ exactly when $P(\overrightarrow{m})$ and $Q(\overrightarrow{m})$ are true. We can do the same using other words of the English language, like *or*, *not* or *implies*.

Of course, in mathematics we have to give such words very precise meanings (or *interpretations*), using for instance truth-tables. These precise meanings are consistent with everyday usage, but sometimes are a little narrower, and so we usually replace the usual words in English with corresponding symbolic counterparts, called *logical connectives*. I am assuming you have seen some basic propositional logic, and are familiar with the usual connectives, such as & for "and", and ¬ for "not" (with their usual meanings), and ∨ for "or" (where $P(\overrightarrow{m}) \vee Q(\overrightarrow{m})$ holds whenever $P(\overrightarrow{m})$ or $Q(\overrightarrow{m})$ hold, or both $P(\overrightarrow{m})$ and $Q(\overrightarrow{m})$ hold). It is worth remembering that you can get all the usual connectives using just *two* basic ones — for instance you can get $P \vee Q$ as $\neg(\neg P \& \neg Q)$, and $P \Rightarrow Q$ (for "P implies Q") as $\neg P \vee Q$.

If we restrict ourselves to logical connectives and *propositional variables* P, Q, etc., which stand for 0-ary relations and can only be interpreted as outright true or false, we get the language of the *propositional calculus*.

I will say more about the role of logical language in Chapter 3.   ⬜

It is now easy to see that just as operations like substitution and primitive recursion lead from recursive, or more particularly primitive recursive, functions to more recursive or primitive recursive functions, a similar thing happens with logical connectives applied to relations.

EXAMPLE 2.2.8 *Show that if P, Q are (primitive) recursive relations, then so are $\neg P$, $P\&Q$ and $P \vee Q$.*

SOLUTION I will give two alternative solutions — one a more formal one, and one using Church's Thesis for relations.

(Formal) We get the characteristic functions of $\neg P$, $P\&Q$ and $P \vee Q$ by substituting known primitive recursive functions:

$$\chi_{\neg P}(\overrightarrow{m}) = 1 \dot{-} \chi_P(\overrightarrow{m})$$
$$\chi_{P\&Q}(\overrightarrow{m}) = \chi_P(\overrightarrow{m}) \times \chi_Q(\overrightarrow{m})$$
$$\chi_{P\vee Q}(\overrightarrow{m}) = \mathrm{sg}[\chi_P(\overrightarrow{m}) + \chi_Q(\overrightarrow{m})].$$

(Using Church's Thesis) We know that $\neg P(\overrightarrow{m})$ holds if and only if $P(\overrightarrow{m})$ is false. Since we can effectively decide whether $P(\overrightarrow{m})$ is true or not (and so whether $P(\overrightarrow{m})$ is false), the result follows by Theorem 2.2.6.

The same sort of argument works for the other two. We just notice that

$$P(\overrightarrow{m})\&Q(\overrightarrow{m}) \text{ holds} \iff P(\overrightarrow{m}) \text{ is true } and \ Q(\overrightarrow{m}) \text{ is true}$$

and

$$P(\overrightarrow{m}) \vee Q(\overrightarrow{m}) \text{ holds} \iff P(\overrightarrow{m}) \text{ is true } or \ Q(\overrightarrow{m}) \text{ is true,}$$

and apply Theorem 2.2.6 again. ⬜

REMARK 2.2.9 You might notice that in this example the formal machinery actually gives us a clearer and more concise solution to the problem. It also gives an immediate proof that the *primitive* recursive relations are closed under applications of the usual logical connectives. On the other hand, the informal solution was perhaps a bit more informative, bringing out the essential triviality of the question! In more complicated examples we will find a suitable balance between the two approaches is what is wanted. ⬜

EXERCISE 2.2.10 *Show that if R and S are primitive recursive sets, then so are $\mathbb{N} - R$, $R \cap S$ and $R \cup S$.*

EXERCISE 2.2.11 *Show that every finite set is primitive recursive.*

EXERCISE 2.2.12 *Let f be a recursive function with infinite range. Show that we can find a one–one recursive function g with $\mathrm{range}(f) = \mathrm{range}(g)$.*

[**Hint:** Describe informally how to list the members of the range of f without repetitions. Use Church's Thesis to extract the required recursive function.]

EXERCISE 2.2.13 *Show that if $R(\overrightarrow{m}, n)$ is a recursive relation, and f is a recursive function, then $R(\overrightarrow{m}, f(n))$ is a recursive relation.*

Here is another useful result, simplifying our usage of the μ-operator.

LEMMA 2.2.14
If $R(\overrightarrow{m}, n)$ is a recursive relation, then f defined by

$$f(\overrightarrow{m}) = \mu n[R(\overrightarrow{m}, n)] \quad (=_{\text{defn}} \text{ the least } n \text{ such that } R(\overrightarrow{m}, n) \text{ holds})$$

is partial recursive.

PROOF To reconcile the definition of f with Definition 2.1.16, just notice

$$f(\overrightarrow{m}) = \mu n[\chi_{\neg R}(\overrightarrow{m}, n) = 0],$$

and use Example 2.2.8 telling us that $\chi_{\neg R}$ is a recursive function. ⬜

EXAMPLE 2.2.15 *Let p_n denote the nth prime number (with $p_0 = 2$). Show that p_n is a recursive function of n.*

SOLUTION p_n is total, since there exist infinitely many primes.
And here is a concise formal description of p_n as a recursive function:

$$p_0 = 2$$
$$p_{n+1} = \mu z[z > p_n \& \Pr(z)].$$

The result follows using the known recursiveness of $>$, p_n and $\Pr$, together with Example 2.2.8, Exercise 2.2.13, and Lemma 2.2.14.

Alternatively, this could be re-expressed as a perfectly good argument using Church's Thesis. ⬜

The computability of p_n is very relevant to its role in codings, as we shall see later. Here is another useful function.

EXERCISE 2.2.16 *Let $(m)_i$ denote the exponent of p_i in the prime factorisation of m (where, for example, $(2^{14}3^7)_1 = 7$).*
Show that $(m)_i$ is a recursive function of m, i.

Sometimes we need a computable coding of all ordered pairs of numbers which is bijective.

EXAMPLE 2.2.17   *Find a recursive pairing function* $\langle \cdot, \cdot \rangle : \mathbb{N} \times \mathbb{N} \overset{1\text{-}1}{\underset{\text{onto}}{\rightarrow}} \mathbb{N}$.

SOLUTION   Use Church's Thesis. Form an array containing all pairs of natural numbers:

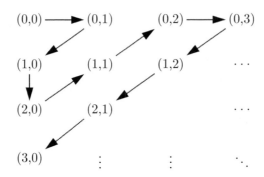

Then list the pairs algorithmically in the order indicated by the arrows, say. This defines the required function.

(Alternatively, if you prefer an explicit definition of the pairing function, $\langle x, y \rangle = 2^x(2y+1) \dot{-} 1$ also does the job, but proceeds through the array a bit more tortuously.)   ⬚

EXERCISE 2.2.18   *Show that one can define inverses* π_1, π_2 *for* $\langle \cdot, \cdot \rangle$, *with*

$$\pi_1(\langle m, n \rangle) = m, \quad \pi_2(\langle m, n \rangle) = n, \qquad \forall m, n \in \mathbb{N},$$

which are also recursive.

You will notice that we can easily define a recursive bijective coding of all $(k+1)$-tuples now, for any given $k+1 > 2$. We just define inductively:

$$\langle n_1, \ldots, n_k, n_{k+1} \rangle = \langle \langle n_1, \ldots, n_k \rangle, n_{k+1} \rangle.$$

EXERCISE 2.2.19 (Bounded Quantifiers)   *Let* $R(\overrightarrow{m}, p)$ *be a primitive recursive relation. Assume new relations* P, Q *to be defined by*

$P(\overrightarrow{m}, n) \iff (\forall p < n)R(\overrightarrow{m}, p)$ (that is, for every $p < n$, $R(\overrightarrow{m}, p)$ is true),
$Q(\overrightarrow{m}, n) \iff (\exists p < n)R(\overrightarrow{m}, p)$ (that is, for some $p < n$, $R(\overrightarrow{m}, p)$ is true).

Show that P *and* Q *are both primitive recursive.*

EXERCISE 2.2.20 (Course-of-Values Recursion) *Let $h(\overrightarrow{m}, n, p)$ be primitive recursive. Let $f(\overrightarrow{m}, n+1) = h(\overrightarrow{m}, n, \widetilde{f}(\overrightarrow{m}, n))$, where*

$$\widetilde{f}(\overrightarrow{m}, n) =_{\text{defn}} p_0^{f(\overrightarrow{m},0)} \times \cdots \times p_n^{f(\overrightarrow{m},n)},$$

and $f(\overrightarrow{m}, 0)$ is primitive recursive. Show that $\widetilde{f}$, and hence f, is in PRIM.

EXERCISE 2.2.21 *Let $h_0(\overrightarrow{n}), \ldots, h_k(\overrightarrow{n})$ be primitive recursive, and let $R_0(\overrightarrow{n}), \ldots, R_k(\overrightarrow{n})$ be primitive recursive relations, exactly one of which holds for any given $\overrightarrow{n}$. Show that if*

$$f(\overrightarrow{n}) =_{\text{defn}} \begin{cases} h_0(\overrightarrow{n}) & \text{if } R_0(\overrightarrow{n}) \\ \vdots & \vdots \\ h_k(\overrightarrow{n}) & \text{if } R_k(\overrightarrow{n}), \end{cases}$$

then f is primitive recursive.

EXERCISE 2.2.22 *A Fibonacci sequence $\{u_n\}_{n \geq 0}$ is given by*

$$u_0 = k_o, \qquad u_1 = k_1, \qquad u_{n+2} = u_{n+1} + u_n.$$

Show that u_n is a primitive recursive function.

2.3 Unlimited Register Machines

Unlimited Register Machines (or URMs) are mathematical abstractions of real-life computers. URMs, more user-friendly than Turing machines, make an ideal introduction to machine models of computability. They were invented some years later than Turing's machines, by Shepherdson and Sturgis, at a time when actual computers already existed. Nowadays they are often called Random Access Machines (RAM).

A URM has *registers* $R_1, R_2, \ldots$ which store natural numbers $r_1, r_2, \ldots$:

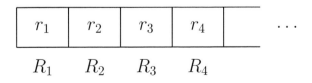

A URM *program* is a finite list of instructions — we follow Nigel Cutland's description — each having one of four basic types:

Type	Symbolism	Effect
zero	$Z(n)$	$r_n = 0$
successor	$S(n)$	$r_n = r_n + 1$
transfer	$T(m, n)$	$r_n = r_m$
jump	$J(m, n, q)$	If $r_n = r_m$ go to instruction q — else go to next instruction

So, for example, instruction S(n) gives a new value for r_n by adding 1 to the existing value of r_n.

Each URM computation using a given program starts with instruction number 1 on the list — and carries out the rest in numerical order *unless* told to jump. A computation will *halt* if it runs out of instructions to obey.

EXAMPLE 2.3.1   *Let P be the URM program:*

> 1. J(2,3,5)
> 2. S(1)
> 3. S(3)
> 4. J(1,1,1)

Describe the computation of P for input $(7, 2)$.

SOLUTION   We just write down the *trace table* of the computation:

Instruction	R_1	R_2	R_3
1	7	2	0
2	7	2	0
3	8	2	0
4	8	2	1
1	8	2	1
2	8	2	1
3	9	2	1
4	9	2	2
1	9	2	2
Halt			

Notice that in the trace table, each instruction in the left column acts on the contents of the registers shown to its left. The instruction J(1,1,1) is called an *unconditional* jump instruction.　　　　　　　　　　　　　　　　□

It is easy to see now that the program P in Example 2.3.1, given any input (n, m), gives output $n + m$ — in other words, P carries out *addition*.

Of course, what we *want* URM programs for is computing functions $\mathbb{N}^k \to \mathbb{N}$ for $k \geq 1$. Example 2.3.1 suggests some basic procedures for doing this.

Input convention: To input a k-tuple $(n_1, \ldots, n_k)$, we start with $n_1, \ldots, n_k$ in registers $R_1, \ldots, R_k$, respectively, and with 0 in all the other registers.

Output convention: If a computation halts, the output is the number in register R_1 — there is no output otherwise.

It is now obvious what we should mean by a URM program *computing* a function.

DEFINITION 2.3.2 *A URM program P **computes** the function $f : \mathbb{N}^k \to \mathbb{N}$ if — for all $(n_1, \ldots, n_k) \in \mathbb{N}^k$ — the computation with input $(n_1, \ldots, n_k)$ using program P halts with output $f(n_1, \ldots, n_k)$.*

*f is **URM-computable** if and only if there is a URM program which computes f.*

So addition: $(m, n) \mapsto m + n$ is URM computable by Example 2.3.1.

Notice that we could get P to compute a function of *any* given $k \geq 1$ variables, just by restricting P to k-ary inputs. We write $\varphi_P^{(k)}$ for the function of k variables computed by a given URM program P.

Here are some very simple URM computable functions:

EXAMPLE 2.3.3 (The Basic URM-Computable Functions)
　　Show that the following functions are URM-computable:
　　(a) **0** $: n \mapsto 0$　(*the zero function*)
　　(b) $' : n \mapsto n' = n + 1$　(*the successor function*)
　　(c) $U_i^k : (n_1, \ldots, n_k) \mapsto n_i$ for $1 \leq i \leq k$ (*the projection functions*)

SOLUTION　All these are computed by one-instruction programs.

　　(a) $\nvdash$ is computed by:　1. Z(1).

(b) $'$ is computed by: 1. S(1).

(c) U_i^k is computed by: 1. T(i,1).   ▯

EXERCISE 2.3.4   *By giving suitable URM programs show that the following functions are URM-computable:*

a) $m \mapsto 3$,

b) $m \mapsto \begin{cases} 0 & \text{if } m = 0 \\ 1 & \text{if } m \neq 0, \end{cases}$

c) $(m, n) \mapsto \begin{cases} 0 & \text{if } m = n \\ 1 & \text{if } m \neq n. \end{cases}$

Flowcharts

It is often hard to envisage exactly what a given program does. It sometimes help to graphically represent its actions in a flowchart.

Take P from Example 2.3.1:

1. J(2,3,5)

2. S(1)

3. S(3)

4. J(1,1,1)

We can represent P by the flowchart:

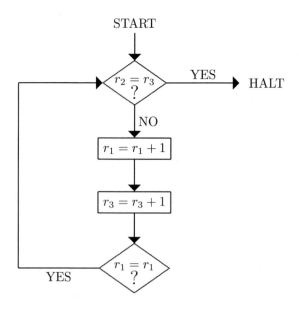

This can be simplified:

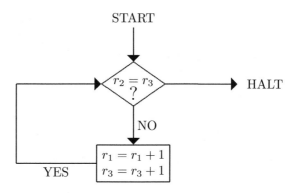

EXERCISE 2.3.5   *Consider the URM program:*

> 1. J(1,4,9)
> 2. S(3)
> 3. J(1,3,7)
> 4. S(2)
> 5. S(3)
> 6. J(1,1,3)
> 7. T(2,1)

a) Draw the flow chart corresponding to this program.

b) Give the trace table of the computation with this program for the following single number inputs: i) 0,   ii) 1,   iii) 4.

c) Say which function $f : \mathbb{N} \to \mathbb{N}$ is computed by this program.

EXERCISE 2.3.6   *Consider the URM program:*

> 1. J(2,3,9)
> 2. J(1,3,9)
> 3. S(3)
> 4. S(4)
> 5. J(2,4,7)
> 6. J(1,1,2)
> 7. Z(4)
> 8. J(1,1,2)
> 9. T(4,1)

a) Draw the flow chart corresponding to this program.

b) Give the trace table of the computation with this program for the following inputs of pairs of numbers: i) (7,3),   ii) (4,2),   iii)   (5,0).

c) Say which function $f : \mathbb{N}^2 \to \mathbb{N}$ is computed by this program.

EXERCISE 2.3.7   *Find URM programs for the following functions:*

a) $(m, n) \mapsto 2m + 5n$,

b) $(m, n) \mapsto m \times n$.

We can of course URM-compute much more complicated functions. How complicated?

Closure properties of the URM-computable functions

One way of generating many examples from a class of functions is to show the class *closed* under lots of operations. For instance in Section 2.1 we got quite complicated recursive functions using *closure* under substitution, primitive recursion and minimalisation.

EXAMPLE 2.3.8   *Show that if f and g are URM-computable functions $\mathbb{N} \to \mathbb{N}$, then so is $f \circ g : n \mapsto f(g(n))$.*

SOLUTION   We need to write down a program $P_{f \circ g}$ which computes $f(g(n))$. The basic idea is that it should first compute $g(n)$, and then feed this value to the URM program for f.

Let P_f and P_g be URM programs for f and g. We would like to take the URM program

$$P_g \qquad\qquad (2.4)$$
$$P_f$$

(the instructions in P_g followed by those in P_f, suitably renumbered) to compute $f \circ g$. But there are three technical problems to deal with:

1. **Problem 1 —**

   P_g with input n may halt via a jump into the *middle* of P_f (or even past its end).

   Action needed — Let $\ell(P) =$ the *length* of P (that is, the number of instructions in P). Define:

DEFINITION 2.3.9   *P is in* **standard form** *if for every J(m,n,q) in P we have $q \leq \ell(P) + 1$.*

We assume from now on that all programs P are in standard form.

2. **Problem 2** —
If we renumber the instructions of P_f in 2.4 in the obvious way, all the jump instructions J(m,n,q) in P_f will need q modifying.

Action needed — Define:

DEFINITION 2.3.10   *The* **join** *of programs P and Q is the program*

$$P$$
$$Q$$

got by writing the instructions of Q after those of P — with each J(m,n,q) in Q replaced by $J(m,n,\ell(P)+q)$.

3. **Problem 3** —
Given a URM program P, let $\rho(P) =$ the largest index k of a register R_k used by P.

Then for P_f to compute $f(g(n))$, it needs to start with the tape looking like:

Action needed — We need to define a "cleaning up" program to insert between P_g and P_f in 2.4, which will remove any accumulated rubbish in registers $R_2, \ldots, R_{\rho(P_f)}$:

DEFINITION 2.3.11 $Z[a; b]$, *for* $b \geq a$, *is the program which cleans up all registers* $R_a, \ldots, R_b$:

$$
\begin{array}{ll}
1. & Z(a) \\
2. & Z(a+1) \\
& \vdots \\
b-a+1. & Z(b)
\end{array}
$$

The solution to Example 2.3.8 is now easy. Let $\rho = \min\{\rho(P_f), \rho(P_g)\}$. Then the program

$$
\begin{array}{c}
P_g \\
Z[2; \rho] \\
P_f
\end{array}
$$

computes the composition $f \circ g$.   ⬜

More generally, we can show:

EXERCISE 2.3.12 *If* $f : \mathbb{N} \to \mathbb{N}$, $g : \mathbb{N} \to \mathbb{N}$ *and* $h : \mathbb{N}^2 \to \mathbb{N}$ *are URM computable then so is the function* $n \mapsto h(f(n), g(n))$.

[**Hint:** In preparation for applying P_h to $(f(n), g(n))$, compute $f(n)$ while storing n in a separate register, then compute $g(n)$ using the stored n while safely storing $f(n)$, and then transfer $f(n), g(n)$ to registers R_1, R_2.]

In fact, defining substitution as in Definition 2.1.1, we can show (although we omit the details of the proof) that:

THEOREM 2.3.13
The URM computable functions are closed under substitution.
That is, if
$$
f(\overrightarrow{m}) = g(h_0(\overrightarrow{m}), \ldots, h_l(\overrightarrow{m})),
$$
where $g, h_0, \ldots, h_l$ *are URM computable functions, then* f *is URM computable too.*

EXERCISE 2.3.14   *(a) Consider the URM program P_f:*

$$
\begin{array}{rl}
1. & T(1,3) \\
2. & T(2,4) \\
3. & J(1,4,9) \\
4. & J(2,3,9) \\
5. & S(3) \\
6. & S(4) \\
7. & S(5) \\
8. & J(1,1,3) \\
9. & T(5,1)
\end{array}
$$

i) Draw the flow chart corresponding to this program.
ii) Give the trace table of the URM computation using P_f for input $(1,1)$.
iii) Find the output of the computation using P_f for input $(1,2)$.
iv) Describe the function $f : \mathbb{N} \times \mathbb{N} \to \mathbb{N}$ computed by P_f.

(b) Devise a URM program P_g to compute the function $g : \mathbb{N} \to \mathbb{N}$ given by

$$g(m) = 3m + 2.$$

By using the above programs P_f and P_g, or otherwise, write down a URM program which computes the function $h : \mathbb{N} \times \mathbb{N} \to \mathbb{N}$ given by

$$h(m) = 3\,|m - n| + 2.$$

By this time it should be no surprise that, with a little more work (which we again omit), we can also show that:

THEOREM 2.3.15
The URM computable functions are closed under primitive recursion.

The next exercise we will find useful later on.

EXERCISE 2.3.16   *If f, g are functions $\mathbb{N} \to \mathbb{N}$, we say that g **dominates** f of for all large enough n (greater than some n_0) we have $g(n) \geq f(n)$.*
   Show that every URM computable function is dominated by a strictly increasing URM computable function.

Since the initial primitive recursive functions are URM computable, Theorems 2.3.13 and 2.3.15 immediately give:

THEOREM 2.3.17

All the primitive recursive functions are URM-computable.

In fact — and this is the main reason Shepherdson and Sturgis first presented URMs in their 1963 paper — it is quite easy to find URM programs to deal with *all three* operations in the Definition 2.1.18 of the recursive functions. And so:

THEOREM 2.3.18

All recursive functions are URM-computable.

We need to know how to code up what URMs do, using recursive functions and relations, before we can prove the converse of Theorem 2.3.18. We leave this until Chapter 4. For the moment, let us be satisfied with the statement of the following remarkable theorem, just one important piece of evidence supporting Church's Thesis.

THEOREM 2.3.19

A function is URM computable if, and only if, it is recursive.

2.4   Turing's Machines

After URMs, Turing machines are frustratingly difficult to program. Even the simplest computational task is a mammoth effort, it seems, for a Turing machine! But underlying this is the great strength of the Turing model of computation, and the reason why it is still the standard one for those wanting a true *measure* of how *difficult* it is to carry out a particular computation. Where URMs do not care, for instance, about the size of numbers encountered in a computation, a Turing program makes us very aware of the work involved in reading very large numbers. The atomic actions of a Turing machine, first described in Turing's 1936 paper, are as basic as one can get.

Apart from this, the work and personality of Alan Turing are integral to the history of science and computing in the last century. This section touches on some of the new thinking that went into revolutionising the way we view our present-day world. You can get invaluable background to the mathematics

which follows from Andrew Hodges' very readable — and much praised —
1983 biography of Turing: *Alan Turing: The Enigma*.

How Turing machines work

The basic hardware for any Turing machine T consists of a tape, sub-
divided into cells, which is infinitely extendable in both directions, and a
reading/writing head:

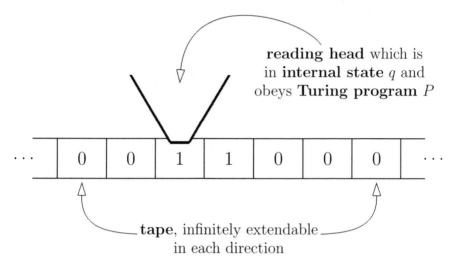

reading head which is
in **internal state** q and
obeys **Turing program** P

tape, infinitely extendable
in each direction

The presence of the reading head is the first sign that we are dealing with
computability in a much more basic way than we did with URMs. URMs may
appear to work like real-life computers, but Turing machines actually *do* it.

It is standard to talk of a *Turing machine* T when we actually mean its
program. Turing programs are not structured as ordered *lists* of instructions,
like URM programs. They use instead *internal states*, whose purpose is not
obvious at first. Turing originally thought of these as corresponding to human
states of mind, and perhaps we should too.

In order to describe a *Turing program* for T we first need the programming
symbols, consisting of:

1. The *tape symbols* $S_0, S_1, \ldots, S_n$ allowed to be written on the tape. Each
   cell contains just one tape symbol. We write S_0 as "0" — interpreted
   as "blank" — and write S_1 as "1". We call the tape symbols other than
   S_0 the *alphabet* Σ of T.

   [**Remark:** We usually just use tape symbols 0 and 1 — but with the
   extra symbols S_2, etc. permitted and renamed as we choose.]

2. An infinite list of *internal states*, $q_0, q_1, \ldots$. At the start of any given
   step of a computation, the reading head is specified to be in one of these
   states.

[**Remark:** Any given Turing program will only use finitely many internal states, but we need to have an unbounded number of states available.]

3. The *action symbols*, used by the program to tell the reading head what to do in relation to its current cell — L (saying "move left one cell"), R ("move right one cell"), 1 ("print 1") and 0 ("erase the current cell").

[**Remark:** Do not worry about the dual usage of 0 and 1 as both tape symbols and action symbols — when we see them in a program, the intended meaning will be clear. But we *never* allow L or R as tape symbols, for reasons that will soon be obvious.]

The program for T will be formed from instructions — called *quadruples* — made up from the programming symbols. Any such quadruple Q must take the form:

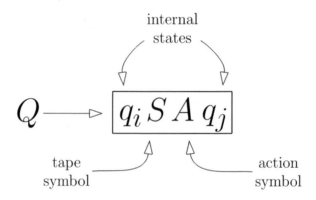

Then $Q = q_i S A q_j$ says: "If T is in state q_i reading tape symbol S, then perform action A and pass into new internal state q_j." If T is in state q_i reading tape symbol S, then we say that Q is *applicable*.

There is now just one small wrinkle in the definition of what a Turing machine is — we have to be sure that its program does not ask for conflicting actions at any step in a computation.

DEFINITION 2.4.1   *(1)   A set X of quadruples is* **consistent** *if*

$$q_i S A q_j, \; q_i S A' q_k \in X \implies A = A' \text{ and } q_j = q_k.$$

(2)   A **Turing machine** *T is a finite consistent set of quadruples.*

REMARK 2.4.2   What happens if we drop the consistency condition in Definition 2.4.1. We will see later that this gives us a way to model *nondeterministic* computations (or computations with guessing).   ▯

How does T compute? Obviously, at each step of a computation, T must obey any applicable quadruple. But we also need some conventions, along the lines of the input and output conventions for computing functions with URMs. We start with the conventions for computing functions $\mathbb{N} \to \mathbb{N}$.

Input convention: To input $n \in \mathbb{N}$, place $n + 1$ consecutive 1's on the tape. Then set the reading head in starting state q_0 reading the leftmost 1.

Output convention: If a computation halts — which only happens when there is no applicable quadruple in T — output the number $\varphi_T(n)$ of 1's left printed on the tape.

REMARK 2.4.3   (1) We input n as $n + 1$ consecutive 1's in the interests of uniformity — this means all computations, even those with input 0, start with the head reading a leftmost 1.

(2) Outputting n as n possibly nonconsecutive 1's is very natural, and certainly the easiest thing in practice. Later, when we want to combine programs, it will be helpful to have the 1's gathered together consecutively before output. But we can achieve that with a suitable "cleaning up" program, like we did for URMs.   ▯

DEFINITION 2.4.4   *A function f is* **Turing computable** *if $f = \varphi_T$ for some Turing machine T.*

EXAMPLE 2.4.5   *Find a Turing machine which computes the successor function $n' = n + 1$.*

SOLUTION   Let T have Turing program $P = \emptyset$ (the *empty program*). Then $\varphi_T(n) = n + 1$ for all $n \in \mathbb{N}$.   ▯

We can visualise this for input 2, say:

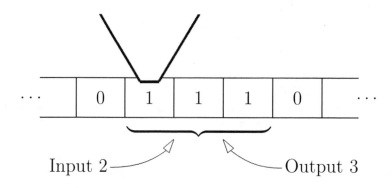

Input 2——————Output 3

EXAMPLE 2.4.6   *Show that the zero function* $\mathbf{0} : n \mapsto 0$ *is Turing computable.*

SOLUTION   Let T have Turing program

$$q_0 1 0 q_1 \quad \leftarrow \quad \text{subroutine for deleting 1's}$$

$$q_1 0 R q_0 \quad \leftarrow \quad \left\{ \begin{array}{l} \text{move right in search of another 1,} \\ \text{in preparation for return to "delete"} \\ \text{subroutine} \end{array} \right.$$

Then $\varphi_T(n) = 0 = \mathbf{0}(n)$ for all $n \in \mathbb{N}$.   ⧠

REMARK 2.4.7   Even simple Turing programs need annotating as above. This is the least we must do for the reader, given that flow charts for Turing programs tend to be too complicated.   ⧠

EXERCISE 2.4.8   *Find a Turing machine which computes the function sg defined by*

$$\text{sg}(n) = \begin{cases} 0 & \text{if } n = 0, \\ 1 & \text{if } n \neq 0, \end{cases}$$

indicating briefly why your program works.

How do we Turing compute functions of *more* than one variable, such as $f(m, n) = m + n$?

> **Extended input/output convention:** We associate with T a partial function of $k > 0$ variables, denoted by $\varphi_T^{(k)}$, by making $\varphi_T^{(k)}(\overrightarrow{n_k})$ be the output got from inputting $n_1, \ldots, n_k$ into T in the form:
>
> $$\underbrace{1 \ldots 1}_{n_1 + 1 \text{ times}} 0 \underbrace{1 \ldots 1}_{n_2 + 1 \text{ times}} 0 \ldots 0 \underbrace{1 \ldots 1}_{n_k + 1 \text{ times}}$$
>
> and setting the reading head in starting state q_0 reading the leftmost 1.

We still write φ_T for $\varphi_T^{(1)}$, of course.

EXAMPLE 2.4.9 *Find a Turing machine T which computes the zero function* $\mathbf{0} : (n_1, \ldots, n_k) \mapsto 0$ *for any $k > 0$.*

SOLUTION We need to adapt the program from Example 2.4.6 to deal with the case when $k > 1$.

Let T be the following Turing program:

$$\left.\begin{array}{r} q_0 1 0 q_1 \\ q_1 0 R q_0 \end{array}\right\} \leftarrow \left\{\begin{array}{l} \text{implement a subroutine to erase the} \\ \text{current block of 1's} \end{array}\right.$$

$$q_0 0 R q_2 \} \leftarrow \left\{\begin{array}{l} \text{when reach end of current block of 1's} \\ \text{move right in search of a new one} \end{array}\right.$$

$$q_2 1 1 q_0 \} \leftarrow \left\{\begin{array}{l} \ldots \text{ and if find one, prepare to return} \\ \text{to the "erase" subroutine} \end{array}\right.$$

Then $\varphi_T^{(k)}(\overrightarrow{n_k}) = 0 = \mathbf{0}(\overrightarrow{n_k})$ for all $\overrightarrow{n_k} \in \mathbb{N}^k$, all $k > 0$.   ☐

I have to say, at this point, that composing Turing programs is much more rewarding than reading them. This is why I have relegated the more interesting programs to the exercises!

EXERCISE 2.4.10 *Show that the addition function $f : (x, y) \mapsto x + y$ is Turing computable.*

EXERCISE 2.4.11 *Write a Turing program for the function $g(x, y) = x \dot{-} y$.*

[My program uses 17 quadruples — you might manage with less.]

EXERCISE 2.4.12   *Show that the projection function* $U_i^k : \overrightarrow{m} \mapsto m_i$ *is Turing computable, for* $1 \leq i \leq k$.

We now know that the initial recursive functions are all Turing computable.

We can also check that the Turing computable functions satisfy just the same closure properties as do the recursive functions and the URMs. We do not need to go into detail about this. But if you do the next exercise, it should give you some confidence that you could do everything if you really had the time and motivation.

EXERCISE 2.4.13   *Let* f *and* g *be Turing computable functions* $\mathbb{N} \to \mathbb{N}$. *Show that the composition* $f \circ g$ *of* f *and* g *is Turing computable.*

[**Hint:** Let f and g be computed by Turing programs P_f and P_g. Merge P_f and P_g into a new program for $f \circ g$ by (a) ("cleaning up") adapting P_g to give its output as a block of consecutive 1's, with the head reading the leftmost 1, and (b) appropriately renumbering the states in P_f.]

The following example will show you how to carry out the cleaning up part. It will also give you an idea of how to compose more complicated Turing programs using extra tape symbols, or markers.

EXAMPLE 2.4.14   *Let* T *be a Turing machine. Find a new program* T^* *which computes exactly like* T, *except that, from input* n, T^* *provides the output in the form of* $\varphi_T(n) + 1$ *consecutive 1's, with the reading head halting on the leftmost 1.*

SOLUTION   The basic idea is to keep track of which cells the reading head visits during the computation — the working space — and to shuffle all the 1's to a single block to the left of this working space.

The first thing we do is to introduce a marker ε for T^* to leave behind in place of the 0's left by T.

To get T^* from T:

1. Replace 0 by ε in all the quadruples of T.

2. Add a quadruple $q_i 0 \varepsilon q_i$ for every internal state q_i in T.

   Result: T^* computes exactly like T, but leaves ε in place of each 0 left by T.

3. (Preparation for the cleaning up subroutine) For each tape symbol $S = \varepsilon$ or 1, each internal state q_i in T, if there is no quadruple of the form $q_i S \cdots$ in T, then add $q_i S S q_M$ to T^*, where M is chosen $>$ any index j of a q_j in T.

4. (Cleaning up subroutine) Include the following quadruples in T^*:

$$
\left.\begin{array}{l}
q_M \varepsilon R q_M \\
q_M 1 R q_M \\
q_M 0 1 q_{M+1}
\end{array}\right\} \quad \leftarrow \quad \left\{\begin{array}{l}\text{move to the right end of the working} \\ \text{space, and add an extra 1}\end{array}\right.
$$

$$
q_{M+1} 1 L q_{M+1} \ \} \quad \leftarrow \quad \left\{\begin{array}{l}\text{move to the left end of the right hand} \\ \text{block of 1's}\end{array}\right.
$$

$$
q_{M+1} \varepsilon 1 q_{M+2} \ \} \quad \leftarrow \quad \{ \text{ if find an } \varepsilon \text{ remaining, change it to a 1}
$$

$$
\left.\begin{array}{l}
q_{M+2} 1 R q_{M+2} \\
q_{M+2} 0 L q_{M+3} \\
q_{M+3} 1 0 q_{M+3} \\
q_{M+3} 0 L q_{M+1}
\end{array}\right\} \quad \leftarrow \quad \left\{\begin{array}{l}\ldots \text{ and move to the right hand 1 and} \\ \text{erase it, preparing to start a new sub-} \\ \text{routine}\end{array}\right.
$$

$$
q_{M+1} 0 R q_{M+4} \ \} \quad \leftarrow \quad \{ \text{ if no } \varepsilon \text{ is left, halt on the leftmost 1.}
$$

It is not difficult to see that T^* does now work.   ⬜

EXERCISE 2.4.15 *Say which function $f(x)$ is computed by the Turing machine below, which uses extra tape symbols ε, η as counters:*

$q_0\, 1\, \varepsilon\, q_1$	$q_2\, \eta\, 1\, q_3$	$q_6\, 1\, L\, q_6$
$q_1\, \varepsilon\, R\, q_1$	$q_3\, 1\, R\, q_4$	$q_6\, 0\, L\, q_7$
$q_1\, 1\, R\, q_1$	$q_4\, 0\, 1\, q_5$	$q_7\, 1\, L\, q_7$
$q_1\, 0\, R\, q_2$	$q_5\, 1\, R\, q_5$	$q_7\, \varepsilon\, 0\, q_7$
$q_2\, 0\, \eta\, q_2$	$q_5\, 0\, \eta\, q_5$	$q_7\, 0\, R\, q_0$
$q_2\, 1\, R\, q_2$	$q_5\, \eta\, L\, q_6$	

Using counters ε, η, or otherwise, find a Turing program for the function $h(x,y) = (x+1)(y+1)$.

Turing machines can be upgraded not just with enlarged languages, but with more tapes and reading heads, for example. None of these modifications changes the class of Turing computable functions we get, but they are interesting to people such as computer scientists, who look at *efficiency* of programs.

Anyway, what is important to us is that we can improve on Exercise 2.4.13, and show, just like we did for URMs, that substitution leads from Turing computable functions to more Turing computable functions. And that the Turing computable functions have *all* the same closure properties as do the URM computable functions. We will leave the interested reader (perhaps you!) to go further into the verification that, once again, the class of computable functions we get is exactly the class of partial recursive functions.

2.5   Church, Turing, and the Equivalence of Models

We have finished our review of some of the most useful and commonly encountered models of computability. And however different they looked to each other, they all led to the same class of functions. We could have looked at other models, such as the lambda calculus, but they would all have turned out to be equivalent. This has become a powerful argument for Church's Thesis, but not the most convincing one.

Turing too had his thesis, independently of Church:

TURING'S THESIS (1936)   *Every function for which there is an intuitively effective process for computing its values can be shown to be Turing computable.*

And as Kleene described in a 1984 interview with William Aspray of the Charles Babbage Institute, Turing's 1936 paper gave "immediate support" to Church's Thesis:

> "because ... Turing's definition of computability was intrinsically plausible, whereas with the other two [recursive functions and lambda computable functions], a person became convinced only after he investigated and found, much by surprise, how much could be done with the definition."

In other words, Turing's Thesis, and its presentation in his paper, was so persuasive, that it made any non-equivalent model of computability impossible to envisage. And that is the real reason why some people go so far as to call Church's Thesis a theorem – for us it is Theorem 2.2.6. Also, in belated recognition of Turing's role, people usually add Turing's name to Church's famous thesis:

CHURCH–TURING THESIS   *f is effectively computable $\iff$ f is partial recursive $\iff$ f is Turing computable.*

All this transforms the way we talk about computability.

Important consequence: Our technical mastery of specific models gives us a strong grasp of the notion of computability. It gives us the security of knowing we can back up intuitive argument with relevant detail when needed. The Church–Turing Thesis allows us to talk of f being "(partial) computable" — or "p.c." — instead of "Turing computable" or "(partial) recursive".

Similarly, the following definition now has real mathematical content:

DEFINITION 2.5.1   *We say a set A is* **computable** *if there is an effective procedure for deciding whether $n \in A$ or $n \notin A$, for each $n \in \mathbb{N}$.*

This is because:

$$A \text{ is computable} \iff \chi_A \text{ is computable} \quad \text{(by definition of } \chi_A\text{)}$$
$$\iff \chi_A \text{ is recursive (or Turing computable)}$$
$$\text{(by the Church–Turing Thesis)}$$

You can see how this works out in practice in the following:

EXAMPLE 2.5.2   *Show that the set* $Ev = \{0, 2, 4, \dots\}$ *is computable.*

SOLUTION   We describe an intuitively satisfying algorithm for deciding whether $n \in Ev$ or not:

Given n, divide n by 2. If the remainder $= 0$, then $n \in Ev$.   ⬚

Notice — If our solution were to be challenged, or we had our own doubts, we could convert the informal solution into a formal one based on the recursive model: $\chi_{Ev}(n) = 1 \dotminus \mathrm{rm}(2, n)$.

EXERCISE 2.5.3   *Show that a set A is computable if, and only if, its complement $\overline{A}$ is computable.*

Chapter 3

Language, Proof and Computable Functions

As well as models of how we compute, there are models of how we accumulate knowledge. In this chapter I am going to briefly review the basic logic needed for the most familiar of these models — the axiomatic theory.

We will also begin to look at the close links between the axiomatic model and our model of computability. When we know more about these two kinds of model, we will see that the parallels between them will help our understanding of both. For instance, in Chapter 8, we will see that one of the most important mathematical results of the last century — Gödel's Incompleteness Theorem — is reducible to a simple fact from computability theory. Remember — it was this theorem which had such negative implications for Hilbert's Programme, in particular in relation to the Provability Theme I mentioned in Chapter 1.

At this point I should mention that Chapters 3 and 8 dealing with logic can be safely missed out, if you are only interested in the computability theory.

3.1 Peano Arithmetic and Its Models

We will start off with some basic questions:

- How can we *capture* a given function $\mathbb{N} \to \mathbb{N}$ in a theory?

- What *sort* of theory is needed to fully describe a given function $\mathbb{N} \to \mathbb{N}$?

- What *class* of functions is describable in a such a theory?

The first two questions are easy to answer. The third has a simple answer, which is harder to verify.

Clearly any theory for talking about functions on the natural numbers must have a language capable of making statements about the natural numbers. But that is not all. When we say a theory $\mathcal{T}$ describes a function f, we do not just mean that the language of $\mathcal{T}$ is rich enough to *define* f in some way. What we understand is that $\mathcal{T}$ has enough theorems to tell us what any particular value $f(n)$ is. So if the class of functions $\mathcal{T}$ describes is to be large, it had better be capable of proving all the usual basic theorems of number theory.

We start with PA, the standard theory of arithmetic, whose basic assumptions — or *axioms* — are essentially due to the Italian logician Giuseppe Peano (1858–1932). Ours is a *first order* theory in that it only allows quantification over variables for individuals — or numbers, as we intend. The symbolic logic on which such theories are based was developed by Gottlob Frege in the latter part of the nineteenth century.

REMARK 3.1.1 If you are not very familiar with axiomatic theories, here are a few pointers as to how they are used and some brief historical background.

Axiomatic mathematics goes back to the ancient Greeks, of course, most influential being Euclid's *Elements*. And when nineteenth century developments in logic and set theory led to seeming contradictions — most famously *Russell's Paradox* (of which more later) — the axiomatic framework was what Hilbert and others saw as the best way of restoring everyday mathematics to a secure basis. The idea was to develop axiomatic theories very formally, with assumptions and ways of deducing from them rigorously formulated so as to ensure no contradictions. Although a theory might be defined with a particular interpretation in mind, the theory itself would involve just formal manipulations, independent of any underlying reality. The full theory might eventually tell us important things about a concrete mathematical structure, but that would only be if the structure turned out to be a valid interpretation — or *model* — of the theory. This explains for instance why we have so-called *numerals* $\bar{m}$ in our theory PA, when it is obvious we are wanting to talk about *numbers* m. We will develop PA formally, but will be very disappointed if the structure $\mathbb{N}$ does not turn out to be a model of PA!

Logicians have a special regard for the modelling potential of axiomatic theories — or *formal systems* as they like to call them. Out in the real world people are not so sure. The early scepticism of Henri Poincaré is specially well known. Here is a quote from his 1902 review of Hilbert's *Foundations of Geometry*:

> The logical point of view alone appears to interest Hilbert. Being given a sequence of propositions, he finds that all follow logically from the first. With the foundations of this first proposition, with its psychological origin, he does not concern himself.

Jacques Hadamard's classic *The Psychology of Invention in the Mathematical Field* (1945) is still an insightful reminder of how unaxiomatic the process of mathematical discovery seems to be. But — as a model of the world which mathematicians would *like* mathematical proofs to inhabit, the axiomatic model still reigns supreme. It is still a very useful model, so long as one does not claim too much for it.   ☐

The language $\mathcal{L}_{\mathrm{PA}}$ of first order arithmetic

Firstly, we want to be able to refer to particular numbers using the language $\mathcal{L}_{\mathrm{PA}}$. The simplest solution is to include just one *constant symbol* $\bar{0}$ (intended interpretation 0, of course).

If we then include a unary function symbol (let's just denote it by $'$) for the successor function $x \mapsto x' = x + 1$, then we can refer to any other natural number n via the *numeral* $\bar{n} = \bar{0} \overbrace{' \cdots '}^{n\ \text{times}}$.

We also include binary function symbols $+$ and $\times$ with the obvious intended interpretations. What we have in mind in setting up this symbolism is the usual structure of the natural numbers, written $\mathfrak{N} = \langle \mathbb{N}, ', +, \times, = \rangle$. We have not included a symbol for the ordering $<$ on numbers, but this is easily defined using the symbols we do have: $x_0 \leq x_1$ stands for the wf $\exists x_3 (x_2 = x_1 + x_3)$.

In addition to these special symbols, there is a certain amount of basic symbolism common to all such theories. Trivially, we allow brackets (,) in our language. We also need enough *variables* to effectively quantify — say the infinite list: $x_0, x_1, \ldots$. We include a *predicate symbol* $=$ for the usual identity relation. And we allow ourselves all the usual logical connectives and quantifiers: $\neg$ (for "not"), $\Rightarrow$ (for "implies"), $\Leftrightarrow$ (for "if and only if"), $\vee$ (for "or"), $\wedge$ (for "and"), $\exists$ (for "there exists") and $\forall$ (for "for all").

We assume some sensible rules for combining these symbols into *formulas* or *wfs* (well-formed formulas) in such a way that they have unambiguous and meaningful interpretations. You can find such rules in any elementary logic text. In particular, you will probably find an inductive definition of the *terms* of $\mathcal{L}_{\mathrm{PA}}$ — these are just objects like $x_2 + (\bar{7} \times x_1')$ formed from the variables and $\bar{0}$ using the function symbols of $\mathcal{L}_{\mathrm{PA}}$. The inductive definition of the wfs will start with the *atomic* wfs — those like $\bar{n} + x_0 = x_1 \times x_1$ with no quantifiers or connectives — and tell you how to sensibly build up to more complex wfs.

REMARK 3.1.2 Your logic text will probably give these rules in the more general context of full *first order logic*. Its language has an infinite set of constants and infinitely many function and predicate symbols of each possible number of arguments. This means the language of any first order theory you want — for instance that of PA — can be got by specialising this general language. ☐

Given such sensible rules for forming wfs, we can use them to write down *axioms* which have meaningful interpretations — and hopefully interpretations which are *valid* in some context we are interested in. Once again, you will have to go to a basic logic text if you want the formal details of how you interpret wfs in a mathematical structure and make precise what one means by the formula being *satisfied* in that structure. We have Alfred Tarski (1902–1983) to thank for making precise our notions of *truth* and *satisfaction* in an interpretation. We use $\models$ to denote "satisfies". So, for instance, $\mathfrak{N} \models \varphi$ means that the wf φ of $\mathcal{L}_{\text{PA}}$ is satisfied in the standard interpretation, and $\models \varphi$ means φ is *universally valid* — that is, satisfied by any interpretation. I will assume that either you know enough logic, or, quite likely, Tarski's definitions have sufficiently entered the culture, for you to deal reasonably confidently with such notation.

EXAMPLE 3.1.3 *Let φ be the wf $\exists x_0 \forall x_1 (x_0 \leq x_1) \Rightarrow \forall x_1 \exists x_0 (x_0 \leq x_1)$. Show that $\models \varphi$.*

SOLUTION We can only have $\not\models \varphi$ if there is some interpretation $\langle \mathbb{M}, \leq \rangle$ of φ in which there is some $k \in \mathbb{M}$, such that for each $a \in \mathbb{M}$ we have $a \not\leq k$. But then there cannot exist an $a \in \mathbb{M}$ such that for *every* $k' \in \mathbb{M}$ $a \leq k'$, since for any such a we would have $a \not\leq k$. So $\models \varphi$, as required. $\quad\square$

EXERCISE 3.1.4 *Let φ be the wf $\forall x_1 \exists x_0 (x_0 \leq x_1) \Rightarrow \exists x_0 \forall x_1 (x_0 \leq x_1)$. Say whether or not (i) φ is universally valid, and (ii) $\langle \mathbb{N}, \leq \rangle \models \varphi$.*

REMARK 3.1.5 The wfs we are specially interested in are those in which all the variables occur quantified — that is, in which there are no *free* variables. These are called *sentences* — the implication being that these are fully formed statements in the everyday sense of the term. They have the important property that they are satisfied in a given interpretation just if they are true statements about it. $\quad\square$

The axioms of first order arithmetic

The axioms of PA fall into two groups. There are those which capture the basic logic of the theory — the *logical axioms* — which are basic to all first order theories. We usually include here some axioms for = capturing the basic properties of the identity relation. We will take:

(E1) $(x_1 = x_2 \Rightarrow (x_1 = x_3 \Rightarrow x_2 = x_3))$, and

(E2) $x_1 = x_2 \Rightarrow (x_1' = x_2')$,

where axiom (E2) gives a very weak form of the substitution rule.

It is *Gödel's Completeness Theorem* which tells us that we can write down a set of axioms — the axioms of basic first order logic in the language of PA — from which all the *logically valid* wfs of PA are derivable. The logically valid wfs are those which are true in any interpretation just by virtue of their logical structure.

We do not really care here what these axioms look like, so long as we know they exist. We assume that any first order theory contains the logical axioms in the appropriate language. The full set of logical axioms give a theory sometimes called *predicate calculus*, or PC for short. In proving theorems in PA, Gödel's Completeness Theorem allows us to use *any* known logically valid formula just as we would an axiom.

REMARK 3.1.6   Why, I hear you ask, do we bother with this theorem of Gödel? Why do we not just take all the logically valid wfs of $\mathcal{L}_{\mathrm{PA}}$ as axioms?

Well the problem is — as we shall see later — that there is no algorithm for deciding if a given wf is logically valid or not. And we usually do ask of any axiomatic theory that we can tell whether a given wf is an axiom of the theory or not!   □

The second group of axioms is meant to capture some basic facts about the structure of the natural numbers. They are chosen in such a way that any other basic fact you can think of can be proved using them.

The special axioms for the first order theory PA for arithmetic

We will take the following axioms (N1) – (N6):

(N1) $(\overline{0} \neq x_1')$

(N2) $(x_1' = x_2' \rightarrow x_1 = x_2)$

(N3) $(x_1 + \overline{0} = x_1)$

(N4) $(x_1 + x_2' = (x_1 + x_2)')$

(N5) $(x_1 \times \overline{0} = \overline{0})$

(N6) $(x_1 \times x_2' = x_1 \times x_2 + x_1)$

together with the axiom *scheme*

(N7)   if $\varphi(x_i)$ is a wf of $\mathcal{L}_{\mathrm{PA}}$, then

$$(\varphi(\overline{0}) \rightarrow ((\forall x_i)(\varphi(x_i) \rightarrow \varphi(x_i')) \rightarrow (\forall x_i)\varphi(x_i)))$$

is an axiom of PA.

We say that (N7) is an *axiom scheme* since corresponding to the infinitely many possible choices of φ there are many different instances of (N7). Actually, the logical axioms were schemes too.

The axioms (N1) – (N6) are based on very elementary facts from arithmetic. Notice that the axioms (N3) – (N6) are just formal versions of the recursive definitions of $+$ and $\times$ which we saw in Chapter 2. Axiom (N7) is called the *induction axiom*, for obvious reasons. In Peano's original formulation, induction appeared as just one axiom — but this needed *second order* quantification over *sets* of numbers, which brings us up against the problem that there is *no* Gödel's Completeness Theorem for second order logic. If we want to axiomatise arithmetic, we do need to use a first order language.

The rules of deduction

We will take the standard rules of first order logic. Writing $\vdash$ in the usual way for "proves", these are:

Modus Ponens $\vdash$ (MP):

$$\varphi, \ (\varphi \Rightarrow \psi) \vdash \psi, \quad \text{and}$$

Generalisation (GEN):

$$\varphi \vdash (\forall x_i)\varphi \qquad \text{(for each } i \geq 0).$$

Of course, all axioms of PA are *theorems*, as are all lines of *proofs*. We inductively define a *proof* in PA to be a finite list of wfs of $\mathcal{L}_{PA}$, each one on the list either being a theorem of PA we already have a proof of or being derived from previous wfs on the list via (MP) or (GEN). Gödel's Completeness Theorem allows us to include any universally valid wf in a proof, since it tells us it is provable in PA.

We write $\vdash_{PA} \varphi$ for "φ is a theorem of PA". And if Γ is a set of wfs of $\mathcal{L}_{PA}$, we write $\Gamma \vdash_{PA} \varphi$ — "Γ proves φ in PA" — if φ can be proved allowing the wfs of Γ as added axioms.

When there is no ambiguity about which theory we are working in we just write $\vdash$ in place of $\vdash_{PA}$.

What do proofs in PA look like?

If we wanted to prove a lot of theorems about numbers, efficiently, we would augment the basic rules of deduction with a range of derivable rules. In modern logic these are often set out in a "sequent calculus", in which the rules are dynamically emphasised in place of the logical axioms. Setting out Modus Ponens as a sequent rule we would have:

$$\frac{\begin{array}{cc} \Gamma & \varphi \Rightarrow \psi \\ \Gamma & \varphi \end{array}}{\begin{array}{cc} \Gamma & \psi \end{array}}$$

How can we prove something like $\bar{0} \neq \bar{7}'$? Or something very basic like $x_1 = x_1$, which we could have included as an axiom, but didn't? The following examples and exercises will give you the flavour of proofs in PA.

EXAMPLE 3.1.7 *Let t be any term of PA. Show that*

$$\vdash_{PA} \ t = x_2 \Rightarrow (t = x_3 \Rightarrow x_2 = x_3).$$

SOLUTION   Here is a proof in PA. We annotate the lines to help the reader keep track:

1. $x_1 = x_2 \Rightarrow (x_1 = x_3 \Rightarrow x_2 = x_3)$ (axiom (E1))

2. $\forall x_1 (x_1 = x_2 \Rightarrow (x_1 = x_3 \Rightarrow x_2 = x_3))$ (GEN applied to 1)

3. $\forall x_1 (x_1 = x_2 \Rightarrow (x_1 = x_3 \Rightarrow x_2 = x_3))$
$\Rightarrow (t = x_2 \Rightarrow (t = x_3 \Rightarrow x_2 = x_3))$ (logically valid)

4. $t = x_2 \Rightarrow (t = x_3 \Rightarrow x_2 = x_3)$ (MP applied to 2,3)

Of course, line 3 means this is not a *full* proof in PA — any more than subsequent "proofs" quoting this theorem will be.   ∎

EXERCISE 3.1.8   *Let t_1, t_2, t_3 be terms of* PA. *Show that*

$$\vdash_{PA} t_1 = t_2 \Rightarrow (t_1 = t_3 \Rightarrow t_2 = t_3).$$

REMARK 3.1.9   We can easily see from Exercise 3.1.8 that the single axioms (E1), (E2) and (N1) – (N6) can all be replaced with corresponding schemes (E1)′, (E2)′ and (N1)′ – (N6)′ in which the variables x_1, x_2, x_3 are replaced by arbitrary terms t_1, t_2, t_3.   ∎

EXERCISE 3.1.10   *Show that the following formulas are theorems of* PA:

$$(E2)' \quad t_1 = t_2 \Rightarrow t_1' = t_2'$$
$$(N1)' \quad 0 \neq (t_1)'$$

where t_1, t_2 are terms of PA.

EXAMPLE 3.1.11   *Show that $\vdash_{PA} t = t$ for any term t of* PA.

SOLUTION   This a proof of $(x_1 = x_1)$ in PA:

1. $x_1 + \bar{0} = x_1$ (N3)

2. $x_1 + \bar{0} = x_1 \Rightarrow (x_1 + \bar{0} = x_1 \Rightarrow x_1 = x_1)$
  (this is just (E1)′ with $t_1 = x_1 + \bar{0}$, $t_2 = x_1$ and $t_3 = x_1$)

3. $x_1 + \bar{0} = x_1 \Rightarrow x_1 = x_1$ (MP applied to 1,2)

4. $x_1 = x_1$ (MP applied to 1,3)

We can now derive $(t = t)$ just as we derived (N1)$'$, etc. ⬜

In particular, Exercise 3.1.11 tells us that if $m = n$ then $\vdash_{\text{PA}} (\bar{m} = \bar{n})$. We also have a converse.

EXERCISE 3.1.12 *Show that if $m \neq n$ then $\vdash_{\text{PA}} \neg(\bar{m} = \bar{n})$.*

[**Hint:** Assuming $m < n = m + k$, say, repeatedly apply (N2)$'$ with logically valid wfs to show $\vdash_{\text{PA}} (\bar{m} = \bar{n}) \Rightarrow (\bar{0} = \bar{k})$. Then use an instance of (N1)$'$ and a logically valid wf to get $\vdash_{\text{PA}} \neg(\bar{m} = \bar{n})$.]

What do models of PA look like?

We have designed PA so that $\mathfrak{N}$ is a model — all the axioms can be interpreted in $\mathfrak{N}$ and are true in that structure. We call $\mathfrak{N}$ the *standard* model of PA. We must now ask to what extent the standard model is captured by our theory. The best outcome would be for $\mathfrak{N}$ to be the *only* model of PA. But this is too much to expect of a first order theory.

Peano's original axioms did have the one model $\mathfrak{N}$, but used second order language to make sure that the domain of any model consisted of natural numbers — that is, interpretations of numerals $\bar{m}$ — only. It is the induction axiom (N7) that has been crucially weakened in the first order version of Peano's axioms. So why on earth have we done that?

The answer is that we *cannot axiomatise* second order logic — there is no Gödel's Completeness Theorem for the basic logic on which second order arithmetic must be based. So axiomatic theories are generally first order. And we modify our expectations of such theories. We ask for *complete* theories, in the sense that they successfully prove all the *first order* statements about their intended models. Specifically, we want PA to be complete in that for every sentence φ of PA, either $\vdash \varphi$ or $\vdash \neg\varphi$. We will see later that even such modest hopes must be dashed.

But first, let us briefly look at the *nonstandard* models of PA. There are some general model-theoretic results which can be used to reveal models for consistent theories. We will list the three most useful such theorems.

THEOREM 3.1.13 (Gödel's Completeness Theorem)
Any consistent first order theory has a model.

This statement of the theorem does give us the version we mentioned earlier — namely, that every logically valid wf is a theorem of PC. To see this, assume φ to be a wf which is not provable in PC. Then PC $\cup \{\neg\varphi\}$ ($=$ the theory with just one special axiom $\neg\varphi$) is consistent, and so has a model $\mathfrak{M}$, say. And since $\mathfrak{M} \not\models \varphi$, φ is not logically valid.

Our second model theoretic result is a powerful tool for finding new models.

THEOREM 3.1.14 (The Compactness Theorem)
Assume that every finite subset of a given set Γ of wfs has a model. Then there exists a model of the whole set Γ.

The third such result gives us small infinite models for certain theories.

THEOREM 3.1.15 (The Löwenheim–Skolem Theorem)
If first order theory T has an infinite model then it has a model which is countable.

Once again, you will need to go to a basic logic text for proofs of these theorems.

REMARK 3.1.16 When we say that a model of a theory is countable, we mean that its domain is countable.

You will probably remember from basic set theory that a set A is *countable* if its members can be listed: $A = \{a_0, a_1, a_2, \ldots\}$, where we allow repetitions in the list. Equivalently, A is countable if $A = \text{range}(f)$, some $f : \mathbb{N} \overset{\text{onto}}{\rightarrow} A$. Cantor's Theorem tells us that *uncountable* sets commonly occur in everyday mathematics — for example $\mathbb{R}$, the set of all real numbers, is not countable.

Cantor introduced infinite numbers — or *cardinals* — and showed how these extended the finite number system, along with its arithmetic, in a very natural way. The first infinite number — the number of elements in any countably infinite set — he called $\aleph_0$. The cardinals are linearly ordered: $0, 1, 2, \ldots, \aleph_0, \aleph_1, \aleph_2, \ldots$ etc, but no one knows what $\aleph_1$ is exactly. The *Continuum Hypothesis* (CH) is the conjecture that $\aleph_1 = 2^{\aleph_0}$ = the number of real numbers — that is, the cardinality of the *continuum*.

Actually, Theorem 3.1.15 is often called the *Downward* Löwenheim–Skolem Theorem — since there is also an *Upward Löwenheim–Skolem Theorem* which says that if T has a model of infinite cardinality κ, then it has models of all cardinalities $\kappa' \geq \kappa$.

You will find plenty of excellent texts covering the technical side of set theory. I would recommend Joseph Warren Dauben's scholarly biography of Cantor — *Georg Cantor: His Mathematics and Philosophy of the Infinite* (Princeton University Press, 1990) — for fascinating historical background to how Cantor initiated controversy and a revolution in mathematical and philosophical thought. ⬜

EXERCISE 3.1.17 *Show that the set $\mathbb{Q}$ of all rational numbers is countable.*

EXERCISE 3.1.18 *Show that if $A \subseteq B$, and B is countable, then A must also be countable.*

EXAMPLE 3.1.19 *Prove Cantor's Theorem — that is, prove that the set of all real numbers is not countable.*

SOLUTION By Exercise 3.1.17, it will be enough to show that the set $[0,1]$ of real numbers between 0 and 1 is uncountable.

To get a contradiction, assume a list: $r_0, r_1, \dots$ of all such real numbers. We can assume each r_i given by an infinite decimal $r_i = 0 \cdot r_i(0) r_i(1) \dots$, say, where each $r_i(j) \in \{0, 1, \dots, 9\}$.

We can then write down a real $r = 0.r(0)r(1) \cdots \in [0,1]$ which is not on our list by defining $r(j) = 9 \dotdiv r_j(j)$ for each $j \in \mathbb{N}$. This, of course, is the contradiction we wanted.   ⬜

EXERCISE 3.1.20 *Show that the set $2^{\mathbb{N}}$ of all subsets of $\mathbb{N}$ is uncountable.*

We already have a countable model of PA — the standard model $\mathfrak{N}$. Do we have any nonstandard models? And if so, how many of these are countable? And what do they look like? The answers to these questions will give us a better idea of how closely first order theories such as PA describe familiar structures like $\mathfrak{N}$.

Exercise 3.1.12 tells us that all models of PA are infinite — they must all contain a "standard part" consisting of the infinite set of interpretations of the numerals $\bar{0}, \bar{1}, \bar{2}, \dots$. It is easy to show that there are models containing nonstandard elements, which are not interpretations of any numeral.

THEOREM 3.1.21
There exists a nonstandard model $\mathfrak{M}$ of PA.

PROOF We first add a new individual constant c to the language $\mathcal{L}_{\mathrm{PA}}$. We then define an extension PA^+ of PA by adding to PA some new special axioms saying that c cannot stand for a natural number:

$$c \neq \bar{0},\ c \neq \bar{1},\ c \neq \bar{2},\ \dots, c \neq \bar{n},\ \dots.$$

> **Claim.** There exists a model $\mathfrak{M}$ of PA$^+$.

PROOF (of claim)   We use the Compactness Theorem.

Let Γ be any finite subset of the axioms of PA$^+$. Choose $\bar{m}$ to be a numeral not appearing in Γ. Then every φ in Γ *either* is an axiom of PA, so true in $\mathfrak{N}$, *or* is of the form $c \neq \bar{n}$, where $m \neq n$ — in which case φ is true in $\mathfrak{N}$, so long as c is interpreted as m. With m so designated, $\mathfrak{N} \models \Gamma$.

But since any such Γ has a model, it follows by the Compactness Theorem that PA$^+$ itself has a model $\mathfrak{M}$, say.   ⬜

But although $\mathfrak{M}$ is also a model of just PA, it cannot be isomorphic to $\mathfrak{N}$. This is because the set of new axioms in their entirety force PA to provide an interpretation for c which cannot be an interpretation of a numeral.   ⬜

So how different to $\mathfrak{N}$ will such a nonstandard model $\mathfrak{M}$ be? Many properties of $\mathfrak{N}$ are guaranteed by PA. For instance, if we define the ordering $\leq$ as described above, then it determines a usual *linear* ordering of $\mathfrak{M}$. ($\leq$ is *linear* if for any a, b in $\mathfrak{M}$ we have either $a \leq b$ or $b \leq a$.) This means that every nonstandard element is *greater* than all the standard ones. And we can verify that every element, except the interpretation of $\bar{0}$, is a successor of another element. We can also use the Löwenheim–Skolem Theorem to provide a *countable* nonstandard model.

At first sight these nonstandard models all look the same — but very big in relation to $\mathfrak{N}$.

REMARK 3.1.22   When we say two models look the same, we really mean that they are *isomorphic*. An *isomorphism* between two structures $\mathfrak{A}$ and $\mathfrak{B}$ is a bijective function $f : \mathfrak{A} \to \mathfrak{B}$ that preserves structure.   ⬜

We first notice that if a model $\mathfrak{M}$ of PA contains one nonstandard element, say γ, then it also contains $\gamma \pm 1, \gamma \pm 2, \ldots$. So one can view the domain of $\mathfrak{M}$ as being isomorphic to an initial ordering $\omega = 0 < 1 < 2 < \ldots$, succeeded by instances of $\omega^* + \omega = \ldots, -2 < -1 < 0 < 1 < 2 < \ldots$. The usual rules of arithmetic, captured by PA, ensure that there is no greatest $\omega^* + \omega$ in the ordering — if γ is a nonstandard element of $\mathfrak{M}$ then 2γ cannot be in the same $\omega^* + \omega$ as γ. Moreover, the $\omega^* + \omega$ instances are *densely* ordered — given γ, λ occupying different copies of $\omega^* + \omega$, $\lfloor \frac{\gamma + \lambda}{2} \rfloor$ will give rise to a new $\omega^* + \omega$ strictly between the original ones.

This is enough to pin down the order-type of any nonstandard model of arithmetic. They all have the same initial ω followed by a dense linear ordering, without greatest or least element, of copies of $\omega^* + \omega$. So the countable nonstandard models all have the *same* order-types. But $+$ and $\times$ may look very different from model to model, due to the abundance of different isomorphic initial segments of the ordering. In fact, there is a theorem of Harvey Friedman which tells us that any countable nonstandard model of PA has lots of proper initial segments isomorphic to it. There are, it turns out, $2^{\aleph_0}$ non-isomorphic countable nonstandard models of arithmetic — that is, as many as there are real numbers.

3.2   What Functions Can We Describe in a Theory?

Theorem 3.2.10 below is what this chapter is all about. It will give yet another, very different, characterisation of the computable functions. The Church–Turing Thesis already gives us the recursive model. Theorem 3.2.10 tells us that every computable function can be fully described in first order arithmetic. We will have to wait until later to see a proper proof that these are the *only* total functions so describable. But you should already have a strong intuition that being able to prove $f(n) = m$ in PA gives us an algorithm for getting the value $f(n)$.

But before we can prove our theorem, we need to make precise what we mean by "describing" a function in a theory. At the same time we will do this for relations and sets.

Remember — the *graph* of a k-ary function f is the relation $\text{graph}(\overrightarrow{m}, n)$ which holds just when $f(\overrightarrow{m}) = n$.

DEFINITION 3.2.1   *(1) We say a k-place relation R is **representable** in PA if there is a wf φ of $\mathcal{L}_{\text{PA}}$ for which*

$$R(\overrightarrow{m}) \text{ holds } \implies \vdash_{\text{PA}} \varphi(\overrightarrow{m}), \quad \text{and} \qquad (*)$$
$$R(\overrightarrow{m}) \text{ is false } \implies \vdash_{\text{PA}} \neg\varphi(\overrightarrow{m}). \qquad (**)$$

*(2) We say a k-place function f is **representable** in PA if its graph is representable in PA.*

*(3) We say a set S is **representable** in PA if the relation "$m \in S$" is representable in PA.*

EXAMPLE 3.2.2   *Show that $=$ is representable in PA.*

SOLUTION   Take $\varphi(x_0, x_1)$ to be the wf $(x_0 = x_1)$. We need to verify both parts $(*)$, $(**)$ of Definition 3.2.1.

$(*)$ If $m_0 = m_1$ then $\bar{m}_0 = \bar{m}_1$ (as terms of PA).

So by Example 3.1.11 we have $\vdash_{\text{PA}} (\bar{m}_0 = \bar{m}_1)$ — that is, $\vdash_{\text{PA}} \varphi(\bar{m}_0, \bar{m}_1)$.

$(**)$ On the other hand, if $m_0 \neq m_1$, then $\vdash_{\text{PA}} \neg(\bar{m}_0 = \bar{m}_1)$ by Exercise 3.1.12. So $\vdash_{\text{PA}} \neg\varphi(\bar{m}_0, \bar{m}_1)$, as required. ⬚

Just as the logical connectives lead from computable relations to more computable relations, it turns out that the representable relations are closed under the use of logical connectives. For example:

EXERCISE 3.2.3 *Show that if the relation $P(\overrightarrow{m})$ is representable in* PA, *then so is $\neg P(\overrightarrow{m})$.*

EXERCISE 3.2.4 *Show that if $P(\overrightarrow{m})$ and $Q(\overrightarrow{m})$ are representable in* PA, *then so are (i) $P(\overrightarrow{m})\&Q(\overrightarrow{m})$, and (ii) $P(\overrightarrow{m}) \vee Q(\overrightarrow{m})$.*

For sets we have:

EXERCISE 3.2.5 *Show that if S and T are representable sets of numbers, then so are (i) $S \cap T$, (ii) $S \cup T$, and (iii) $\overline{S} = \mathbb{N} - S$.*

EXERCISE 3.2.6 *Show that a set S is representable if and only if its characteristic function χ_S is representable in* PA.

The next example provides the first step towards showing all the recursive functions — and hence all the computable functions — are representable.

EXAMPLE 3.2.7 *Show that the zero function $\mathbf{0}$ is representable in* PA.

SOLUTION Take $\varphi(x_0, x_1)$ to be the wf $(x_0 = \bar{0})$, and show that it represents the graph of $\mathbf{0}$.

$(*)$ If $\mathbf{0}(m_0) = m_1$, we have $m_1 = 0$. So $\vdash_{\text{PA}} \bar{m}_0 = \bar{0}$ — that is $\vdash_{\text{PA}} \varphi(\bar{m}_0, \bar{m}_1)$.

$(**)$ And if $\mathbf{0}(m_0) \neq m_1$, then $m_1 \neq 0$. So $\vdash_{\text{PA}} \neg(\bar{m}_0 = \bar{0})$ — which is $\vdash_{\text{PA}} \neg\varphi(\bar{m}_0, \bar{m}_1)$. ⬚

It is not too difficult to show that the other initial recursive functions are representable too.

EXERCISE 3.2.8   *Show that the successor function m' is representable.*

EXERCISE 3.2.9   *Show that the projection functions U_i^n, $1 \le i \le n$, are representable in* PA.

[**Hint:** Take $\varphi(x_0, x_1, \ldots, x_n)$ to be the wf $(x_n = x_{i-1})$.]

THEOREM 3.2.10
  All recursive functions are representable in PA.

PROOF   This is by induction on the number of applications of rules (d) (substitution), (e) (primitive recursion) and (f) (minimalisation).

This is not the place to go into detail about this induction — you are likely to find that even your recommended logic text is reticent about certain technical aspects — but I will give enough to give you a feel for the structure of the induction, and some confidence that with enough time and application you could fill in the rest. A good reference for the gory details is Elliott Mendelson's *Introduction to Mathematical Logic*, any edition.

We already have the representability of the initial functions from Example 3.2.7 (the zero function), and Exercises 3.2.8 and 3.2.9 (the successor and projection functions).

We need to show that rules (d), (e) and (f) lead from representable functions to more representable functions.

I will give just part of the verification for rule (d), substitution, in a simplified form:

Assume that f is defined by $f(m) = h(g(m))$, where

$$h \text{ is represented by } \varphi(x_0, x_1), \quad \text{and}$$
$$g \text{ is represented by } \psi(x_0, x_1).$$

We show that θ represents f where

$$\theta(x_0, x_1) =_{\text{defn}} \exists z \, [\psi(x_0, z) \& \varphi(z, x_1)].$$

Part (*) of the definition of "represents":   Say $f(m) = n$.
  Then there exists a p such that $g(m) = p$ and $h(p) = n$. For this p

$$\vdash_{\text{PA}} \psi(\bar{m}, \bar{p}) \quad \text{and} \quad \vdash_{\text{PA}} \varphi(\bar{p}, \bar{n}),$$

since ψ, φ represent g, h.

We can then get $\vdash_{\text{PA}} \psi(\bar{m}, \bar{p}) \& \varphi(\bar{p}, \bar{n})$, using an instance of the logically valid wf $(\mathcal{A} \Rightarrow (\mathcal{B} \Rightarrow \mathcal{A}\&\mathcal{B}))$, and modus ponens twice.

Since $\chi(t) \Rightarrow \exists x\, \chi(x)$ is logically valid, so provable in PA for any wf χ and term t, we get

$$\vdash_{\mathrm{PA}} \psi(\bar{m}, \bar{p}) \& \varphi(\bar{p}, \bar{n}) \Rightarrow \exists z\, [\psi(\bar{m}, z) \& \varphi(z, \bar{n})].$$

So by MP, $\vdash_{\mathrm{PA}} \exists z\, (\psi(\bar{m}, z) \& \varphi(z, \bar{n}))$ — that is, $\vdash_{\mathrm{PA}} \theta(\bar{m}, \bar{n})$.

The verification of part (**) of the definition of represents is left as an exercise for the reader. ▢

REMARK 3.2.11 Actually, things are not quite as simple as I have pretended above. Sometimes one needs to reveal more of the inductive structure in a proof than is actually wanted for the result. In this case the proof of Theorem 3.2.10 usually uses a stronger definition of representability of functions, along the following lines:

DEFINITION 3.2.12 $\varphi(x_0, \ldots, x_k)$ **functionally represents** *a k-ary function $f(\overline{m})$ in* PA *if for all $\overline{m}$*

(i) $f(\overline{m}) = n$ *implies that* $\vdash_{\mathrm{PA}} \varphi(\overline{m}, \bar{n})$, *and*

(ii) $\vdash_{\mathrm{PA}} \varphi(\overline{m}, x_k) \Rightarrow (x_k = \overline{f(\overline{m})}$.

Representability and functional representability are in fact equivalent. It is the fact that a wf φ may represent f in PA, but fail to functionally represent φ, that gives Definition 3.2.12 a role in the above proof. ▢

EXERCISE 3.2.13 *Show that if φ functionally represents f in* PA, *then φ also represents f in φ.*

EXERCISE 3.2.14 *Let the k-ary function f be represented in* PA *by a wf θ. Show that there exists a wf φ which functionally represents f.*
[**Hint:** Take φ to be the wf $\theta(\overline{x}, x_k) \& \forall z (z \neq x_k \Rightarrow \neg\theta(\overline{x}, z)).$]

It is now easy to get a similar result to Theorem 3.2.10 for sets and relations.

COROLLARY 3.2.15
All recursive sets and relations are representable in PA.

PROOF Let S be a recursive set, so that by definition χ_S is recursive. Then χ_S is representable in PA by Theorem 3.2.10. And hence S is representable by Exercise 3.2.6.

The proof for relations is similar.   ⬚

Finally, by Church's Thesis, we have the result we aimed for in this chapter:

COROLLARY 3.2.16

Every computable function, relation or set over ℕ *is representable in* PA.

And as we said before, we will soon be able to turn this into another *characterisation* of the computable functions. The computable functions turn out to be *exactly* those functions we can fully describe in a first order theory.

Another important role for Theorem 3.2.10 is in proving Gödel's Incompleteness Theorem. But before going on to that, we will need to look at coding techniques. And after seeing some more concepts and results from computability, we will be ready to view Gödel's Theorem as just one facet of a basic computability theoretic phenomenon.

Chapter 4

Coding, Self-Reference and the Universal Turing Machine

In the twentieth century, science had to face up in earnest to the problematic relationship between the local and the global.

In mathematics this surfaced in various guises. At the turn of the century there was the foundational crisis coming out of unregulated self-reference — most famously via Russell's Paradox in the newly developing set theory. By the 1930s Hilbert's formalism set the stage for self-reference on a more secure but limited basis — using Gödel numberings and variants of that technique — with no less startling results. In the physical sciences quantum theory threw up the Einstein–Rosen–Podolsky thought experiment, radically challenging people's preconceptions about causal locality. While by the second half of the century, chaos theory was telling us how the most everyday phenomena — even a dripping tap — could present us with practical incomputability, framed by strangely emergent *patterns* of events.

This chapter will show you how coding can turn a Turing machine into the input of a Turing computation. Most importantly, I will describe how to build Turing's *universal* machine, whose consequences go far beyond the early aims of its inventor. Later, in Part II, I will return to the wider scientific context, and describe a model of computationally complex environments based on Turing machines which live in the real world.

4.1 Russell's Paradox

Using Frege's logic and Cantor's set theory, the philosopher Bertrand Russell (1872–1970) described a devastatingly simple example of a statement which could be neither true nor false.

> **RUSSELL'S PARADOX (1901)** Let S be the set of all sets which do not belong to themselves — that is, $S = \{X \mid X \notin X\}$. Is S a member of itself — or in symbols, is $S \in S$?

Clearly, if we answer "Yes" to the question, we have a contradiction, because then S does not satisfy its own defining property. But if we answer "No", then S *does* satisfy the condition for it to be in S! We have a model of truth which fails to define truth in accordance with the model. It looks like the model is not consistent.

The most noticeably unusual feature of Russell's example is the use of *self reference* — in this case defining a set in such a way that it could conceivably belong to itself. In a post-Freudian world it is hard for us to accept that self reference in itself is disallowed. What Gödel did was to describe how a theory can prove things about itself in a quite legitimate way, by the use of suitable *codings*. What Gödel had not at first expected was that the paradox would re-emerge (in his Incompleteness Theorem), but in the form of a limitation on provability rather than on truth. This time there is no paradox, since we have fewer preconceptions about provability than we do about truth.

Our first application of Gödel's coding technique will be to Turing machines.

4.2 Gödel Numberings

We will now use a Gödel-style coding to get a computable list of Turing programs. This will mean defining a function gn from Turing programs to $\mathbb{N}$, in such a way that with a little adjustment we can get a computable inverse gn^{-1}. We will then be able to get Turing machines which can call up other Turing programs, much as a modern computer does, by computing with Gödel numbers.

Our coding is very crude, and we care more about transparency than efficiency. The computability of the inverse gn^{-1} depends on our being able to uniquely factorise numbers into their prime divisors.

Gödel numbers for Turing machines

We start off by coding the tape symbols, action symbols, and internal states. Let

$$gn(L) = 2 \quad \text{(the } 0^{\text{th}} \text{ prime number } p_0\text{)}$$
$$gn(R) = 3$$
$$gn(q_i) = p_{2+2i} \quad \text{(the } (2 + 2i)^{\text{th}} \text{ prime number)}$$
$$gn(S_i) = p_{2+2i+1}.$$

We can then code the quadruples. If $Q = q_i S A q_j$, let

$$gn(Q) = 2^{gn(q_i)} \times 3^{gn(S)} \times 5^{gn(A)} \times 7^{gn(q_j)}$$

And finally, if $P = \{Q_0, Q_1, \ldots, Q_k\}$ is a Turing program, let

$$gn(P) = 2^{gn(Q_0)} \times 3^{gn(Q_1)} \times \cdots \times p_k^{gn(Q_k)}$$

be a Gödel number for P. The value gn(P) obviously depends on the order in which its quadruples are listed, so a program P may have more than one Gödel number. This does not matter, and nor does the fact that not every number is a Gödel number.

What *is* important for the following definition is that we can computably *tell* what a number n Gödel numbers, if anything. To compute $gn^{-1}(n)$ one just needs to completely factorise n, including all exponents appearing, and to read off what, if anything, is Gödel numbered, according to the above recipe.

DEFINITION 4.2.1 *(1) The* **eth** **Turing machine** P_e *is defined by*

$$P_e = \begin{cases} P & \text{if } gn^{-1}(e) \downarrow = \text{some Turing program } P \\ \emptyset \text{ (the empty program)} & \text{otherwise.} \end{cases}$$

We say P_e *has* **index** e.
(2) We can now simplify our earlier notation $\varphi_T^{(k)}$ *via an index of T:*

$$\varphi_e^{(k)} = \text{ the } k\text{-place partial function computed by } P_e.$$

We call $\varphi_e = \varphi_e^{(1)}$ *the* **eth** **partial computable (p.c.) function**.

REMARK 4.2.2 So every number e is the index of *some* Turing machine T, and given e we can effectively find T. □

Adding just a little more to what we know about the list $\varphi_0, \varphi_1, \ldots$, we get:

THEOREM 4.2.3 (The Enumeration Theorem)
$\varphi_z(x)$ *is a p.c. function of x and z, such that for each e, φ_e is the eth (unary) p.c. function.*

PROOF We only need to describe how to compute $\varphi_z(x)$:

1. Find P_z.

2. Give P_z input x and wait for the computation to terminate.

3. In which case, set the output $= \varphi_z(x)$.

The rest of the theorem follows from Definition 4.2.1.   ⬜

It would be much nicer if we could get a similarly well-behaved list of all *total* computable functions. Unfortunately, not ...

EXERCISE 4.2.4 *Show that there is no list* $\{f_e\}_{e \geq 0}$ *of all (total) computable functions for which* $f_z(x)$ *is a computable function of* x *and* z.

[**Hint:** Consider the function $h : x \mapsto f_x(x) + 1$.]

It is a short step from this to the existence of quite simple incomputable objects, as we shall see in the next chapter. For instance, it will be easy to see that we cannot computably decide whether an arbitrary Turing computation ever terminates or not.

We often need to obtain indices of p.c. functions, sometimes with particular properties, sometimes by modifying given indices or Turing machines. Here is a very simple, and aptly named example.

EXERCISE 4.2.5 (The Padding Lemma)   *Show that if* f *is a given p.c. function, there exist infinitely many indices* i *for* f *(that is, numbers* i *for which* $f = \varphi_i$*).*

Often in the background is a practical version of the Church–Turing Thesis. This tells us not just that we can computably get the index of any intuitively acceptable algorithm. It says that if we are computably given a list of algorithms, then we can compute the corresponding list of indices.

For example, say we have a p.c. function $f(x, y) = \varphi_e^{(2)}(x, y)$. For any fixed x, the program P_e immediately gives an algorithm for computing $f(x, y)$ as a function of y, and its index $g(x)$ we can compute from e and x.

It is customary in text books to dignify this observation as a theorem (originally due to Stephen Kleene):

THEOREM 4.2.6 (The s-m-n Theorem)
If $f(x, y)$ *is p.c., there is a computable* g *for which* $f(x, y) = \varphi_{g(x)}(y)$.

In applying this and its variants later on, we will rely on your assimilation of the intuition. Even though I will try not to mention the s-m-n Theorem again, I am sure you would like to know the origins of its name — in Kleene's classic *Introduction to Metamathematics*, my $g(x)$ appears in more general form as $S_n^m(z, y_1, \ldots, y_m)$.

EXERCISE 4.2.7 *Let* $m, n \geq 1$. *Show that there exists a computable function* S_n^m *such that for all* $x, y_1, \ldots, y_m, z_1, \ldots, z_n$

$$\varphi_{S_n^m(x, y_1, \ldots, y_m)}^{(n)}(z_1, \ldots, z_n) = \varphi_x^{(n+m)}(y_1, \ldots, y_m, z_1, \ldots, z_n).$$

But let us get back to more important matters.

4.3 A Universal Turing Machine

What was so special about Turing's 1936 paper was not so much the invention of the Turing machine — Emil Post had earlier devised a good machine-based model of computability — but the description of a universal machine that could perform any computational task without special hardware. As well as playing a role in the development of the stored-program computer, it led on to a deeper understanding of what could and could not be computed.

THEOREM 4.3.1 (Turing, 1936)
There exists a Turing machine U — *the* **Universal Turing Machine** — *which if given input* (e, x) *simulates the* e^{th} *Turing machine with input* x.
That is, $\varphi_U^{(2)}(e, x) = \varphi_e(x)$.

PROOF Using Turing's Thesis, choose U which computes $\varphi_z(x)$. ∎

REMARK 4.3.2 I would love to tell here the full story behind the Universal Turing Machine.

One would have to include in it Charles Babbage and his struggle to build his analytic engine, which would — if it had not hopelessly over-stretched nineteenth century engineering capabilities and Babbage's savings — have been a true computing machine, complete with cogs. His faithful follower Ada, Countess of Lovelace, is credited with the first computer program, which she wrote to enable the analytic engine to compute Bernoulli numbers. There is even a computer programming language which was named "Ada" after her by the U.S. Department of Defense. Babbage died an embittered and disappointed man, and his work was pretty much forgotten until quite recently. Despite Turing's awareness of Babbage, his universal machine had a different genesis.

If you want to know more, I very much recommend Doron Swade's *The Cogwheel Brain: Charles Babbage and the Quest to Build the First Computer,* and Benjamin Woolley's *The Bride of Science: Romance, Reason, and Byron's Daughter.*

I have already mentioned Martin Davis' *The Universal Computer: The Road from Leibniz to Turing*, and this is very good on the role Turing's work played in the history of the computer. And for the technically inclined, there is a whole book of articles on *The Universal Turing Machine: a half century survey* from 1988, edited by Rolf Herken.   ⬜

4.4   The Fixed Point Theorem

While we are talking about indices for computable functions, here is another remarkable result. Its statement and proof look simple, but it is not so easy to penetrate their underlying meaning. We will see later what a useful — if slightly mysterious — tool it provides.

THEOREM 4.4.1 (The Kleene Fixed Point Theorem)
If f is a computable function, there exists a $k \in \mathbb{N}$ for which

$$\varphi_{f(k)} = \varphi_k.$$

PROOF We start by using the practical version of the Church–Turing Thesis to get a computable h such that

$$\varphi_{h(x)} = \varphi_{\varphi_x(x)}. \qquad (4.1)$$

[For each x, $h(x)$ is the Gödel number of the Turing machine which if given input z, first seeks to compute $\varphi_x(x)$ — and then if $\varphi_x(x) \downarrow$ computes $\varphi_{\varphi_x(x)}(z)$.]

Then $f \circ h$ has an index — say $f(h(x)) = \varphi_e(x)$, each x. Finally, taking $x = e$ in $\varphi_{f(h(x))} = \varphi_{\varphi_e(x)}$, we get

$$\varphi_{f(h(e))} = \varphi_{\varphi_e(e)} = \varphi_{h(e)}$$

from 4.1. So we can take our fixed point to be $k = h(e)$.   ⬜

There are all sorts of variations and generalizations of this result — sometimes also called the *Recursion Theorem*. Here is one we will find useful later. We can get the proof by making the obvious modifications in the one above.

EXERCISE 4.4.2 (The Fixed Point Theorem with Parameters)   *Show that if $f(x, \overrightarrow{y})$ is a computable function, there exists a computable $k(\overrightarrow{y})$ such that $\varphi_{f(k(\overrightarrow{y}), \overrightarrow{y})} = \varphi_{k(\overrightarrow{y})}$.*

4.5 Computable Approximations

The list $\{\varphi_i\}_{i \in \mathbb{N}}$ of all p.c. functions has nice properties given by the Enumeration Theorem. There are other computable features we would like the list to have. For instance, it would be very useful to have an algorithm for testing whether $\varphi_e(x) = y$ or not.

QUESTION: Is $\varphi_e(x) \downarrow$ or $\varphi_e(x) = y$ a **computable** *relation (of e, x, y) ?*

Even though we cannot answer this important question for the moment, we do have:

EXERCISE 4.5.1 *Show that if $\varphi_e(x) \downarrow$ is a computable relation, then so is $\varphi_e(x) = y$.*

And we can define closely related *computable approximations* to the above relations:

DEFINITION 4.5.2 *We write*

$$\varphi_{e,s} = y \Leftrightarrow_{\text{defn}} x, y, e < s \text{ and } y \text{ is the output of } \varphi_e(x) \text{ in } < s$$
$$\text{computational steps of } P_e.$$

The bound on the number of computational steps is the one which ensures computability by truncating infinite searches. The bound $x, y, e < s$ is just to give — this will be convenient later on — only finitely many e, x, y for a given s for which $\varphi_{e,s} = y$ holds.

The important facts to check are:

COROLLARY 4.5.3
(a) $\varphi_{e,s}(x) \downarrow$, $\varphi_{e,s}(x) = y$ are both computable relations (of e, s, x, y).
(b) $\varphi_e(x) \downarrow \Leftrightarrow (\exists s)\varphi_{e,s}(x) \downarrow$ and $\varphi_e(x) = y \Leftrightarrow (\exists s)\varphi_{e,s}(x) = y$.

PROOF I will just do $\varphi_{e,s}(x) = y$.

For (a): To computably verify if $\varphi_{e,s}(x) = y$, first check $x, y, e < s$ — then take P_e, give it input x, and follow through the resulting computation to see if we get output y *within s steps* of the computation.

(b) follows straight from Definition 4.5.2.   ⬜

EXERCISE 4.5.4   *Verify Corollary 4.5.3 for $\varphi_{e,s}(x) \downarrow$.*

We will see later that $\varphi_e(x) \downarrow$ is *not* a computable relation. But first let us put this incomputability phenomenon in context.

Chapter 5

Enumerability and Computability

We now have some good models of computability, and they all turn out to be equivalent. These models enable us to *talk about* computability with some confidence that what we are saying can be rigorously justified. Many natural examples of everyday sets, functions and relations can be easily fitted into these models. But not *all* such examples — as we come to suspect in what follows. Before this chapter is finished we will have found that mathematics, at least, is riddled with easily described, but incomputable, objects.

More and more we will rely on arguments based on the Church–Turing Thesis. As our confidence increases, we will increasingly dispense with the safety net provided by *explicit* reference to our models of computation.

5.1 Basic Notions

Let us look at some examples of naturally arising sets.

EXAMPLE 5.1.1 *Let $S = \{x \in \mathbb{N} \mid x \text{ is a prime number}\}$. Discuss the computability of S.*

SOLUTION In this case S is easily seen to be computable. Appealing to the Church–Turing Thesis, we could just observe that the well-known "sieve" of Eratosthenes of Cyrene, from the third century B.C., provides a simple algorithm for identifying prime numbers. Or, more formally, we could recall our earlier proof of the primitive recursiveness of the relation $\Pr(n)$. ▯

Here is something a little more subtle:

EXAMPLE 5.1.2 Let $S = \{x \in \mathbb{N} \mid \exists$ a sequence of at least x 7's in the decimal expansion for $\pi\}$. Is S computable?

SOLUTION   Again the answer is "yes". The argument falls into two cases:
Case 1: We have $S = \{0, 1, \ldots, n\}$ for some $n \geq 0$. In which case we can either notice that its characteristic function is given primitive recursively by $\chi_S(x) = \overline{\mathrm{sg}}(x \dot{-} n)$, or we can use the fact, seen earlier, that any finite set is computable. Or —

Case 2: We have $S = \mathbb{N}$, an infinite computable set (with characteristic function the constant function 1). ⬚

Now let us make a small modification to the definition of S.

EXAMPLE 5.1.3 Let $S = \{x \in \mathbb{N} \mid \exists$ a sequence of exactly x 7's (and no more) in the decimal expansion for $\pi\}$. Discuss the computability of S.

SOLUTION   It is now not clear whether S is computable or not, without knowing more about the decimal expansion of π. But it is notoriously difficult to identify patterns in the decimal expansion of π, and we do not know any established facts about π which would help us here.

On the other hand, π is what Alan Turing called a *computable real* — we can effectively write down the decimal for π as far as we like, using for instance one of the well-known infinite series for π known to converge rapidly. This means that although we cannot necessarily tell whether a given x is in S, we *can* effectively *enumerate*, or *list*, all the members of S (although perhaps not in order of magnitude). The algorithm for enumerating S would be to examine progressively more of the decimal for π, and each time we come across a block of exactly x 7's, to enumerate x into S. ⬚

This leads us to a new notion:

DEFINITION 5.1.4 $A \subseteq \mathbb{N}$ *is* **computably enumerable** *(c.e.) if there is an effective process for enumerating all the members of A.*

Or more formally: A *is c.e.* $\iff$ $A = \emptyset$, *or there is a computable function f such that* $A = \{f(0), f(1), \ldots\} = \mathrm{range}(f)$.

Notice that the more formal definition does seem to capture the content of the intuitive one. It will come in useful in proving some of the more technical facts about c.e. sets.

The first question which comes to mind is:

> **How does this new notion of A being computably enumerable relate to A being computable?**

This was a question first considered by Emil Post — born in Poland on 11 February 1897, brought by his parents to the United States when he was seven, and whose mathematical career was cut short by a heart-attack, seemingly brought on by electric shock treatment for a depressive mental illness, at the early age of fifty-seven. His first answer, in a seminal 1944 paper in the *Bulletin of the American Mathematical Society*, is not unexpected.

THEOREM 5.1.5

If $A \subseteq \mathbb{N}$ is computable, then A is also computably enumerable.

PROOF Say A is computable, so that we can effectively decide whether $x \in A$ for any given $x \in \mathbb{N}$.

Then we can effectively enumerate the members of A by asking, in turn, "Is $0 \in A$?", "Is $1 \in A$?", ..., and each time we get a "yes" to "Is $x \in A$?", enumerating x into A. □

EXERCISE 5.1.6 *Express the proof of Theorem 5.1.5 within the recursive model. That is, show that if A is recursive, then either $A = \emptyset$, or there is a recursive function f such that $A = \text{range}(f)$.*

We are not ready to say yet whether the converse of Theorem 5.1.5 holds. But, following Post, we can show that if A is c.e., then it is *very close* to being computable.

THEOREM 5.1.7 (The Complementation Theorem)

$A \subseteq \mathbb{N}$ is computable $\iff$ both A and $\overline{A}$ are computably enumerable.

PROOF ($\Rightarrow$) Say A, and hence $\overline{A}$, is computable. Then both A and $\overline{A}$ are c.e. by Theorem 5.1.5 above.

($\Leftarrow$) Assume that A and $\overline{A}$ are c.e. We describe an algorithm for effectively deciding whether a given $n \in A$ or not:

1. Set in motion effective enumerations of A and $\overline{A}$, say:

$$f(0), f(1), \cdots \in A \qquad \text{and} \qquad g(0), g(1), \cdots \in \overline{A}.$$

2. Search both enumerations until we find a number $y(n)$, say, such that $n = f(y(n))$ or $n = g(y(n))$. (Notice that $y(n)$ does exist, since $\mathbb{N} = A \cup \overline{A}$.)

Then $n \in A \iff n = f(y(n))$.   ▯

EXERCISE 5.1.8   *If the f in Definition 5.1.4 is recursive, then we say that A is **recursively enumerable**. Show, without using the Church–Turing Thesis, that a set A is recursive if and only if both A and $\overline{A}$ are recursively enumerable.*

The following example shows how we can combine the formal and informal methods:

EXAMPLE 5.1.9   *A function f is said to be **increasing** if $f(x) < f(x+1)$ for all $x \in \mathbb{N}$. Show that an infinite set $A \subseteq \mathbb{N}$ is computable if and only if it is the range of an increasing computable function.*

SOLUTION   ($\Rightarrow$) Let A be computable. Define a function f by

$$f(0) = \mu y\,[y \in A],$$
$$f(n+1) = \mu y\,[y \in A \& y > f(n)].$$

Then f is an increasing computable function, and $A = \text{range}(f)$.
  ($\Leftarrow$) Say we have $A = \text{range}(f)$, with f increasing and computable. Then

$$x \in A \iff (\exists y \le x)[f(y) = x],$$

showing A to be computable.   ▯

EXERCISE 5.1.10   *Show that every infinite c.e. set contains an infinite computable subset.*

EXERCISE 5.1.11   *Show that an infinite set is c.e. if and only if it is the range of a one–one computable function.*

We can show that the set of all c.e. sets (written $\mathcal{E}$) has certain *closure properties* with respect to the set theoretical operations $\cup$ and $\cap$. In Chapter 12 we will find that the resulting *lattice* structure has some very interesting properties.

EXAMPLE 5.1.12 (The Union Theorem) *Show that if A and B are c.e. sets of numbers, then so is $A \cup B$.*

SOLUTION Let A, B be c.e. We can assume that neither set is empty — since if, say, $A = \emptyset$, then $A \cup B = B$, and the result follows immediately.

Assume $A = \text{range}(f)$, $B = \text{range}(g)$, with f, g computable. Then

$$f(0), g(0), f(1), g(1), \ldots$$

gives an effective enumeration of $A \cup B$. ☐

EXERCISE 5.1.13 *Show that if A and B are c.e. sets, then so is $A \cap B$.*

5.2 The Normal Form Theorem

We find c.e. sets turning up in all sorts of contexts. In this section we trace some of the different guises they can appear in.

We first need some new notation and terminology. These are based on the following equivalence:

$$P_e \text{ halts on input } x \iff \varphi_e(x) \downarrow \iff x \in \text{dom}(\varphi_e).$$

The following definitions will be connected up with the notion of a set being c.e. in the Theorem 5.2.10 below.

DEFINITION 5.2.1 *(1) We write $W_e = \text{dom}(\varphi_e)$, for each $e \in \mathbb{N}$.*
*(2) We say that P_e has **halting set** W_e. Then the **halting problem** for the eth Turing machine is said to be **solvable** if W_e is a computable set.*

Intuitively, the halting problem for a Turing machine T is solvable if and only if we can tell effectively whether or not T halts on any given input x.

EXERCISE 5.2.2 (The Padding Lemma for C.E. Sets)   *Let T be a Turing machine with halting set A. Show that A has infinitely many indices (that is, numbers e such that $A = W_e$).*

EXERCISE 5.2.3 (The Fixed Point Theorem for C.E. Sets)   *Given any computable function $f : \mathbb{N} \to \mathbb{N}$, show that there exists a $k \in \mathbb{N}$ such that*

$$W_{f(k)} = W_k.$$

DEFINITION 5.2.4 (Level 1 of the Arithmetical Hierarchy)
     (1) If for all $x \in \mathbb{N}$ we have $x \in A \iff (\exists y)R(x, y)$, for some computable relation R, then we say that A is a Σ_1^0 set, and write $A \in \Sigma_1^0$.
     (2) If we have $x \in A \iff (\forall y)R(x, y)$, some computable R, then we say that A is a Π_1^0 set, and write $A \in \Pi_1^0$.
     (3) If $A \in \Sigma_1^0 \cap \Pi_1^0$, then we say that A is a Δ_1^0 set, and write $A \in \Delta_1^0$.

REMARK 5.2.5   This definition is part of a much more general definition of the *arithmetical hierarchy* (in Chapter 10), which classifies relations (in the language of first-order arithmetic) in terms of their quantifier forms.

     The subscript 1 in the above notation just says there is one quantifier involved. In actual fact we can replace y by $\overrightarrow{y}$ in the above definition, so allowing a whole block of quantifiers. This can be shown equivalent to the definition given by using the coding function $\langle . \rangle$ to write, for example:

$$(\exists \overrightarrow{y})R(x, \overrightarrow{y}) \iff (\exists z)[z = \langle \overrightarrow{y} \rangle \mathbin{\&} R(x, \overrightarrow{y})].$$

     The superscript 0 (which is sometimes omitted when there is no risk of confusion) tells us that all quantification is over natural numbers (rather than over *sets* of natural numbers, for example).   ▯

EXAMPLE 5.2.6   *Given e, show that $\{x \mid \varphi_{e,s}(x) \downarrow\}$ is a Σ_1^0 set.*

SOLUTION   The equivalence

$$\varphi_{e,s}(x) \downarrow \iff \exists s[\varphi_{e,s}(x) \downarrow]$$

immediately gives the result.   ▯

     This example suggests the following computable approximations to the halting set W_e, using those for the p.c. function φ_e.

DEFINITION 5.2.7   *Write $W_{e,s} = \mathrm{dom}(\varphi_{e,s})$, for each $e, s \in \mathbb{N}$.*

It is then straightforward to verify the sort of properties we want of such approximations:

EXERCISE 5.2.8   *Show (i) that $W_{e,s}$ is a finite computable set, and (ii) that $W_e = \cup_{s \geq 0} W_{e,s}$.*
[**Hint:** Use Corollary 4.5.3.]

The following exercise gives an alternative characterisation of the Π_1^0 sets.

EXERCISE 5.2.9   *Show that $A \subseteq \mathbb{N}$ is a Σ_1^0 set if and only if its complement $\overline{A}$ is a Π_1^0 set.*

More generally, we will say that a relation $R^*(\overrightarrow{x})$ is Σ_1^0 if we have

$$R^*(\overrightarrow{x}) \iff (\exists \overrightarrow{y})R(\overrightarrow{x}, \overrightarrow{y}),$$

for some computable relation R. And then we can say that R^* is a Π_1^0 relation if and only if its negation $\neg R^*$ is Σ_1^0.

We have already seen that logical connectives such as $\neg$, $\&$ and $\vee$ lead from computable relations to more computable relations. What happens when we use quantifiers? As we shall see in the next theorem, it is at most one existential quantifier which separates a c.e. set from a corresponding computable relation. We are now ready to prove:

THEOREM 5.2.10 (The Normal Form Theorem for C.E. Sets)
The following are equivalent:
(1) A is c.e.,
(2) $A \in \Sigma_1^0$,
(3) $A = W_e$ for some $e \in \mathbb{N}$ (so that $W_e = $ the eth c.e. set).

PROOF $\boxed{(1) \Rightarrow (2)}$ Assume A to be c.e.

If $A = \emptyset$, we have $x \in A \iff (\exists s)(x = x + 1)$.
Assume that $A = \mathrm{range}(f)$ for some computable function f. Then

$$x \in A \iff (\exists s)[f(s) = x],$$

where, of course, $f(s) = x$ is a computable relation of s and x.

$\boxed{(2) \Rightarrow (3)}$ Assume that we have

$$x \in A \iff (\exists s)R(x, s),$$

with R computable. Define

$$\psi(x) = \begin{cases} 0 & \text{if } (\exists s)R(x, s), \\ \text{undefined} & \text{otherwise.} \end{cases}$$

It is easy to see, using the Church–Turing Thesis, that ψ is p.c.
[*Algorithm for* ψ: Given x, check in turn which of $R(x, 0), R(x, 1), \ldots$ holds.
If we come across an s for which $R(x, s)$ holds, set $\psi(x) = 0$ — but if the
search does not terminate, we have $\psi(x) \uparrow$, as required.]

So by the *Enumeration Theorem*, $\psi = \phi_e$ for some e. But then $x \in A \iff$
$\psi(x) \downarrow \iff \phi_e(x) \downarrow$, giving $A = \text{dom}(\phi_e) = W_e$.

$\boxed{(3) \Rightarrow (1)}$ Assume $A = W_e \neq \emptyset$.

Let $p \in A$ (say $p = $ the least member of A), and define for each $x, s \in \mathbb{N}$

$$f(\langle x, s \rangle) = \begin{cases} x & \text{if } \phi_{e,s}(x) \downarrow \\ p & \text{otherwise.} \end{cases}$$

Then since $\langle ., . \rangle : \mathbb{N}^2 \to \mathbb{N}$ is computable and surjective, and since $\phi_{e,s}(x) \downarrow$ is
a computable relation, this defines f as a (total) computable function. Also

$$x \in A \iff \phi_e(x) \downarrow \iff (\exists s)[\phi_{e,s}(x) \downarrow] \iff f(\langle x, s \rangle) = x \text{ (some } s) ,$$

so that $A = \text{range}(f)$, and hence is c.e.   ▯

An immediate corollary of the proof is that $A \subseteq \mathbb{N}$ is c.e. if and only if
it is the domain of some p.c. function. There are a number of other simple
applications of this very useful theorem.

EXERCISE 5.2.11 *Show that $A \subseteq \mathbb{N}$ is c.e. if and only if it is the range
of a p.c. function.*

EXERCISE 5.2.12 *Show that $A \in \Delta_1^0$ if and only if it is computable.*

EXERCISE 5.2.13 *Show that if P and Q are Σ_1^0 relations, then so are
$P \& Q$ and $P \vee Q$. Deduce the results of Example 5.1.12 and Exercise 5.1.13.*

EXERCISE 5.2.14 *Show that if $X = \{n_0, n_1, \ldots, n_k\}$ is a finite set of natural numbers, then one can effectively find a Turing machine T for which $\mathrm{Dom}(\varphi_T) = X$.*

Deduce that for each such X one can effectively find an index i for X as a computably enumerable set: $X = W_i$.

EXERCISE 5.2.15 *Show that the following sets are all c.e.:*

    (i) $K_1 = \{x \mid W_x \neq \emptyset\}$,

    (ii) $\mathrm{Range}(\varphi_e) = \{y \mid (\exists x)[\varphi_e(x) = y]\}$,

    (iii) $\mathrm{Graph}(\varphi_e) = \{\langle x, y \rangle \mid \varphi_e(x) = y\}$.

EXERCISE 5.2.16 *Show that $\psi : \mathbb{N} \to \mathbb{N}$ is p.c. if and only if $\mathrm{Graph}(f) = \{\langle x, y \rangle \mid f(x) = y\}$ is c.e.*

EXERCISE 5.2.17 *Prove that if ψ is a p.c. function and A is c.e., then (i) $\psi(A)$, and (ii) $\psi^{-1}(A)$ are also c.e.*

EXERCISE 5.2.18 *Let $X = W_e$, $Y = W_f$ be c.e. sets, where we write $X^s = W_{e,s}$, $Y^s = W_{f,s}$. (We call $\{X^s\}_{s \geq 0}$, $\{Y^s\}_{s \geq 0}$* **standard c.e. approximating sequences** *for X and Y.)*

Let $X \setminus Y = \{z \mid (\exists s)(z \in X^s - Y^s)\}$ and $X \searrow Y = (X \setminus Y) \cap Y$. Show that:

(i) Both $X \setminus Y$ and $X \searrow Y$ are c.e. sets.

(ii) $X \setminus Y = (X - Y) \cup (X \searrow Y)$.

(iii) If $X \searrow Y$ is finite then $X - Y$ is c.e.

(iv) Assuming that numbers are enumerated into **at most one** *of $X^s = W_{e,s}$, $Y^s = W_{f,s}$ at any given stage s, and taking $\hat{X} = X \setminus Y$, $\hat{Y} = Y \setminus X$, prove the* **Reduction Principle** *for X, Y:*

Given c.e. sets X and Y, there exist c.e. sets $\hat{X} \subseteq X$ and $\hat{Y} \subseteq Y$ such that $\hat{X} \cap \hat{Y} = \emptyset$ and $\hat{X} \cup \hat{Y} = X \cup Y$.

EXERCISE 5.2.19 *A set D is the* **difference of c.e. sets** *(d.c.e.) if and only if $D = A - B$ where A and B are c.e.*

(i) Show that the set of all d.c.e. sets is closed under the formation of intersections.

(ii) Show that if $C_n = \{e \mid |W_e| = n\}$, then C_n is d.c.e. for all $n \geq 0$.

We end this section with a c.e. counterpart to the universal Turing machine.

THEOREM 5.2.20 (The Enumeration Theorem for C.E. Sets)
There is a c.e. set K_0 such that for each $e \in \mathbb{N}$ we have the eth c.e. set
$W_e = \{x \mid \langle x, e \rangle \in K_0\}$.

PROOF   We start by defining K_0 in the obvious way:

DEFINITION 5.2.21   *Define* $K_0 = \{\langle x, e \rangle \mid x \in W_e\}$.

Then:

$$\langle x, e \rangle \in K_0 \iff x \in W_y \iff (\exists s) \underbrace{[x \in W_{y,s}]}_{\text{computable relation}} \ ,$$

from which we get $K_0 \in \Sigma_1^0$, giving K_0 c.e. by the Normal Form Theorem.   □

5.3   Incomputable Sets and the Unsolvability of the Halting Problem for Turing Machines

The reader who knows a little about countable and uncountable sets will already be able to deduce the existence of incomputable sets from the Enumeration Theorem for p.c. functions. From a set-theoretic point of view, it turns out that *almost all* sets of numbers are not computable.

We say that a set A is *countable* if it is empty or its members can be listed $a_0, a_1, \dots$ (allowing repetitions). It is easy to see that a subset of a countable set is also countable. Clearly there are at most countably many computable sets, since all their characteristic functions will appear somewhere in the list $\varphi_0, \varphi_1, \dots$. On the other hand Cantor's Theorem tells us that no such list can contain *all* sets $\subseteq \mathbb{N}$.

Actually, the reader who has seen a proof of Cantor's Theorem already, will probably have seen it in the form of a proof that the set of all real numbers is uncountable. All will become clear via the following constructive version of that famous proof. This will also tell us that not only are most sets incomputable, but some of those incomputable sets are very closely connected with everyday mathematics.

THEOREM 5.3.1

There exists a computably enumerable set which is not computable.

PROOF We define a set K which (1) is c.e., but (2) whose complement is not c.e., and which is hence not computable by the Complementation Theorem.

DEFINITION 5.3.2 *Define $K = \{x \mid x \in W_x\}$.*

Then (1) K is Σ_1^0 and hence c.e., since

$$x \in K \iff x \in W_y \iff (\exists s) \underbrace{[x \in W_x]}_{\text{computable relation}} .$$

For (2), assume that $\overline{K}$ is c.e. Then $= W_e$, some $e \in \mathbb{N}$, giving

$$x \in W_e \iff x \in \overline{K} \iff x \notin K \iff x \notin W_x.$$

Taking $x = e$, we get a contradiction. □

One can visualise this proof as a proof by *diagonalisation*, whereby one proceeds down the leading diagonal of an infinite array of questions:

$$\boxed{0 \in W_0?} \qquad 1 \in W_0? \qquad 2 \in W_0? \qquad \dots$$

$$0 \in W_1? \qquad \boxed{1 \in W_1?} \qquad 2 \in W_1? \qquad \dots$$

$$0 \in W_2? \qquad 1 \in W_2? \qquad \boxed{2 \in W_2?} \qquad \dots$$

$$\vdots \qquad\qquad \vdots \qquad\qquad \vdots \qquad\qquad \ddots$$

making $\overline{K} \neq W_i$ by forcing different answers to the questions "$i \in \overline{K}$?" and "$i \in W_i$?" for each $i \in \mathbb{N}$.

EXERCISE 5.3.3 *Show that the lattice $\mathcal{E}$ of all c.e. sets is not closed under the formation of complements.*

EXERCISE 5.3.4 *Show that there is a partial recursive function ψ, such that if we drop the requirement in the definition of $\mu y[\psi(x, y) = 0]$ that if*

$\mu y[\psi(x, y) = 0] = z$ then $\psi(x, y) \downarrow$ for all $y \leq z$, then the function f defined by $f(x) = \mu y[\psi(x, y) = 0]$ is not partial recursive.

Deduce that the partial recursive functions are not closed under such an unrestricted μ-operator.

[**Hint:** Define

$$\psi(x, y) = \begin{cases} 0 & \text{if } y = 1, \text{ or if } y = 0 \text{ and } x \in K, \\ \text{undefined} & \text{otherwise,} \end{cases}$$

and show that for each x $\chi_{\overline{K}}(x) = \mu y[\psi(x, y) = 0]$.]

EXERCISE 5.3.5   *We say that φ_e has a* **computable completion** *if and only if there is a computable (total) function f such that*

$$\forall \in \mathbb{N} [\varphi_e(x) \downarrow \Rightarrow f(x) = \varphi_e(x)].$$

Show that:

(i) There is a partial function ψ which is not computable but has a computable completion.

(ii) There is no computable function f such that for all $e \in \mathbb{N}$ $\varphi_{f(e)}$ is a computable completion of φ_e.

Theorem 5.3.1 actually opens the floodgates to a whole host of incomputable sets and unsolvable problems. For instance, we have Turing's famous results from 1936–37:

COROLLARY 5.3.6

There exists a Turing machine with an unsolvable halting problem.

PROOF   Since K is c.e., we have by the Normal Form Theorem that $K = W_e$ for some $e \in \mathbb{N}$. And since W_e is not computable, the halting problem for the Turing machine with program P_e is unsolvable.   ☐

COROLLARY 5.3.7

The halting problem for the universal Turing machine U is unsolvable.

PROOF   Let P_e be as in Corollary 5.3.6.

Then by the definition of the universal Turing machine U, we have that $x \in W_e \iff U$ halts on input (e, x). But then, the solvability of the halting problem for U would lead to that of the halting problem for P_e — and this would contradict Corollary 5.3.6. □

EXERCISE 5.3.8 *The* **Printing Problem** *for a Turing machine T and a symbol S_k is the problem of determining, for any given input x, whether T ever prints the symbol S_k.*

Find a Turing machine T for which the printing problem is unsolvable.

5.4   The Busy Beaver Function

Here is a particularly nice example of an incomputable function, due to the distinguished combinatoricist Tibor Radó. It needs our knowledge of unlimited register machines.

I will first describe the idea behind the example. It uses a common trick for getting a function obviously outside of a given set of functions.

DEFINITION 5.4.1   *Let $f, g : \mathbb{N} \to \mathbb{N}$.*

*(1) We say g **dominates** f if for some $n_0 \in \mathbb{N}$*

$$n > n_0 \implies g(n) \geq f(n).$$

*(2) If S is a set of functions, we say that g **dominates** S if g dominates $f(n) + 1$ for every function $f \in S$.*

So, for example $n \mapsto 2^n$ dominates the set of all polynomial functions — and $n \mapsto n!$ dominates $\{n \mapsto k^n \mid k \in \mathbb{N}\}$ = the set of exponential functions.

Now notice:

IMPORTANT FACT: *If g dominates S, then $g \notin S$.*

What made the Ackermann function we saw in Subsection 2.1 not primitive recursive? It was that it dominated the set of all primitive recursive functions.

The aim now is to produce a function B which dominates all computable functions. We need the following:

> **LEMMA 5.4.2**
> *There are only finitely many functions (partial or total) computed by a URM program with at most n instructions.*

PROOF   We just observe:

> *(1) A program P with at most n instructions uses at most 2n registers.*

Since $Z(k), S(k)$ use just one register, and $T(k, l), J(k, l, q)$ use at most two. And:

> *(2) If P is a URM program which uses 2n different registers, then P is equivalent to a program P^* which uses only $R_1, \ldots, R_{2n}$.*

If P uses $R_{l_1}, \ldots, R_{l_{2n}}$ with $l_1 = 1 < l_2 < \cdots < l_{2n}$, we can just replace each reference to an R_{l_t} in P by a reference to R_t — giving a new program P^* with $\varphi_P^{(1)} = \varphi_{P^*}^{(1)}$.

And hence:

> *(3) Such a P^* chooses its instructions from a finite set.*

Namely, the set:

$$\{Z(k) \mid 1 \le k \le 2n\} \cup \{S(k) \mid 1 \le k \le 2n\}$$
$$\cup \{T(k, l) \mid 1 \le k \le 2n \text{ and } 1 \le l \le 2n\}$$
$$\cup \{J(k, l, q) \mid 1 \le k \le 2n,\ 1 \le l \le 2n,\ 1 \le q \le n+1\}$$

which contains $4n(n^2 + 2n + 1)$ instructions.

It is now easy to see that there are at most $[4n(n^2 + 2n + 1)]^n$ instructions for P^*, which gives the lemma.   □

We can now define:

DEFINITION 5.4.3 (The Busy Beaver Function)   *Let*

> $B(n) =$ *the maximum output, for input* 0*, of any URM program with at most* n *instructions.*

The computability of B looks quite approachable. We easily get $B(1) = 1$, $B(2) = 2$, etc. — but by the time we get to $B(10) \geq 39$, we begin to have problems replacing the "$\geq$" with "$=$".

EXERCISE 5.4.4   *Verify that* $B(10) \geq 39$.

EXERCISE 5.4.5   *Show that if the halting problem for URMs is solvable, then B is computable.*

Well, of course, B is *not* computable — it's the unsolvability of the halting problem for URMs which makes our beaver so busy! So let us get on and prove the incomputability of B. We need some notation and three simple facts.

We write $P[n]$ for the URM program with $\leq n$ instructions for which $\varphi_{P[n]}^{(1)}(0) = B(n)$.

LEMMA 5.4.6
B *is a strictly increasing function.*

PROOF   We need to show that for all $n \in \mathbb{N}$, $B(n) < B(n+1)$.

The program

$$P[n]$$
$$S(1)$$

has at most $n + 1$ instructions — and output $B(n) + 1$ for input 0.
So $B(n+1) \geq B(n) + 1 > B(n)$.   ⬜

LEMMA 5.4.7
For all $n \geq 1$, we have $B(n + 5) \geq 2n$.

PROOF Consider:

$$
\begin{array}{rl}
1. & S(1) \\
\vdots & \vdots \\
n. & S(1) \\
n+1. & T(1,2) \\
n+2. & J(2,3,n+6) \\
n+3. & S(1) \\
n+4. & S(3) \\
n+5. & J(1,1,n+2)
\end{array}
$$

It has $n+5$ instructions, and output $2n$ for input 0. So $B(n+5) \geq 2n$. ☐

LEMMA 5.4.8
Every URM computable function is dominated by a strictly increasing URM computable function.

PROOF Let f be URM computable. Define g by

$$
g(0) = f(0) + 1
$$
$$
g(n+1) = \max\{g(n), f(n+1)\} + 1.
$$

Since the URM computable functions are closed under primitive recursion, we have g URM computable.

Also $g(n) > f(n)$ for all n, and $g(n+1) > g(n)$. ☐

We can now prove:

THEOREM 5.4.9 (Tibor Radó, 1962)
B dominates every URM computable function.

PROOF Say f is URM computable.

By Lemma 5.4.8, we can take a URM computable strictly increasing g which dominates f.

Aim — $\boxed{\text{Show } B \text{ dominates } g.}$

Let P_g compute g, with k_0, say, $= \ell(P_g)$.

First step — | For each n, devise a URM program giving $g(B(n))$ from input 0.

Let P_g^+ be obtained from P_g by adding $\rho(P[n])$ to every register number other than 1 occurring in the instructions in P_g.

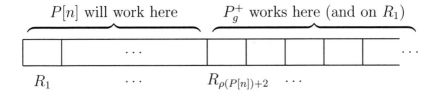

$$P[n] \text{ will work here} \qquad P_g^+ \text{ works here (and on } R_1)$$

$R_1 \qquad \qquad R_{\rho(P[n])+2}$

Then

$$P[n]$$
$$P_g^+$$

has at most $n + k_0$ instructions, and output $g(B(n))$ for input 0.

So — | $B(n + k_0) \geq g(B(n))$ | for all $n \geq 1$.

Hence — using the fact that g, B are increasing —

$$B(n + k_0 + 6) > B(n + k_0 + 5)$$
$$\geq g(B(n + 5))$$
$$\geq g(2n) \qquad \text{(by Lemma 5.4.7)}$$
$$> g(n + k_0 + 6) \qquad \text{for } n > k_0 + 6.$$

This means $B(m) > g(m)$ for $m > 2(k_0 + 6)$.
So B dominates g and, hence, f also. ☐

It immediately follows that B dominates the set of all URM computable functions So:

COROLLARY 5.4.10
The function B is not URM computable.

So — using the Church–Turing Thesis — we get another example of a function which is not computable. We did have to work a little harder to get it, but the reward is that it is a particularly natural example. Of course to a nineteenth century mathematician, there would not be much "natural" about the function B. We will see in the next chapter how one needs to work even harder to satisfy more basic demands for "natural" incomputable functions.

To end on a technical note — did you spot how we used Lemma 5.4.7 to avoid using Gödel numbers?

Chapter 6

The Search for Natural Examples of Incomputable Sets

The proof of Corollary 5.3.7 is a simple example of how one proves problems to be unsolvable in practice. Almost always one *reduces* some known unsolvable problem to the one one is interested in, to get the required unsolvability. It is a very powerful method, and is used not just in classical computability theory but in situations arising in computer science with more restrictive notions of computability.

However, it is dependent on a small number of special incomputable sets, and this has a downside. It restricts the *range* of unsolvable problems discovered, and we shall say more about this later. For instance, in Chapter 7 we will see that the incomputability of K gives us simple proofs of the undecidability of various basic mathematical theories, but in such a way that these theories can be viewed as being little more than notational variants of K.

So although we quickly discover a number of naturally arising unsolvable problems — such as that of telling whether a given computer program halts on a given input or (as we shall see) that of deciding whether a given statement is logically sound or not — they all turn out to be essentially *the same*.

This suggests some fundamental questions concerning incomputability in mathematics and real life:

> **How rich a variety of unsolvable problems is there? Does incomputability impinge on everyday life? And if so, can we find an informative theory of incomputability?**

These questions can be seen to underlie many bitter debates in science and mathematics, and prominent figures can be found ranged on both sides of this controversy. It has to be said that there are as yet no generally agreed answers to these questions, but quite a lot of pointers to positive ones. We will of course concentrate on the mathematical evidence in this book. But there exist fascinating discussions concerning extensions of the Church–Turing Thesis to the material Universe (see Section I.8 of Volume I of Odifreddi's book on *Classical Recursion Theory*) and of incomputability in Nature (see, for example, Roger Penrose's *The Emperor's New Mind*).

Even more divisive is the debate as to how the human mind *relates* to practical incomputability. The unavoidable limitations on computers and axiomatic theories suggest that mathematics — and life in general — may be an essentially *creative* activity which transcends what computers do.

To what extent can the human mind be likened to a Turing machine or a large computer?

The basic inspiration for Alan Turing's computing machines was, of course, the human mind, with things like "states of mind" feeding into the way he described the way his machines worked. Turing made clear in a number of places which side of the argument he was on! On the other, we feel like our mental processes are not entirely mechanical, in the sense that a Turing machine is. And various people have explicated these feelings to a point where it can be argued convincingly that these feelings have more than purely subjective content. For instance, there is the famous and influential book of Jacques Hadamard on *The Psychology of Invention in the Mathematical Field*, or the philosophically remarkable *Proofs and Refutations: The Logic of Mathematical Discovery* by Imre Lakatos. In science, Karl Popper effectively demolished the inductive model of scientific discovery — as was accomplished, more debatably, by Thomas Kuhn at the social level. This raises the question of how to model the way theories are hypothesised, via a process which seems neither random nor simply mechanical.

A purely mathematical answer to the question is very difficult. Roger Penrose (in his *Shadows of the Mind*) has argued (unsuccessfully it seems) that the overview we have of Gödel's Incompleteness Theorem for axiomatic arithmetic (see Chapter 8) shows that the human mind is not constrained by that theorem. But it is hard to be clear what it is that the human mind may be doing that Turing machines are incapable of. Obviously it will help to know more about both the physical and the logical structures involved. What is really needed is an *alternative* mathematical model to that of the Turing machine, and providing this must be one of the main aims of computability theory. Some speculations in this direction are provided in the 2003 paper by myself and George Odifreddi on "Incomputability in Nature".

6.1　The Ubiquitous Creative Sets

We now return to the mathematical development which will feed into eventual answers to such deep and intractable questions.

So how is it that our first incomputable set K turns out to be so typical of all the other unsolvable problems we know of? We need to define a type of set, due to Post, which is a direct abstraction from that feature of Gödel's Incompleteness Theorem on which Penrose based his argument for the essential creativeness of human thought.

DEFINITION 6.1.1   *We say $A \subseteq \mathbb{N}$ is* **creative** *if and only if*
(i) A is c.e., and
(ii) there is a computable function f such that for each e

$$W_e \subset \overline{A} \Rightarrow f(e) \in \overline{A} - W_e.$$

If A satisfies (i) and (ii), we call f the **creative function for** A.

An important property of creative sets is given by:

EXERCISE 6.1.2   *Show that if A is creative then A is not computable.*

The following theorem will be no surprise now.

THEOREM 6.1.3
Creative sets do exist. In particular K is creative.

PROOF K is known to be c.e. by Theorem 5.3.1.
We take the creative function for K to be the identity function $f : x \mapsto x$.
Assume $W_e \subseteq \overline{K}$. Then $e \notin W_e$, since otherwise we would have $e \in K \cap W_e$, a contradiction. So $f(e) = e \in \overline{K} - W_e$, as required. $\qquad\square$

Here is another property of creative sets which parallels the situation with axiomatic theories (see Chapter 8).

EXAMPLE 6.1.4   *Let C be a creative set. Show that*
(i) There is an algorithm whereby, given n members of $\overline{C}$, some $n \geq 0$, one can computably find $n + 1$ members of $\overline{C}$.
(ii) And hence $\overline{C}$ contains an infinite c.e. subset.

SOLUTION   (i) Let $y_1, y_2, \ldots, y_n$ be a (possibly empty) list of members of $\overline{C}$, and let C have corresponding creative function f.

By 5.2.14 we can find an i for which $W_i = \{y_1, y_2, \ldots, y_n\}$. Since $W_i \subseteq \overline{C}$, we have $f(i) \in \overline{C} - W_i$. But then $\{y_1, \ldots, y_n, f(i)\} \subseteq \overline{C}$ where $y_j \neq f(i)$, any $j = 1, \ldots, n$.

(ii) Enumerate an infinite c.e. subset $A = \{f(i_0), \ldots, f(i_n), \ldots\}$ of $\overline{C}$ as follows:

- Compute i_0 such that $W_{i_0} = \emptyset$, and enumerate i_0 into A.
- Assume n numbers enumerated into A already, with corresponding index computed. Say $\{f(i_0), \ldots, f(i_{n-1})\} = W_{i_n} \subseteq \overline{C}$. Enumerate $f(i_n)$ into A. □

We will see in Chapters 7 and 8 that the property of creativeness is shared by all the naturally arising unsolvable problems discovered by Church, Turing, and others in the 1930s and after. But first we will see that there are other ways in which incomputable sets can arise. The basic technique will still be diagonalisation, so long as we allow that description of a non-linear "diagonal" whose precise positioning is incomputable and chosen by us!

6.2   Some Less Natural Examples of Incomputable Sets

As we have seen, the complements of creative sets contain infinite c.e. subsets. So if we could find an incomputable c.e. set whose complement does not have that property, it would certainly be very different to K. This, of course, was Post's thinking when he formulated the following:

DEFINITION 6.2.1 (Post, 1944)   A c.e. set S is said to be **simple** *if and only if*
  (i) $\overline{S}$ *is infinite, and*
  (ii) $(\forall e)[W_e$ *infinite* $\Rightarrow W_e \cap S \neq \emptyset]$.

This is easily visualised:

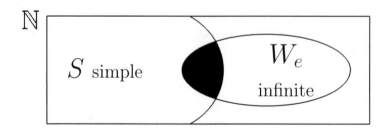

Note: Another way of saying a set S has infinite complement is to say it is *coinfinite*. Similarly we talk about cofinite or co-c.e. sets, and so on.

COROLLARY 6.2.2 (of the definition)

If S is simple, then S is not computable.

PROOF   Assume, in order to get a contradiction, that S is simple and computable. Then $\overline{S}$ is infinite (by (i)) and computable. Which means that $\overline{S}$ is an infinite c.e. set, and so is equal to an infinite W_e, some $e \in \mathbb{N}$ — contradicting condition (ii) for S to be simple.   ▯

Simple sets are not naturally occurring sets in the way that creative sets are. We have to *construct* them. The proof of the following theorem will look rather different from anything we have seen before. It has some of the features of what we will later call a *priority argument*.

THEOREM 6.2.3

There exists a simple set A.

PROOF   The idea behind the proof is that for each e we wait for some stage s at which some x is enumerated into $W_{e,s}$. We then enumerate x into A to make $A \cap W_e \neq \emptyset$. However, we have to be a bit more choosy about which x we use if we are to have $\overline{A}$ infinite. And we can only use computable information in choosing x if we are to end up with a set A which is *seen* to be computably enumerable.

So — we computably enumerate the members of A at stages $0, 1, \ldots, s, \ldots$ so as to satisfy the *requirements:*

$$\mathcal{N}_e : \qquad |A \cap \{0, 1, \ldots, 2e\}| \leq e$$
$$\mathcal{P}_e : \qquad \text{If } W_e \text{ is infinite, then } W_e \cap A \neq \emptyset,$$

for all $e \in \mathbb{N}$. It is easy to see that if the requirements are satisfied, then A is simple. For part (i) of Definition 6.2.1, we just notice that if $\mathcal{N}_e$ holds then $|\overline{A} \cap \{0, 1, \ldots, 2e\}| > e$. So if $\mathcal{N}_e$ holds for infinitely many e, then $\overline{A}$ must be infinite. And $\mathcal{P}_e$ is just part (ii) of Definition 6.2.1.

The algorithm for enumerating A is:

(1)   For each as yet *unsatisfied* $\mathcal{P}_e$, *wait* for a stage s at which there is a number $x \in W_{e,s}$ with $x > 2e$.

(2)   If such an x appears, enumerate x into A, at which stage $\mathcal{P}_e$ becomes *satisfied*.

It is usual to call such an algorithm a *construction* of A and to describe the proof that the construction works as the *verification*. In this case the verification consists of an observation that A is c.e., by the effectiveness of the algorithm for enumerating the members of A, and proofs of two simple lemmas.

LEMMA 6.2.4
$\mathcal{P}_e$ *is satisfied for each* $e \in \mathbb{N}$.

PROOF OF LEMMA 6.2.4   Assume that W_e is infinite. Let s be the least stage at which we get some $x \in W_{e,s}$ with $x > 2e$. By the construction at stage s, $\mathcal{P}_e$ is not already satisfied, which means that it becomes so via part (2) of the construction at stage s via $x \in W_e \cap A$.   ☐

LEMMA 6.2.5
$\mathcal{N}_e$ *is satisfied for each* $e \in \mathbb{N}$.

PROOF OF LEMMA 6.2.5   Since we can only enumerate a number x into A with $x \leq 2e$ *on behalf of* some W_i with $i < e$ — and for each such i *at most one* such x is enumerated — the lemma follows immediately.   ☐

So the theorem is proved.   ☐

See Example 12.2.6 later on for a shorter proof — but one which does not give such an approachable introduction to priority arguments.

We saw earlier (Union Theorem 5.1.12 and Exercise 5.1.13) that the c.e. sets are closed under formations of unions and intersections. Is this true of the set of all simple sets?

EXAMPLE 6.2.6   *(i) Show that if A is a simple set and W is an infinite c.e. set then $A \cap W$ is an infinite c.e. set.*
(ii) Deduce that if A and B are simple then $A \cap B$ is simple.

SOLUTION   (i) Let A be simple and W be infinite c.e. Then $A \cap W$ is c.e. by Exercise 5.1.13.

Now say $A \cap W$ is finite. Then $X = W - (A \cap W)$ is also infinite and c.e., and $X \subseteq \overline{A}$, contradicting A being simple.

(ii) Say A, B are simple (and so c.e.). Then $A \cap B$ must be c.e.

Now let W be infinite and c.e. Then, $A \cap W$ is an infinite c.e. set by part (i). But then $B \cap (A \cap W)$ is infinite (by (i) again). So $(A \cap B) \cap W \neq \emptyset$.

Finally, $\overline{A \cap B}$ is infinite since $\overline{A \cap B} \supseteq$ the infinite set $\overline{A}$. □

On the other hand:

EXERCISE 6.2.7 *(i) Assuming the existence of a simple set, show that there exists a simple set $S \supseteq Ev$ (the set of all even numbers).*

(ii) Show that there exist simple sets A and B such that $A \cup B$ is not simple.

EXERCISE 6.2.8 *Show that the complement of a creative set contains an infinite computable subset.*

Hence show that each creative set is contained in some simple set.

Here are some more interesting facts concerning unions of c.e. sets. Let $W_e \searrow W_f$ be defined as in Exercise 5.2.18.

EXERCISE 6.2.9 *Let $A = W_f$ be any incomputable c.e. set $\subseteq \mathbb{N}$. Show that there exist disjoint c.e. sets B_0, B_1 such that $A = B_0 \cup B_1$, and for every c.e. set W_e, if $W_e \searrow A$ is infinite, then $W_e \cap B_i \neq \emptyset$ for each $i = 0, 1$.*

Deduce that if either $W_e - B_0$ or $W_e - B_1$ is c.e. then $W_e - A$ is c.e.

If $A = B_0 \cup B_1$ is the disjoint union of c.e. sets B_0, B_1 we say B_0, B_1 *split* A and write $A = B_0 \sqcup B_1$. A is *splittable* if there is such a splitting $A = B_0 \sqcup B_1$.

EXERCISE 6.2.10 (Friedberg Splitting Theorem) *Given any incomputable c.e. $A \subseteq \mathbb{N}$, use Exercise 6.2.9 to show that there exist disjoint c.e. sets B_0, B_1 such that:*

(i) $A = B_0 \sqcup B_1$, and

(ii) B_0 and B_1 are not computable.

[**Hint:** Take B_0 and B_1 as in Exercise 6.2.9, and consider $W = \mathbb{N} - B_i$, $i = 0, 1$.]

Are there incomputable c.e. sets other than creative and simple sets? Of course! For instance:

EXERCISE 6.2.11 (Shore, 1978) *We say that a c.e. set B is **nowhere simple** if for every c.e. $W \supseteq B$ with $W - B$ infinite, there exists an infinite* *c.e.* $U \subseteq W - B$.

   Show that any given incomputable c.e. set A can be split into two nowhere simple sets.

[**Hint:** Let B_0, B_1 be as in Exercise 6.2.9. Given W c.e., use $W = (W - B_i) \cup (W \cap B_i)$, for $i = 0, 1$, to show that if either of $W \cap B_0$ or $W \cap B_1$ is finite, then $W - A$ is c.e.

   We will see later that the nowhere simple sets are not creative and contain more diverse information content than the creative sets do.

   The theory of $\mathcal{E}$, the lattice of all c.e. sets under set inclusion, is a very rich and mathematically interesting one. For further reading we recommend Robert I. Soare's 1987 book on *Recursively Enumerable Sets and Degrees*.

6.3   Hilbert's Tenth Problem and the Search for Really Natural Examples

   The unsolvable problems found during the 1930s, such as those embodied in the Halting Problem and Church's Theorem 9.2.7, grew out of very basic questions. But although they do seem very natural, there is a big difference between these and the sort of set described in Example 5.1.3.

   The typical researcher, in mathematics or its applications, deals with real numbers, sets of natural numbers, maybe even sets of real numbers, and abstractions based on such things. The computer scientist is certainly interested in questions regarding the termination of computations. But there are other more practical constraints on what one can compute, which are not obviously to do with full-blown incomputability. What evidence is there that incomputability arises independently of the theoretical framework of computability theory?

   One of the oldest and most basic areas of mathematics is number theory. Around 250 AD the Greek mathematician Diophantus, sometimes known as "the father of algebra", could be found working in Alexandria, the main scientific centre of the old Greek world. It was here he is believed to have written his celebrated *Arithmetica*, in which he developed notation and techniques for discussing solutions to polynomial equations.

PROBLEM: **Given any polynomial equation in one or more variables, with integer coefficients, find a solution *consisting entirely of integers* — that is, solve any given *Diophantine equation*.**

Diophantus was actually interested in rational solutions to polynomial equations with rational coefficients, but it is easy to see that this problem is reducible to the one above. Also, if one knows that an integer solution to a given Diophantine equation does exist, then one can find it by systematically trying out candidate solutions until we discover the equation to be satisfied. This is the basis for one of the most celebrated mathematical problems of the twentieth century:

HILBERT'S TENTH PROBLEM: **To find a general way of telling** *effectively* **whether a given Diophantine equation has a solution or not.**

Its solution needed twentieth century mathematics, and was not at all what Hilbert had expected. As we saw in Chapter 1, Hilbert had the utmost faith in the onward progress of mathematics, and despite the misgivings of leading number-theorists, could only envisage a positive solution — that is, the discovery of a suitable algorithm.

A few simple examples convince one that a general algorithm to deal with every kind of Diophantine equation is not going to be easy.

For equations of the form $ax + by = c$, we have one of the oldest algorithms of all, that of Euclid. Like Diophantus, Euclid also worked in Alexandria, but much earlier, around 300 BC. His algorithm tells us that a solution to the linear Diophantine equation $ax + by = c$ exists exactly when the highest common factor of a and b divides c.

Equations like $3x^2 - 5y^2 + 2xy = 0$ can also be dealt with, using the sort of methods for quadratics initiated by Diophantus. But the algorithm is very different from that for linear equations. And our experience is that many different algorithms are needed to deal with the various special cases for which techniques exist.

At one time it was even suspected that an incomputable set could arise from Fermat's Last Theorem, it seemed so difficult to tell for which natural numbers n the equation $x^n + y^n = z^n$ has integer solutions x, y, z. Of course, we now have Wiles' Theorem, which tells us that n must be 1 or 2, but the proof of the theorem does not suggest the existence of a uniformly applicable algorithm of the sort hoped for by Hilbert. And for good reason!

Let us start to describe the technical framework which led to incomputable sets invading the heartland of classical mathematics. We first need to relate Hilbert's tenth problem to sets of natural numbers.

DEFINITION 6.3.1   *A set $A \subseteq \mathbb{N}$ is* **Diophantine** *if*

$$A = \{x \in \mathbb{N} \mid (\exists y_1, \ldots, y_n \in \mathbb{N})[p_A(x, y_1, \ldots, y_n) = 0]\}$$

for some polynomial $p_A(x, y_1, \ldots, y_n)$ (with integer coefficients).

It is clear from the definition that any Diophantine set is Σ_1^0 and so computably enumerable.

All sorts of c.e. sets turn out to be Diophantine. For instance, say

$$A = \{1, 3, 4, 5, 7, 8, 9, 11, \ldots\} = \{x \in \mathbb{N} \mid x \neq 4k + 2, \text{ any } k \in \mathbb{N}.$$

Then $A = \{x \in \mathbb{N} \mid x = y_1^2 - y_2^2, \text{ some } y_1, y_2 \in \mathbb{N}\}$. So A is Diophantine with $p_A(x, y_1, y_2) = y_1^2 - y_2^2 - x$.

Such facts led the logician Martin Davis, in his 1950 doctoral dissertation, to make a brave conjecture, whose verification would provide the key to a *negative* solution to Hilbert's tenth problem.

DAVIS' STRATEGY: Show that *every* **computably enumerable set is Diophantine.**

Let's see how this could lead to a solution to Hilbert's problem.

A consequence would be that the set K, for example, from Definition 5.3.2 would be Diophantine, with

$$K = \{x \in \mathbb{N} \mid (\exists y_1, \ldots, y_n \in \mathbb{N})[p_K(x, y_1, \ldots, y_n) = 0]\}$$

for suitable p_K.

But since K is not computable, there cannot be any general algorithm for deciding whether the Diophantine equations in the list:

$$p_K(0, y_1, \ldots, y_n) = 0$$
$$p_K(1, y_1, \ldots, y_n) = 0$$
$$p_K(2, y_1, \ldots, y_n) = 0$$
$$\vdots$$

have solutions in $\mathbb{N}$. On the other hand, a positive solution to Hilbert's tenth problem *would* have to provide such an algorithm.

The painstaking verification of the diophantine nature of larger and larger classes of c.e. sets, most of it carried out by the American mathematician

Julia Robinson, eventually hit a serious obstacle — that of showing that c.e. sets that increase *exponentially* (such as $2, 4, 8, 2^4, 2^5, \ldots$) are Diophantine.

In 1960 the stage was set for the final ingredient in a negative solution. Julia Robinson, the philosopher Hilary Putnam (whose contribution is sometimes overlooked), and Martin Davis showed that all that was needed was to show that there exists *one* exponentially increasing set which is Diophantine. This final ingredient was provided by some more very old mathematics, and an ingenious piece of number theory by a twenty-two year old unknown Russian mathematician named Yuri Matiasevich.

In 1202 Leonardo Pisano of Pisa, better known by his nickname Fibonacci, published his *Liber Abaci*, which played an important role in popularising the Hindu-Arabic place-valued decimal system and the use of Arabic numerals in Europe. Also to be found there were Fibonacci sequences, a typical such sequence being

$$1, 1, 2, 3, 5, 8, 13, 21, 34, \ldots, a_n, a_{n+1}, a_{n+2} = a_{n+1} + a_n, \ldots.$$

What is relevant here is that despite its initial appearances of being relatively slow growing, this sequence does in fact increase exponentially — for large n $\frac{1}{\sqrt{5}}[\frac{1}{2}(1 + \sqrt{5})]^n$ is a close approximation.

THEOREM 6.3.2 (Yuri Matiasevich, 1970)
The Fibonacci sequence is Diophantine.

PROOF   Let FIB(n) denote the nth Fibonacci number.   Then here are Matiasevich's equations:

$$u + w - \text{FIB}(2u) - 2 = 0$$
$$l - 2\text{FIB}(2u) - 2a - 1 = 0$$
$$l^2 - lz - z^2 - 1 = 0$$
$$g - bl^2 = 0$$
$$g^2 - gh - h^2 - 1 = 0$$
$$m - c(2h + g) - 3 = 0$$
$$m - fl - 2 = 0$$
$$x^2 - mxy + y^2 - l = 0$$
$$(d - l)l + u - x - 1 = 0$$
$$x - \text{FIB}(2u) - (2h + g)(l - 1) = 0.$$

Of course, we expected to see just one Diophantine equation to which to apply the existential quantifiers, in place of the given system. We get it by squaring and then summing all the left-hand sides, and equating the result to 0.   ▯

COROLLARY 6.3.3 (Davis, Matiasevich, Putnam, Robinson)
(i) Every computably enumerable set is Diophantine.
(ii) There is no positive solution to Hilbert's Tenth Problem.

We have omitted much of the detailed proof of this justly famous result. A good source for the interested reader is the 1993 MIT Press English translation of Matiasevich's book *Hilbert's Tenth Problem*. This contains a fascinating foreword by Martin Davis, in which he gives a personal account of the solving of the problem, including his predicting of the emergence of a "clever young Russian" who would provide that final ingredient.

The proof attracted a huge amount of attention. Julia Robinson went on, in 1982, to become the first woman President of the American Mathematical Society. There is a lot more I could tell you, and if you are interested I very much recommend you to investigate the story further. A good place to begin would be with the celebrated 1970 biography of Hilbert himself, by Julia Robinson's sister Constance Reid.

REMARK 6.3.4   The equations of Matiasevich given above involve fourteen unknowns. Matiasevich and Robinson, collaborating at a distance, also found that there was a number N such that *any* computably enumerable set could be presented as a Diophantine set involving less than N unknowns. A lot of work went into pinning down minimal values. The best result found for Matiasevich's equations was an equivalent set of equations with just *three* unknowns. And the number N was eventually brought down as far as *nine*. ▯

There are all sorts of consequences of Corollary 6.3.3, mainly in giving unsolvability of a wide range of other problems in mathematics and beyond. But not all consequences are negative.

A long-standing problem in number theory had been the search for a polynomial with integer coefficients whose positive values were the prime numbers. The negative solution to Hilbert's Tenth Problem gives, with a little manipulation:

COROLLARY 6.3.5
There exists a polynomial $p(x_1, \ldots, x_n)$ whose positive values are exactly the prime numbers.

This result does not tell us how to directly find such a polynomial. But new knowledge, however theoretical, may have the potential to change the world we live in.

Just a few years passed by before this prediction was explicitly confirmed by Jones, Sato, Wada and Wiens in 1976. The polynomial enumerating the prime numbers they gave was:

$$(k+2)\{1 - [wz + h + j - q]^2$$
$$- [(gk + 2g + k + 1)(h + j) + h - z]^2$$
$$- [2n + p + q + z - e]^2$$
$$- [16(k+1)^3(k+2)(n+1)^2 + 1 - f^2]^2$$
$$- [e^3(e+2)(a+1)^2 + 1 - o^2]^2$$
$$- [(a^2 - 1)y^2 + 1 - x^2]^2 - [16r^2y^4(a^2 - 1) + 1 - u^2]^2$$
$$- [((a + u^2(u^2 - a))^2 - 1)(n + 4dy)^2 + 1 - (x + cu)^2]^2$$
$$- [n + l + v - y]^2 - [(a^2 - 1)l^2 + 1 - m^2]^2$$
$$- [ai + k + 1 - l - i]^2$$
$$- [p + l(a - n - 1) + b(2an + 2a - n^2 - 2n - 2) - m]^2$$
$$- [q + y(a - p - 1) + s(2ap + 2a - p^2 - 2p - 2) - x]^2$$
$$- [z + pl(a - p) + t(2ap - p^2 - 1) - pm]^2\}.$$

Their polynomial looks a little strange, since it appears to be the product of two factors. It works because we are only interested in its positive values, and we only get positive values when $k + 2$ is prime and the second factor has value 1.

REMARK 6.3.6 So how does Corollary 6.3.3 help us with the search for natural examples of incomputable sets? What it does is shows us that everyday mathematics, such as elementary arithmetic, leads us unavoidably to incomputable sets. These incomputable sets may not have any recognisably special mathematical properties, and this may prevent our coming up with *particular* Diophantine sets for which we have methods of proving them incomputable. But if we go back to the dictionary meaning of *natural* — "existing in or caused by nature" (*The New Oxford Dictionary of English*, 1998 edition) — then the inextricable intertwining of incomputability with such basic mathematics, intimately related to natural phenomena, is a strong indicator of naturalness in the wider sense. □

In the next chapter we will start to develop a useful theory of incomputability. This analysis of the incomputable can be thought of as paralleling that of Cantor for the transfinite.

Chapter 7

Comparing Computability and the Ubiquity of Creative Sets

Some people argue that because all the sets we encounter in everyday life are finite, we should exclude mathematics in which infinite sets appear. This is not a popular view! The main reason for this is that however finite the material universe appears, its mathematics inevitably leads one to the consideration of infinite sets and approximation processes involving algorithmic repetition. One can, with great difficulty, reconstruct a large part of classical mathematics on a finitary basis. But in doing this we just mimic in our mathematics the concrete processes by which nature itself is constrained. And because of this, such mathematics does nothing to divert us from the overwhelming impression that we do need the mathematics of the infinite to fully understand our own Universe. In practice, of course, it isn't philosophy which widens our mathematical horizons. From Newton onwards, it has been the pure *explanatory power* of infinitary mathematics which has left us little alternative.

In the same way, there is much we will never understand about the world we live in without a mathematics which takes due account of how incomputable phenomena arise and interrelate. We have already seen how the iteration of very simple rules can lead to incomputable sets — even diophantine functions, using basic arithmetical operations, can have incomputable ranges. Such iterations are abundant in nature, where we find a level of incomputability associated with chaotic phenomena built on computable local events (e.g., problems in predicting the weather). Fractals provide a more graphic expression of the mathematics. For instance, Roger Penrose has asked about the computability of familiar objects like the Mandelbrot and Julia sets.

In this chapter we set out to answer such basic questions as: Are there different sorts of incomputable sets? How do we mathematically *model* situations involving interactions between incomputable phenomena?

7.1 Many–One Reducibility

We start off with a very natural way of *comparing* the computability of different — possibly incomputable — sets of numbers A and B.

DEFINITION 7.1.1 (Post, 1944)　*We say B is* **many–one reducible** *(or* **m-reducible***) to A (written $B \leq_m A$) if and only if there is a computable function f such that for all $x \in \mathbb{N}$*

$$x \in B \iff f(x) \in A.$$

Here is the picture which goes with this definition:

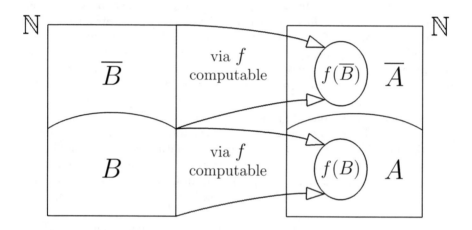

It is clear from Definition 7.1.1 — and from the picture — that

$$B \leq_m A \iff \overline{B} \leq_m \overline{A}.$$

$\leq_m$ is a very natural reducibility, and is an abstraction of many commonly occurring reductions between what look like very different problems. For instance we will see later how translations between theories can be viewed as many–one reductions.

$\leq_m$ behaves somewhat like you would expect an ordering to:

THEOREM 7.1.2
The ordering $\leq_m$ is (1) reflexive and (2) transitive.

PROOF　(1) $A \leq_m A$ via $f(x) = x$.

(2) Say

$$A \leq_m B \quad \text{via} \quad f,$$
$$B \leq_m C \quad \text{via} \quad g.$$

Then $x \in A \iff f(x) \in B \iff g(f(x)) \in C$. So $A \leq_m C$ via the computable function $g \circ f$. ◻

Intuitively, $B \leq_m A$ says that B is "at least as computable" as A is. This intuition is borne out in the following:

THEOREM 7.1.3
(1) If $B \leq_m A$ and A is computable, then B is also computable.
(2) If $B \leq_m A$ and A is c.e., then B is also c.e.

PROOF Say $B \leq_m A$ via f computable.
(1) Then $\chi_B = \chi_A \circ f$.
(2) Use the Normal Form Theorem.
Assume $A \in \Sigma_1^0$, with $x \in A \iff (\exists y) R(x, y)$, some computable R.
Then $x \in B \iff (\exists y) R(f(x), y)$, giving $B \in \Sigma_1^0$ and so c.e. ◻

The following example and exercises give us a first taste of the expressive power of m-reducibility. What a useful and natural notion it is!

EXAMPLE 7.1.4 Let $K_0 = \{\langle x, y \rangle \mid x \in W_y\}$. *Show that $K \leq_m K_0$, and hence K_0 is an incomputable c.e. set.*

SOLUTION We have already seen in the proof of the Enumeration Theorem for c.e. sets that K_0 is c.e.

We also have $x \in K \iff x \in W_x \iff \langle x, x \rangle \in K_0$, each $x \in \mathbb{N}$ — and so $K \leq_m K_0$ via $x \mapsto \langle x, x \rangle$ (computable since $\langle \cdot, \cdot \rangle$ is).

But by Theorem 5.3.1, K is not computable. So by Theorem 7.1.3, K_0 is not computable. ◻

EXERCISE 7.1.5 Show that $\varphi_y(x) \downarrow$ is an incomputable relation of x, y.
Deduce that $\varphi_y(x) = z$ is an incomputable relation of x, y, z.

EXERCISE 7.1.6 Let $A \subseteq \mathbb{N}$. Show that A is c.e. $\iff A \leq_m K_0$.

If you make the m-reduction f one–one, you get a more restrictive form of m-reducibility called *one–one reducibility*:

EXERCISE 7.1.7 *We say B is **one–one reducible** (or **1-reducible**) to A (written $B \leq_1 A$) if there is a one–one computable function f such that for all $x \in \mathbb{N}$*

$$x \in B \iff f(x) \in A.$$

Show that A is c.e. $\iff A \leq_1 K_0$.

Many more of the results I will give for m-reducibility extend without much extra work to 1-reducibility — but without much point. I will rarely mention 1-reducibility again. It is easy to see that $\leq_1$ and $\leq_m$ are different reducibilities, though.

EXERCISE 7.1.8 *Give an example of c.e. sets A, B such that $B \leq_m A$ but $B \not\leq_1 A$. Can you also make B not computable here?*
[**Hint:** Consider sets B such that $B \leq_1$ or $\leq_m$ a simple set S.]

We can get some particularly interesting incomputable sets to apply $\leq_m$ to by looking at sets of indices of p.c. functions or of c.e. sets.

DEFINITION 7.1.9 *Let $\mathcal{A}$ be a set of p.c. functions — or of c.e. sets. Then the **index set** of $\mathcal{A}$ is the set A of all indices of members of $\mathcal{A}$.*

For instance — Tot $=_{\text{defn}} \{e \mid \varphi_e$ is total$\}$ is an index set — as is $K_1 =_{\text{defn}} \{e \mid W_e \neq \emptyset\}$.

However, not every set of indices is an index set! Clearly, if A is an index set of some set $\mathcal{A}$ of c.e. sets, then A satisfies

$$[x \in A \,\&\, \varphi_x = \varphi_y] \implies y \in A, \quad \text{for all } x, y \in \mathbb{N}.$$

Notice of course that an index set of a set of p.c. functions is also an index set of c.e. sets.

EXAMPLE 7.1.10 *Show that K is not an index set.*

SOLUTION Let $f(n) =$ the index of $\{n\}$, each n.
Then by the Fixed Point Theorem we get $W_{f(e)} = W_e$, some e. So

$$e \in W_e = W_{f(e)} = \{e\}.$$

But taking some different index e' of $\{e\}$, we have $e' \notin W_{e'} = \{e\}$. We then have $e \in K$ and $W_e = W_{e'}$ — but $e' \notin K$. □

The following theorem tells us that index sets are indeed incomputable, except for two trivial exceptions.

THEOREM 7.1.11 (Rice's Theorem)

If A is an index set — $\neq \emptyset$ or $\mathbb{N}$ — then $K \leq_m A$ or $K \leq_m \overline{A}$.

Hence — every nontrivial index set is incomputable.

PROOF Say A is an index set of c.e. sets with $e \in A$ and $e' \in \overline{A}$.

Case I: $\emptyset$ has no index in A.

Then we can use the practical Church–Turing Thesis to get a computable function f such that

$$W_{f(x)} = \begin{cases} W_e & \text{if } x \in K \\ \emptyset & \text{otherwise.} \end{cases}$$

Then $x \in K \iff f(x) \in A$ — so $K \leq_m A$ via f.

Case II: $\emptyset$ has no index in $\overline{A}$.

Then a similar argument, with $e', \overline{K}$ in place of e, K, gives $K \leq_m \overline{A}$.

In either case the incomputability of K gives that of A by a now familiar argument. □

Magical as the above result may seem, its proof is firmly rooted in the unsolvability of the halting problem. Intuitively, it says that because we cannot computably decide whether a given computation ever halts, we cannot distinguish between machines which halt on *some* input and those which halt on *no* input — and if we cannot even do that, then there cannot be *any* computable dichotomy of the (indices of) machines.

EXERCISE 7.1.12 *Show directly — that is, without using Rice's Theorem — that K_1 is not computable.*

Anyway, we now have a rich variety of incomputable sets. As well as the index sets we have seen — such as Tot and K_1 — we have for instance

$$\text{Fin} =_{\text{defn}} \{x \mid W_x \text{ is finite}\},$$
$$\text{Inf} =_{\text{defn}} \mathbb{N} - \text{Fin} = \{x \mid W_x \text{ is infinite}\},$$
$$\text{Cof} =_{\text{defn}} \{x \mid W_x \text{ is cofinite}\},$$

a far from exhaustive list, of course. But having fleshed out the information content underlying $\leq_m$, let us get back to using $\leq_m$ to structure it.

7.2   The Non-Computable Universe and Many–One Degrees

It is usual to gather together collections of sets which cannot be distinguished from each other using only many–one reducibility.

DEFINITION 7.2.1   *We write $A \equiv_m B$ (A is **many–one equivalent to** B) if $A \leq_m B$ and $B \leq_m A$.*

The crucial fact is:

LEMMA 7.2.2
  $\equiv_m$ *is an equivalence relation.*

PROOF   We easily show:

(1) (reflexivity) $\forall A \subseteq \mathbb{N}$, $A \leq_m A$ by the reflexivity of $\leq_m$.

(2) (symmetric) $\forall A, B \subseteq \mathbb{N}$, $A \equiv_m B \iff A \leq_m B\,\&\,B \leq_m A \iff B \equiv_m A$.

(3) (transitivity) $A \equiv_m B\,\&\,B \equiv_m C \Rightarrow A \leq_m B\,\&\,B \leq_m C \Rightarrow A \leq_m C$ by transitivity of $\leq_m$. Similarly, we get $C \leq_m A$. So $A \equiv_m C$.   ☐

This now means that everything interesting about the ordering $\leq_m$ can be discovered by focusing on the structure it *induces* on the equivalence classes under $\equiv_m$.

DEFINITION 7.2.3   *(1) An equivalence class under $\equiv_m$ is called an m-degree (or **many–one degree**).*
  We write $\mathbf{a}_m = \deg_m(A) = \{X \subseteq \mathbb{N} \mid A \equiv_m X\}$ — and $\boldsymbol{\mathcal{D}}_m =$ the set of all m-degrees.

  (2) We write $\mathbf{b}_m \leq \mathbf{a}_m$ if and only if $B \leq_m A$ for some $A \in \mathbf{a}_m$, $B \in \mathbf{b}_m$.

REMARK 7.2.4   (1) Say $A_0 \equiv_m A_1$ and $B_0 \equiv_m B_1$. Then $B_0 \leq_m A_0 \Rightarrow B_1 \leq_m A_1$ (using transitivity of $\leq_m$ twice). We can picture this:

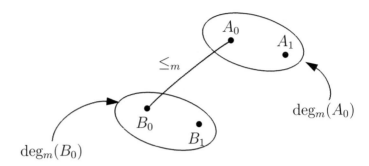

So — | $\mathbf{a}_m \leq \mathbf{b}_m$ is *well defined.* |

(2) Notice that if $X \leq_m \emptyset$ then $X = \emptyset$, giving $\deg_m(\emptyset) = \{\emptyset\}$. Similarly we get $\deg_m(\mathbb{N}) = \{\mathbb{N}\}$, facts which have more to do with the particularities of m-reducibility than with computability.

So — | We exclude $\emptyset$ and $\mathbb{N}$ as members of m-degrees. | ⧠

There are mathematical benefits from the way we have defined $\leq$ on $\mathcal{D}_m$. The ordering $\leq$ looks even more like the orderings we are used to than $\leq_m$ does.

PROPOSITION 7.2.5
 Let $\mathbf{a}_m, \mathbf{b}_m, \mathbf{c}_m \in \mathcal{D}_m$. Then $\leq$ *satisfies:*
  (1) ($\leq$ reflexive) $\mathbf{a}_m \leq \mathbf{a}_m$.
  (2) ($\leq$ transitive) $\mathbf{a}_m \leq \mathbf{b}_m \ \& \ \mathbf{b}_m \leq \mathbf{c}_m \implies \mathbf{a}_m \leq \mathbf{c}_m$.
  (3) ($\leq$ antisymmetric) $\mathbf{a}_m \leq \mathbf{b}_m \ \& \ \mathbf{b}_m \leq \mathbf{a}_m \implies \mathbf{a}_m = \mathbf{b}_m$.
  The properties (1)–(3) make $\leq$ what we call a **partial ordering (p.o.)** *on* $\mathcal{D}_m$.

PROOF   I will just give the proof of (3):

 Say $\mathbf{a}_m = \deg_m(A) \leq \mathbf{b}_m = \deg_m(B)$ and $\mathbf{b}_m \leq \mathbf{a}_m$.
 Then $A \leq_m B$ and $B \leq_m A$ (by Definition 7.2.3) — and hence $A \equiv_m B$ (by Definition 7.2.1).
 So $\mathbf{a}_m = \deg_m(A) = \deg_m(B) = \mathbf{b}_m$.   ⧠

EXERCISE 7.2.6 *Complete the proof of Proposition 7.2.5 by showing that $\leq$ is reflexive and transitive on $\boldsymbol{D}_m$.*

Here are some exercises which bring out some of the variety one can expect among partial orderings.

EXERCISE 7.2.7 *Show that the usual linear ordering $\leq$ on the set $\mathbb{R}$ of real numbers is a partial ordering.*

EXERCISE 7.2.8 *Let A be a set, and let $2^A = \{X \mid X \subseteq A\} =$ the set of all subsets of A. Show that $\subseteq$ is a partial ordering on 2^A.*

DEFINITION 7.2.9 *A partially ordered set L is a **lattice** if every pair $a, b \in L$ has both a least upper bound $lub\{a, b\} = a \vee b$ and greatest lower bound $glb\{a, b\} = a \wedge b$.*

It is easy to see that the set 2^A in Exercise 7.2.8 partially ordered by $\subseteq$ is a lattice.

EXERCISE 7.2.10 *Show that $\mathcal{E}$ with the ordering relation $\subseteq$ forms a lattice — called the **lattice of c.e. sets**.*

Notice that the lattice $\langle \mathcal{E}, \subseteq \rangle$ has a *least* element $\emptyset$, and a *greatest* element $\mathbb{N}$. Also:

EXERCISE 7.2.11 *Show that the lattice $\langle \mathcal{E}, \subseteq \rangle$ is **distributive** — that is, for all $A, B, C \in \mathcal{E}$ we have*

$$(A \cup B) \cap C = (A \cap C) \cup (B \cap C),$$
$$(A \cap B) \cup C = (A \cup C) \cap (B \cup C).$$

DEFINITION 7.2.12 *We say a lattice L, with least and greatest elements 0, 1, is **complemented** if every $a \in L$ has a **complement** $\bar{a} \in L$ such that $a \vee \bar{a} = 1$ and $a \wedge \bar{a} = 0$.*

*We call a complemented distributive lattice a **Boolean algebra**.*

EXERCISE 7.2.13   *Show that* $\mathcal{E}$, *the lattice of c.e. sets, is not a Boolean algebra.*

It is also straightforward to verify that the ordering of 2^A via $\subseteq$ has least element $\emptyset$ and greatest element A, and is a Boolean algebra.

QUESTION: Does $\mathcal{D}_m$ **have a least or greatest element?**

Well — the answer is "yes" and "no".

EXERCISE 7.2.14   *Let* $A \subseteq \mathbb{N}$ *be computable. Show that for each non-empty* $B \subsetneq \mathbb{N}$, *we have* $A \leq_m B$.
[**Hint:** Choose members $p_1 \in B$, $p_2 \in \overline{B}$ and define a computable f such that $f(A) \subseteq \{p_1\}$ and $f(\overline{A}) \subseteq \{p_2\}$.]

COROLLARY 7.2.15 (of the exercise)
$\mathcal{D}_m$ *has a least member* $\mathbf{0}_m$ *consisting of all computable sets (other than* $\emptyset$ *and* $\mathbb{N}$).

PROOF   We start off by defining

$$\mathbf{0}_m = \{A \subset \mathbb{N} \mid A \text{ is computable and} \neq \emptyset \text{ or } \mathbb{N}\}.$$

Then (a) $A, B \in \mathbf{0}_m \implies A \leq_m B$ and $B \leq_m A$ (by Exercise 7.2.14).
So $\mathbf{0}_m$ is certainly *contained in* some m-degree.

And (b) if $A \equiv_m B \in \mathbf{0}_m$, then $A \leq_m B$ computable, and so A must be computable by Theorem 7.1.3 — and since $A \neq \emptyset$ or $\mathbb{N}$, we end up with $A \in \mathbf{0}_m$.
Hence $\mathbf{0}_m \subseteq \deg_m(B)$.

Then (a) and (b) together show that $\mathbf{0}_m$ is itself an m-degree.

Finally (c) we show $\mathbf{0}_m$ to be *least* in $\mathcal{D}_m$.
Given any m-degree $\mathbf{a}_m = \deg_m(A)$, we have (by Exercise 7.2.14) that $B \leq_m A$ for each $B \in \mathbf{0}_m$.
But then — by the definition of $\leq$ on $\mathcal{D}_m$ — we have $\mathbf{0}_m \leq \mathbf{a}_m$.   ☐

We will see later that $\mathcal{D}_m$ has no greatest element. But there are important substructures of $\mathcal{D}_m$ which *do*.

DEFINITION 7.2.16 *(1) We say* $\mathbf{a}_m$ *is* **computably enumerable** *if there exists a c.e.* $A \in \mathbf{a}_m$.

(2) We write $\mathcal{E}_m$ *for the set of all c.e. m-degrees.*

Then:

THEOREM 7.2.17

Let $\mathbf{0}'_m =_{\text{defn}} \deg_m(K_0)$. *Then the following are equivalent:*

(a) $\mathbf{a}_m \leq \mathbf{0}'_m$,

(b) $\mathbf{a}_m$ *is c.e.,*

(c) **Every** $A \in \mathbf{a}_m$ *is c.e.*

Hence — there is a **greatest** *c.e. m-degree* $\mathbf{0}'_m > \mathbf{0}_m$.

PROOF   I will just do (a) $\Longleftrightarrow$ (b):

From Exercise 7.1.6 we have

$$A \text{ is c.e. } \Longleftrightarrow A \leq_m K_0. \tag{7.1}$$

The task — typically in such proofs — is to translate a fact about the reducibility into a degree-theoretic one —

$$\begin{aligned}
\mathbf{a}_m \text{ is c.e. } &\Longleftrightarrow \exists A \text{ c.e. } \in \mathbf{a}_m \\
&\Longleftrightarrow \exists A \in \mathbf{a}_m \text{ with } A \leq_m K_0, \text{ by Equation (7.1)} \\
&\Longleftrightarrow \mathbf{a}_m \leq \deg_m(K_0) = \mathbf{0}'_m.
\end{aligned}$$

$\square$

EXERCISE 7.2.18   *Complete the proof of Theorem 7.2.17.*

Here is another example of such a translation:

EXERCISE 7.2.19   *(a) Let* $A \oplus B =_{\text{defn}} \{2x \mid x \in A\} \cup \{2x + 1 \mid x \in B\}$ *be the* **computable join** *of A and B.*

Show that for all $A, B \subseteq \mathbb{N}$

i) $A \leq_m A \oplus B$ *and* $B \leq_m A \oplus B$, *and*

ii) If $A \leq_m C$ *and* $B \leq_m C$ *then* $A \oplus B \leq_m C$.

(b) Defining the **join** $\mathbf{a}_m \cup \mathbf{b}_m$ *of* $\mathbf{a}_m = \deg_m(A)$, $\mathbf{b}_m = \deg_m(B)$ *by*

$$\mathbf{a}_m \cup \mathbf{b}_m =_{\text{defn}} \deg_m(A \oplus B),$$

deduce that $\mathbf{a}_m \cup \mathbf{b}_m = \mathrm{lub}\,\{\mathbf{a}_m, \mathbf{b}_m\}$, *the least upper bound of* $\{\mathbf{a}_m, \mathbf{b}_m\}$ *with respect to* $\leq$ *on* $\boldsymbol{\mathcal{D}}_m$.

Here is a good place to reveal the picture of $\boldsymbol{\mathcal{D}}_m$ I am aiming to build up — where the ordering goes up the page. There are a number of ingredients of the picture — such as the "fatness" of the ordering, and the exact contents of $\mathbf{0}'_m$ — still to establish.

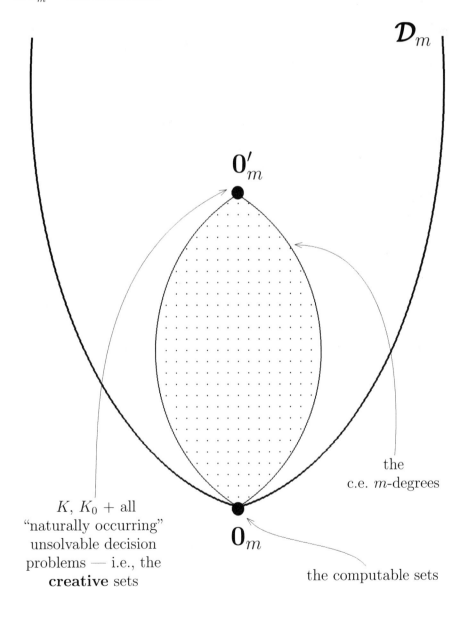

7.3   Creative Sets Revisited

Here is another key ingredient in the above picture:

THEOREM 7.3.1
  Let A be any c.e. set. If there is a creative set $C \leq_m A$, then A is also creative.

PROOF   Let f be the creative function for C — so $f(e) \in \overline{C} - W_e$ for all $W_e \subseteq \overline{C}$ — and assume $C \leq_m A$ via g computable.

Given W_e, we know $g^{-1}(W_e)$ is c.e., by Exercise 5.2.17. And we get — computably in e — an index $i(e)$:

$$W_{i(e)} = g^{-1}(W_e),$$

using the practical version of the Church–Turing Thesis. Now here is the picture:

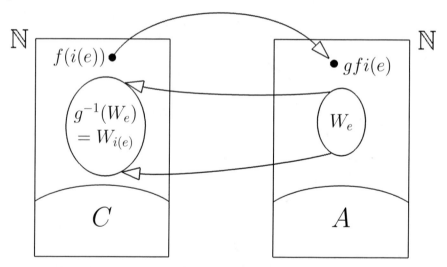

Then

$$W_e \subseteq \overline{A} \implies W_{i(e)} \subseteq \overline{C}$$
$$\implies f(i(e)) \in \overline{C} - W_{i(e)}$$
$$\implies gfi(e) \in \overline{A} - W_e.$$

So A is creative with creative function $g \circ f \circ i$.   ⬜

This theorem tells us a lot about the information content of $0'_m$.

COROLLARY 7.3.2

K_0 — and **every** $A \in 0'_m$ — *is creative.*

PROOF K_0 is creative since K is creative, and $K \leq_m K_0$. And every $A \in 0'_m$ is $\geq_m K_0$, and so is also creative.   ⬜

This means that the halting problem for the universal Turing machine U is closely connected with a creative set — since K_0 is in a natural sense the halting set for U.

We also get to know more about the *structure* of $\mathcal{D}_m$.

COROLLARY 7.3.3

There exist c.e. m-degrees $\mathbf{a}_m$ *with* $\mathbf{0}_m < \mathbf{a}_m < \mathbf{0}'_m$.

PROOF   We know that there exist simple sets, and simple sets were designed to be incomputable c.e. sets which are not creative. So taking $\mathbf{a}_m = \deg_m(A)$, with A simple, we get the corollary.   ⬜

Of course, we do not know yet whether $K \in 0'_m$. The next theorem fills that gap.

DEFINITION 7.3.4   *If every c.e.* $X \leq_m A$, *we say* A *is* m-**complete**.

THEOREM 7.3.5

Every creative set C *is m-complete.*

PROOF   Let C be creative with creative function f, and let A be any c.e. set. Then there is a computable g for which

$$W_{g(x,y)} = \begin{cases} \{f(x)\} & \text{if } y \in A, \\ \emptyset & \text{if } y \notin A \end{cases}$$

This is because one can computably enumerate such a $W_{g(x,y)}$ by (1) enumerating nothing into $W_{g(x,y)}$ until y enters A, in which case (2) just enumerate $f(x)$ into $W_{g(x,y)}$ — and the index $g(x,y)$ is computably obtainable from x and y by the practical Church–Turing Thesis.

It then follows from the Fixed Point Theorem with parameters that there is a computable k such that $W_{k(y)} = W_{g(k(y),y)}$. So

$$y \in A \implies fk(y) \in W_{k(y)} \implies W_{k(y)} \not\subseteq \overline{C} \implies fk(y) \in C, \quad \text{and}$$
$$y \notin A \implies W_{k(y)} = \emptyset \implies fk(y) \in \overline{C} \quad (\text{since } \emptyset \subseteq \overline{C}).$$

It follows that $A \leq_m C$ via fk. $\qquad\qquad$ □

So we finally have the following quite beautiful result, which does much to bolster our view of $\leq_m$ as a very *natural* reducibility.

COROLLARY 7.3.6 (John Myhill, 1955)
The set of all creative sets is exactly $= \mathbf{0}'_m$.

PROOF By Corollary 7.3.2, $\mathbf{0}'_m \subseteq$ the set of all creative sets.
By Theorem 7.3.5, the set of all creative sets $\subseteq \mathbf{0}'_m$. $\qquad$ □

REMARK 7.3.7 So where has all this got us to? Having proclaimed in the introduction to this chapter the importance of the incomputable in the real world, and developed a first model of this incomputable universe, we seem to be finding that all the *mathematically* natural examples of incomputable sets occupy a small corner of our model. Increasingly we find that the apparent variety of unsolvable problems found by Church, Turing and others in the thirties is subsumed by the creative sets — and these are to be found exclusively in $\mathbf{0}'_m$.

Although this is not unexpected — most of what is "natural" according to everyday terminology is not unique mathematically — it is an undeniable obstacle to the recognition of incomputability. And the circumstantial evidence — the uncovering of the rich degree structure for the c.e. degrees, and its correspondingly rich information content — still leaves a gap for those mathematicians dissatisfied with the available natural examples to worry (themselves and us) about.

The next chapter puts this in context. Gödel's Incompleteness Theorem tells us that even if the things we desire do exist, proving it may be very hard indeed. $\qquad\qquad$ □

Finally — just to make things worse — it is time to notice the inadequacy of m-reducibility to base our model of the incomputable universe on. Despite the naturalness of $\leq_m$ and its usefulness in abstracting some very common reductions between problems, it is just not *general* enough to model the algorithmic content of computationally complex environments. For instance, a basic intuition is that if a set A can be computed, then so can its complement $\overline{A}$. However:

EXAMPLE 7.3.8   *Show that, in general, $\overline{A} \not\equiv_m A$.*

SOLUTION   Take $A = K$.

Then $\overline{K} \not\leq_m K$ — since otherwise we would have $\overline{K}$ c.e. by Theorem 7.1.3.

And this would give K computable by the Complementation Theorem, contradicting what we found in Theorem 5.3.1.   ⧠

In Part II we will modify our model, using the notion of an *oracle Turing machine*. But first let us take what is probably the most celebrated mathematical theorem of the 20th century and put it in a computability-theoretic context.

Chapter 8

Gödel's Incompleteness Theorem

Gödel's Incompleteness Theorem (actually there are two of them) is a theorem about axiomatic theories — first order axiomatic theories. It tells us that if such a theory has enough in the way of language and axioms — enough, that is, to capture basic high-school arithmetic — then it cannot prove all the true statements in that language.

The theorem does have general implications. In mathematics, it torpedoed Hilbert's Programme for capturing all of true mathematics in suitable axiomatic theories. And it reflects badly on reductionism more generally. But having said that it is not as deep as some people will try to tell you. The real world is not reducible to an axiom system. The theorem does not explain quantum ambiguity, or consolidate a post-modern worldview in which truth itself becomes relative. And in this chapter, I will present it as just another aspect of some basic computability theory.

8.1 Semi-Representability and C.E. Sets

Gödel's Theorem applies in the first instance to first order Peano arithmetic, although it will be easy to see it actually applies to any theory satisfying a small number of basic requirements. We already know the link between computability and PA — the computable functions, sets and relations are exactly the ones *representable* in PA. What about the c.e. sets?

> **CRUCIAL QUESTION:**   What can we say about the c.e. sets within the theory PA? Or within any first order theory T?

Let us start by doing what Gödel did — computably assign Gödel numbers to the theory PA in the same way we did it for Turing machines in Section 4.2. Gödel originally wanted this "arithmetisation of arithmetic" to help him prove facts about PA — such as its consistency — within PA. But as we shall see, it led to unexpected results!

Again, we can obtain the Gödel numbering by first assigning primes to the individual symbols. Then if $\psi = s_0 s_1 \ldots s_n$ is sequence of symbols of $\mathcal{L}_{PA}$, we can define $gn(\psi) = 2^{gn(s_0)} \times 3^{gn(s_1)} \times \cdots \times p_n^{gn(s_n)}$. And we can do the same sort of thing to get $gn(\overrightarrow{\psi})$ for $\overrightarrow{\psi}$ a list of wfs of PA. Just as before, the unique factorisation theorem will ensure that gn has a computable inverse, and because we can computably tell, for instance, whether ψ is a wf, or whether $\overrightarrow{\psi}$ is a proof in PA, we have lots of corresponding computable relations, such as:

$$Form(m) \iff_{\text{defn}} gn^{-1}(m) \text{ is a wf of PA}$$

$$Ax(m) \iff_{\text{defn}} gn^{-1}(m) \text{ is an axiom of PA}$$

$$MP(m,n,p) \iff_{\text{defn}} Form(m), Form(n), Form(p) \text{ and } gn^{-1}(p) \text{ is}$$
$$\text{derived from } gn^{-1}(m) \text{ and } gn^{-1}(n) \text{ via MP}$$

$$Gen(m,n) \iff_{\text{defn}} Form(m), Form(n) \text{ and } gn^{-1}(n) \text{ is derived from}$$
$$gn^{-1}(m) \text{ via Gen}$$

$$Proof(m) \iff_{\text{defn}} gn^{-1}(m) \text{ is a proof of PA}$$

Although these are straightforward consequences of the Church–Turing Thesis, I will make explicit the argument for the last of these.

EXAMPLE 8.1.1 *Assuming that $Form(m)$, $Ax(m)$, $MP(m,n,p)$ and $Gen(m,n)$ are computable relations, verify that $Proof(m)$ is computable.*

SOLUTION Let ℓh be the computable *length function* defined by $\ell h(m) =_{\text{defn}}$ the largest i such that $p_i | m$ — for example $\ell h(6) = \ell h(2 \times 3) = 1$. Then:

$$Proof(m) \iff gn^{-1}((m)_0) \, gn^{-1}((m)_1) \ldots gn^{-1}((m)_{\ell h(m)}) \text{ is a proof in PA.}$$

So to computably decide whether $Proof(m)$ holds or not:

(1) Check that each of $Form((m)_0), \ldots, Form((m)_{\ell h(m)})$ hold (if not $Proof(m)$ fails), and then

(2) For each $(m)_i$, check that
either $Ax((m)_i)$
or for some $p, q < i$, $MP((m)_p, (m)_q, (m)_i)$
or $Gen((m)_p, (m)_i)$.   ▯

What about the set $T_{PA} =_{\text{defn}} \{m \, | \vdash_{PA} gn^{-1}(m)\}$? Is the set of all Gödel numbers of theorems of PA computable? This much is clear:

THEOREM 8.1.2
The set T_{PA} of all Gödel numbers of theorems of PA is c.e.

PROOF Since

$$m \in T_{\text{PA}} \iff \exists p\, [Proof(p)\, \&\, (p)_{\ell h(p)} = m],$$

we have $T_{\text{PA}} \in \Sigma_1^0$. ⬜

This result easily extends to any theory which has a computable set of axioms and computable rules of deduction. Given a first order theory T we can Gödel number it just as we did for PA — in fact we can do this for *any* such theory using a Gödel numbering of first order logic, and from now on we will assume we have such a coding. We can then define Ax_T, $Proof_T$, T_T in the obvious way. Then a first order T will be *computably axiomatisable* — or just *axiomatisable* — if it has a set of axioms for which Ax_T is computable.

EXERCISE 8.1.3 Let T be a computably axiomatisable first order theory. Show that the set T_T of all Gödel numbers of theorems of T is c.e.

There is also a converse due to William Craig:

EXERCISE 8.1.4 Let T be a first order theory such that T_T is c.e. Show that T is computably axiomatisable.
[**Hint:** Cf. Example 5.1.9 and Exercise 5.1.10.]

This is a good point at which to return to our crucial question — since an answer will tell us what PA can say about T_{PA}.

We do need some consistency assumptions, since it is well known that an inconsistent theory will tell us anything we want!

DEFINITION 8.1.5 (1) A theory T is **consistent** if for each wf φ of the theory, either not-$\vdash_T \varphi$, or not-$\vdash_T \neg\varphi$.

(2) A theory T — in the language of PA — is ω-**consistent** if for each wf $\varphi(x_i)$ of $\mathcal{L}_{\text{PA}}$, if $\vdash_T (\exists x_i)\neg\varphi(x_i)$, then not-$\vdash_T \varphi(\bar{m})$ for some $m \in \mathbb{N}$.

EXERCISE 8.1.6 Show that if a first order T is **not** consistent, then $\vdash_T \psi$, for each wf ψ of $\mathcal{L}_T$.

EXERCISE 8.1.7 *Show that if T is ω-consistent, then T is consistent.*

We do not expect *every* acceptable theory to be ω-consistent. But we do believe PA to have the property.

EXAMPLE 8.1.8 *Show that if the standard model $\mathfrak{N}$ is a model of PA, then PA is ω-consistent.*

SOLUTION   Say $\mathfrak{N} \models$ PA, and that $\vdash_{PA} \varphi(\bar{m})$ for every $m \in \mathbb{N}$.

Then $\mathfrak{N} \models \neg(\exists x_i)\neg\varphi(x_i)$. So — since $\mathfrak{N} \models (\exists x_i)\neg\varphi(x_i)$ cannot hold — we cannot have $\vdash_{PA} (\exists x_i)\neg\varphi(x_i)$.   ⬚

We need one more ingredient for our answer to that crucial question.

DEFINITION 8.1.9   *We say $S \subseteq \mathbb{N}$ is **semi-representable** in PA if there exists a wf φ of $\mathcal{L}_{PA}$ such that*

$$m \in S \iff \vdash_{PA} \varphi(\bar{m}),$$

for each $m \in \mathbb{N}$.

THEOREM 8.1.10 (The Semi-Representability Theorem)
Assume that PA is ω-consistent. Then for each $S \subseteq \mathbb{N}$ the following are equivalent:

  (a) S is c.e.

  (b) S is semi-representable in PA.

  (c) $S \leq_m T_{PA}$.

PROOF $\boxed{(a) \Rightarrow (b)}$ Assume S to be c.e., and hence Σ_1^0.

And let $m \in S \iff \exists n R(m, n)$, with R computable.

Assume R is represented in PA by $\psi(x_0, x_1)$, say — and let $\varphi(x_0) =_{\text{defn}} (\exists x_1)\psi(x_0, x_1)$.

We show that $m \in S \iff \vdash_{PA} \varphi(\bar{m})$.

To show ($\Rightarrow$): Say $m \in S$, so that $(\exists n)R(m,n)$ — say $R(m,n_0)$ holds. Then — since ψ represents R — we have $\vdash_{\text{PA}} \psi(\bar{m},\bar{n})$.

Then $\vdash_{\text{PA}} \psi(\bar{m},\bar{n}) \implies (\exists x_1)\psi(\bar{m},x_1)$ (since all logically valid wfs are provable) — and MP gives $\vdash_{\text{PA}} \psi(\bar{m},x_1)$.

That is, $\vdash_{\text{PA}} \varphi(\bar{m})$.

To show ($\Leftarrow$): Say $\vdash_{\text{PA}} \varphi(\bar{m})$ — that is $\vdash_{\text{PA}} (\exists x_1)\psi(\bar{m},x_1)$.

Hence $\vdash_{\text{PA}} (\exists x_1)\neg(\neg\psi(\bar{m},x_1))$.

So by ω-consistency, $\vdash_{\text{PA}} \neg\psi(\bar{m},\bar{n})$ fails for some n — which means $R(m,n)$, since ψ represents R.

But then $(\exists n)R(m,n)$ — giving $m \in S$, as required.

$\boxed{(b) \Rightarrow (c)}$ Assume S semi-representable via φ — that is, $m \in S \iff \vdash_{\text{PA}} \varphi(\bar{m})$, each $m \in \mathbb{N}$. So

$$m \in S \iff gn(\varphi(\bar{m})) \in T_{\text{PA}}.$$

Hence $S \leq_m T_{\text{PA}}$ via $gn(\varphi(\bar{m}))$.

$\boxed{(c) \Rightarrow (a)}$ By Theorem 8.1.2, T_{PA} is c.e. So if $S \leq_m T_{\text{PA}}$, then S is also c.e., by Theorem 7.1.3. □

It immediately follows that *any* first order axiomatisable theory T is computably reducible to PA — so in a sense PA is a *universal* theory.

EXERCISE 8.1.11 *Show that if T is an axiomatisable first order theory, then $T_T \leq_m T_{\text{PA}}$.*

On the other hand, PA is not all that special — the Semi-Representability Theorem holds for a wide range of theories.

EXERCISE 8.1.12 *If T is any axiomatisable first order theory in the language of PA, show that:*

(i) If $S \subseteq \mathbb{N}$ is semi-representable in T then S is c.e. , and

(ii) If T is ω-consistent, and every computable relation is representable in T, then every c.e. S is semi-representable in T.

To end this section — remember that Corollary 3.2.15 tells us that all computable sets are representable in PA. You might have already worked out an informal proof of a converse. We can now do this more formally.

EXERCISE 8.1.13 *Show that if PA is consistent, and S is representable in PA, then both S and $\overline{S}$ are semi-representable in PA. Deduce that the sets representable in PA are exactly the computable sets.*

8.2   Incomputability and Gödel's Theorem

If we now feed an incomputable c.e. set into PA we get — with some help from the Semi-Representability Theorem — Gödel's famous incompleteness result. But first let us be more precise about what we mean by (in)completeness of a theory.

DEFINITION 8.2.1   *A theory T is **complete** if for every sentence φ of T, **either** $\vdash_T \varphi$ **or** $\vdash_T \neg\varphi$ (but not both).*

Then:

LEMMA 8.2.2

(1) Let T be any axiomatisable first order theory in which K is semi-representable. Then T is not complete.

(2) In particular — assuming ω-consistency — PA is not complete.

PROOF   (1) Let φ semi-represent K in our axiomatisable T — so T must be consistent.

Since $\overline{K}$ is not c.e., it cannot be semi-representable in T by Exercise 8.1.12. So

$$\text{not } -(\, n \in \overline{K} \iff \vdash_T \neg\varphi(\bar{n})\,).$$

But if $\vdash_T \neg\varphi(\bar{n})$ then $n \in \overline{K}$ by consistency of T.

So there must be some $n \in \overline{K}$ for which $\vdash_T \neg\varphi(\bar{n})$ is not true — and, of course, $\vdash_T \varphi(\bar{n})$ also fails since $n \notin K$ and φ semi-represents K in T.

(2) now follows immediately by the Semi-Representability Theorem.   ◻

REMARK 8.2.3   Notice — although $\varphi(\bar{n})$ is an undecidable sentence of the theory T or PA — *we* have no problem deciding the truth or otherwise of $\varphi(\bar{n})$, since we know that $n \notin K$. This feature — which for some people shows that humans are cleverer than axiomatic theories — will carry over to Gödel's Theorem and all the results building on Lemma 8.2.2.   ◻

There are various directions in which we can develop this result — for example, we can vary the conditions on T and show we still get incompleteness:

EXERCISE 8.2.4 *Let $\mathcal{T}$ be an ω-consistent computably axiomatisable theory in which every computable relation is representable.*

Show that $\mathcal{T}$ is incomplete.

[**Hint:** Use Exercise 8.1.12.]

Or we can show that it applies to a whole range of theories relatable to PA. We will have to designate one of these as "Gödel's Incompleteness Theorem" — although of course all these results flow directly from Gödel's early 1930s papers.

DEFINITION 8.2.5 *Let $\mathcal{T}$ be a theory and Σ be a set of wfs of $\mathcal{L}_{\mathcal{T}}$.*

*(1) We call $\mathcal{T}' = \mathcal{T} \cup \Sigma$ — the theory $\mathcal{T}'$ got by adding the wfs in Σ to $\mathcal{T}$ as extra axioms — an **extension** of $\mathcal{T}$.*

*(2) Further — $\mathcal{T}'$ is a **finite extension** if Σ is finite.*

THEOREM 8.2.6 (Gödel's First Incompleteness Theorem)
There is no ω-consistent axiomatisable extension of PA which is complete.

PROOF Let $\mathcal{T}$ be an ω-consistent axiomatisable extension of PA.

Clearly any wf provable in PA is also provable in $\mathcal{T}$. In particular — if φ represents a relation R in PA, then φ represents R in $\mathcal{T}$. Hence every computable relation is representable in $\mathcal{T}$ (as well as in PA).

The result follows by Exercise 8.2.4. ∎

REMARK 8.2.7 (1) Barkley Rosser has shown how we can replace ω-consistency with just consistency in all the previous results — including Gödel's Theorem — although doing that does take a little extra work.

(2) We can formulate "PA is consistent" as a number theoretic statement, say:

$$Consis_{\mathrm{PA}} \iff_{\text{defn}} \text{not } \vdash_{\mathrm{PA}} \neg(\bar{0} = \bar{0}) \iff gn(\neg(\bar{0} = \bar{0})) \notin \mathcal{T}_{\mathrm{PA}}.$$

And this means we can follow Gödel in trying — and failing — to provide Hilbert with the proof of consistency of PA within PA that he so much wanted:

GÖDEL'S SECOND INCOMPETENESS THEOREM:
Not ⊢_PA *Consis*_PA — that is, PA is unable to prove its own consistency.

Again, the proof involves some extra non-computability theoretic work —
mainly in proving within PA that *Consis*_PA implies a cleverly chosen, known
unprovable φ.   ⬜

EXERCISE 8.2.8 *Given the standard model $\mathfrak{N}$ for PA, show that PA **does**
have a compete extension.*
[**Hint:** Define

$$\Sigma = \{\varphi \mid \varphi \text{ is a sentence of PA and } \mathfrak{N} \models \varphi\},$$

and show that the theory Th($\mathfrak{N}$) = PA ∪ Σ is a complete extension of PA.]

Th($\mathfrak{N}$) is sometimes called (the theory of) *true first order arithmetic*. It is of
course very far from being computably axiomatisable, and its set of theorems
is so far from being Σ_1^0 (see Exercise 9.1.3) that it cannot be *described* in the
language of PA using *any number of quantifiers*.

REMARK 8.2.9 (1) Arithmetic is basic to mathematics, and this is why
Hilbert was so keen to achieve a proof of its consistency — and using the
methods of elementary arithmetic itself. This is why Gödel's Incompleteness
Theorem is so far-reaching in its consequences. Arithmetic is an ingredient of
very many other important mathematical theories — and so these, with some
minor adjustments to the language, can be viewed as *extensions* of PA, and
so subject to Theorem 8.2.6.

For instance, we can define the numbers and the operations and relations
of elementary arithmetic within the language of set theory. And in such a
way that we can derive all the facts of elementary arithmetic — as captured
in PA — within the standard first order theory for sets, usually termed ZFC
(consisting of the Zermelo axioms as augmented by Abraham Frankel, plus
the Axiom of Choice). So ZFC is incomplete.

(2) So, is Hilbert's Programme completely killed off by Gödel's Theorem?
No — we can still argue that all the mathematics we *really need to know* is
computably derivable, via a suitable axiom system, say.

We have seen how elusive are natural examples of incomputable objects.
Can we find examples of true mathematical statements the average scientist
would be interested in which are not provable in PA or ZFC?

It turns out that not only is most mathematics of interest to non-logicians
provable in nothing more complicated than PA or ZFC — it is provable in

quite weak sub-systems or *fragments* of such theories. And we can typically measure the strength of such fragments according to how complicated the wf φ is allowed to be in the induction axiom. Many such results come out of a very active research area — called *reverse mathematics* — devoted to classifying mathematical theorems according to how weak a fragment can prove them. If you are interested, there is an impressive book by Stephen G. Simpson on *Subsystems of Second Order Arithmetic* with a lot on reverse mathematics.

On the other hand, there have been quite dramatic discoveries of basic statements undecidable in each of ZFC and PA. Most famously, Cantor's old question — Hilbert's First Problem — asking exactly how many real numbers there are, cannot be settled in ZFC. More precisely, the Continuum Hypothesis (CH) we earlier saw in Chapter 3 is *independent* of ZFC. We will benefit from some of the technical fallout from Paul Cohen's 1963 proof of that fact in our Chapter 13, *Forcing and Category*.

More down-to-earth — there are quite simple statements in number theory (which can be easily described to a non-logician) that have proofs which cannot be carried out in PA. The first of these appeared — the Paris–Harrington version of Ramsey's Theorem — in 1976. There have been many more, some of the most natural and interesting being due to the remarkable logician Harvey Friedman.

However, arguably one of the oldest and best theorems unprovable in PA is Goodstein's Theorem, dating back to 1944 — and I say this not just because R.L. Goodstein (1912 – 1985) was my thesis supervisor — or because he was the first mathematical logician to hold a chair at a British university!   ☐

Let us finish with a brief description of Goodstein's Theorem. For a particularly nice computability theoretic proof of its independence from PA, you should go to Adam Cichon's 1983 paper in the *Proceedings of the American Mathematical Society*.

GOODSTEIN'S PROCESS: Given a natural number N:

1) Write it in base x — i.e., write N as a sum of powers of x, and then the exponents also, and the exponents of exponents, etc.

2) Increase the base of the representation by 1, then

3) Subtract 1 from the new number thus obtained.

4) Repeat the procedure 1) – 3), successively increasing the base by 1 and subtracting 1.

The process is said to **terminate** if the number 0 is eventually reached.

Struggling to get some insight into what is happening? Let us see if an example will help.

EXAMPLE 8.2.10   For $x = 2$, $N = 25$ we get step 1):

$$25 = 2^{2^{2^1}} + 2^{2^1 + 2^0} + 2^0$$

Then 2) changing the base to 3 gives $3^{3^{3^1}} + 3^{3^1 + 3^0} + 3^0$.
And 3) subtracting 1 gives $3^{3^3} + 3^{3+1} = 7625597485068$.

Then Goodstein's Theorem states that *this process always terminates —* and the theorem of Kirby and Paris from 1982 tells us that *this cannot be proved in PA.*

You will notice that the Goodstein process terminates rather sluggishly, taking in on the way some ludicrously large numbers! The next number reached in Example 8.2.10 is already huge — over 150 digits — and when we get to base 5 we arrive at a number of 2216 digits. Even $N = 4$ sets in motion a process which takes around 10^{129} iterations to terminate!

The way such a complex phenomenon is based on an iteration of a simple mechanical process has something in common with the genesis of fractals and chaotic natural phenomena. Interestingly, Paris and Tavakol have brought this out in a 1993 paper ("Goodstein Algorithm as a Super-Transient Dynamical System") in *Physics Letters A*, describing how the number 0 emerges from instances of the Goodstein process in the same way as strange attractors manifest themselves in turbulent environments.

This brings us back to computability. It is not just the proof theoretic capabilities of PA which are constrained by its mechanical structure and reliance on induction, as we shall see in the next chapter.

Chapter 9

Decidable and Undecidable Theories

We have seen that any attempt to capture mathematics in Hilbert's proof-theoretical framework is doomed.

What is worse, we now notice, is that provability — in PA and a wide range of theories related to PA — is necessarily *unpredictable*. That is, we cannot computably decide whether a given sentence is going to be provable or not. And this is a big problem when we have limited time to wait for our theory to deliver a decision on a given φ, and the theory is incomplete — and so does not guarantee to tell us *anything* about φ!

9.1 PA is Undecidable

Decidability is for theories what *computability* is for sets.

DEFINITION 9.1.1 *Let* $\mathcal{T}$ *be a first order theory. If* $T_{\mathcal{T}}$ *is computable, then we say that* $\mathcal{T}$ *is* **decidable**.

Intuitively, $\mathcal{T}$ is decidable if we can computably decide, for each φ, whether $\vdash_{\mathcal{T}} \varphi$ or not.

EXERCISE 9.1.2 *Show that if* $\mathcal{T}$ *is decidable, then* $\mathcal{T}$ *is computably axiomatisable.*

EXERCISE 9.1.3 *Show that the theory* $\mathrm{Th}(\mathfrak{N})$ *of true first order arithmetic is not computably axiomatisable, and hence is not decidable.*
Also show that the set of theorems of $\mathrm{Th}(\mathfrak{N})$ *is not c.e.*

From the Semi-Representability Theorem we have that $K \leq_m T_{\mathrm{PA}}$. Then the incomputability of K immediately tells us that T_{PA} is not computable — and so PA is undecidable. In fact we can prove something a lot stronger.

> **DEFINITION 9.1.4** *A first order theory* T *is said to be* **creative** *if* T_T
> *is creative.*

Then:

> **THEOREM 9.1.5**
> *PA is creative — and hence PA is not decidable.*

EXERCISE 9.1.6 *Show that if an incomputable set S is semi-representable in* T, *then* T *is undecidable.*

9.2 Other Undecidable Theories and Their Many–One Equivalence

We can now — not very surprisingly — get a range of undecidable theories in the same way that we showed PA undecidable.

EXERCISE 9.2.1 *Let* T *be any axiomatisable* ω-*consistent first order theory in the language of PA.*

Show that if every computable relation is representable in T, *then* T *is creative — and hence is undecidable.*

EXERCISE 9.2.2 *Show that every* ω-*consistent axiomatisable extension of PA is creative.*

Once again, of course, we can replace ω-consistency with consistency in the above results.

Now let us see something less expected, and in some ways even more basic than the incompleteness and undecidability of PA.

Surely we can trim PA down to a decidable theory by disposing of its special axioms? Actually — *no*: A remarkable theorem of Alonzo Church shows that just the richness of language can be enough to make a theory undecidable, and even creative. What is so fundamental about this result is that it eventually leads to the undecidability of first order logic itself. Its proof needs the following:

LEMMA 9.2.3
Let T' be a finite extension of T. Then $T_{T'} \leq_m T_T$.

PROOF Let $T' = T' \cup \Sigma$, where $\Sigma = \{\varphi_1, \ldots, \varphi_n\}$ is a finite set of sentences φ_i of $\mathcal{L}_T$. Then for each $m \in \mathbb{N}$:

$$
\begin{aligned}
m \in T_{T'} &\Longleftrightarrow \vdash_{T'} gn^{-1}(m) \\
&\Longleftrightarrow \varphi_1, \ldots, \varphi_n \vdash_T gn^{-1}(m) \\
&\Longleftrightarrow \varphi_1 \& \ldots \& \varphi_n \vdash_T gn^{-1}(m) \\
&\Longleftrightarrow \vdash_T \varphi_1 \& \ldots \& \varphi_n \Rightarrow gn^{-1}(m) \quad \text{(by the deduction theorem)} \\
&\Longleftrightarrow \underbrace{gn(\varphi_1 \& \ldots \& \varphi_n \Rightarrow gn^{-1}(m))}_{\text{computable function of } m} \in T_T.
\end{aligned}
$$

So $T_{T'} \leq_m T_T$, as required. $\qquad\Box$

This result is not yet relevant to interesting subtheories of PA, since the axiom (N7) of PA is in fact a *scheme*. This prevents our viewing PA as a finite extension of the more interesting of its subtheories. We are rescued by an early piece of reverse mathematics due to Raphael Robinson (the husband of Julia Robinson, who played such an important role in the solving of Hilbert's Tenth Problem).

RAPHAEL ROBINSON'S THEOREM
   There is a **finitely** axiomatisable theory Q in the language of PA such that:
   (a) All computable functions are representable in Q, and
   (b) For each wf φ of PA, $\vdash_Q \varphi \implies \vdash_{PA} \varphi$ (so that if PA is consistent, or ω-consistent, so is Q).

The idea behind the proof is that if one checks the proof that all computable functions are representable in PA, one sees that just a few instances of the induction axiom (N7) appear. In fact one can absorb the role of (N7) into the single axiom

$$
x_1 \neq \bar{0} \rightarrow \exists x_2 \, (x_1 = x_2'),
$$

added to the usual axioms (N1) – (N6) of PA.

COROLLARY 9.2.4

If PA is (ω-)consistent, then Q is incomplete and undecidable — and even creative.

PROOF　As for PA we get every c.e. set semi-representable in Q — from which everything else follows in just the same way as for PA.　　　　□

We cannot yet use Lemma 9.2.3 to show first order logic undecidable, since PC has a different language from that of PA. But:

EXERCISE 9.2.5　*Let PC^- be PC, with language restricted to that of PA. Then PC^- is undecidable and creative.*

Here is the final ingredient linking the undecidability of a PA-like theory to that of predicate calculus.

LEMMA 9.2.6

If φ is a wf of PA, then $\vdash_{PC^-} \varphi \iff \vdash_{PC} \varphi$.

　　Hence $T_{PC^-} \leq_m T_{PC}$.

PROOF　Let φ be a wf of PA. Then Gödel's Completeness Theorem applies to both PC and PC^- — giving

$$\vdash_{PC^-} \varphi \iff \varphi \text{ is logically valid} \iff \vdash_{PC} \varphi$$

So — remembering $Form(m)$ holds iff $gn^{-1}(m)$ is a wf of PA, and that $Form$ is a computable relation — we have for each $m \in \mathbb{N}$ that

$$m \in T_{PC^-} \iff Form(m) \And m \in T_{PC} \iff \chi_{Form}(m) \times m \in T_{PC},$$

giving $T_{PC^-} \leq_m T_{PC}$ via $m \mapsto \chi_{Form}(m) \times m$.　　　　□

Then

THEOREM 9.2.7 (Church's Theorem)

PC is not decidable.

　　Hence there is no computer program for deciding of a given wf whether or not it is logically valid.

PROOF Immediate from $T_{\mathrm{PC}-} \leq_m T_{\mathrm{PC}}$ and — Exercise 9.2.5 — the undecidability of $T_{\mathrm{PC}-}$. ⧠

REMARK 9.2.8 What is so remarkable about this theorem is that it is counter-intuitive — part of being intelligent, we believe, is being able to mentally cut through the illogicality of the world around us. This is not a personal conceit on our part — it is a comfortable belief in ourselves as members of that post-enlightenment scientific community which can at least tell what is logical and what is not — and without using anything beyond what a computer might be capable of. But Church's Theorem says there is nothing simple even about logicality. It shows us computability, and the reductive structures it is built on, breaking down in a quite unexpected way. ⧠

And there are other striking consequences:

COROLLARY 9.2.9

PC is creative.

PROOF Since the creative set K is semi-representable in $\mathcal{Q}$, we get

$$K \leq_m T_{\mathcal{Q}} \leq_m T_{\mathrm{PC}-} \text{ (by Lemma 9.2.3) } \leq_m T_{\mathrm{PC}}.$$

The result follows by Theorem 7.3.1. ⧠

REMARK 9.2.10 So we now find that all the undecidable theories, all the incomputable sets, and all the unsolvable problems coming out of the work of Gödel, Turing, Church and Kleene in the 1930s, are only superficially different. They are all captured via a creative set, and so all are many–one equivalent and occupy the same many–one degree $\mathbf{0}'_m$. We already suspect that there are very different incomputable c.e. sets — simple sets for example — whose diverse information content entails a rich degree-theoretic structure. We also suspect that there is naturally arising incomputability which is no more mathematically special than is an arbitrary diophantine set.

But Theorem 9.2.9 is certainly one for the sceptics. We do not even have K semi-representable in PC, but the theory still ends up creative. In fact given a few more tricks for linking PA to theories of other structures — for instance, by *defining* aspects of one structure in terms of another — we can get all sorts of superficially different theories which also turn out to be creative. ⧠

EXERCISE 9.2.11 *Let T be a first-order theory, all of whose theorems in the language of PA are true in the standard model $\mathfrak{N}$ of PA. Show that T has a finite extension in which all the computable functions are representable.*

Deduce that T is undecidable.

[**Hint**: Consider the two cases depending on whether T is computably axiomatisable or not.]

EXERCISE 9.2.12 (Tarski, Mostowski & Robinson, 1953)   *We say a first order theory T is **strongly undecidable** if it is finitely axiomatisable, and every theory T' in the language of T that is consistent with T — that is, such that $T \cup T'$ is consistent — is undecidable.*

Show that Q is strongly undecidable.

EXERCISE 9.2.13   *We say $S \subseteq \mathbb{N}$ **separates** $A, B \subseteq \mathbb{N}$ if $A \subseteq S$ and $B \subseteq \overline{S}$. Then $A, B \subseteq \mathbb{N}$ are **computably inseparable** if no computable S separates A from B. A theory T is said to be **computably inseparable** if T_T is computably inseparable from $\{gn(\varphi) \mid \vdash_T \neg\varphi\}$.*

Show that PA is computably inseparable.

Deduce that there exist disjoint c.e. sets A and B which are computably inseparable.

[**Hint:** Show that if a c.e. set S separates PA, then $K \leq_m S$ giving S creative.]

9.3   Some Decidable Theories

Please do not leave this chapter thinking that *any* useful mathematical theory is incomplete and undecidable! Just to redress the balance, I ought to finish with some examples of decidable theories which — although they avoid saying much about arithmetic — still capture some significant mathematics.

Let us first observe the following:

PROPOSITION 9.3.1
Any complete axiomatisable theory is decidable.

Informally you can see how to decide whether a wf φ is provable in the theory or not — just enumerate the wfs provable, and those whose negations are provable, until we observe φ appear in one of these computable lists.

The next exercise asks for a more detailed proof, and shows how incompleteness of PA is a simple consequence of its undecidability.

EXERCISE 9.3.2   *Let T be a computably axiomatisable first order theory.*

(i) Show that $S \leq_m T_T$ where

$$S = \{n \mid Form_T(n) \ \& \ \vdash_T \neg gn^{-1}(n)\}.$$

(ii) Deduce that S is c.e.

(iii) Show that if T is complete then

$$\mathbb{N} - T_T = S \cup \{m \mid \neg Form_T(m)\},$$

and hence that T is decidable.

Hence show that Gödel's Incompleteness Theorem for PA can be deduced from the fact that PA is undecidable.

EXERCISE 9.3.3 *Let T be axiomatisable, and let $S \subseteq \mathbb{N}$ be simple. Show that if $T_T \leq_m S$ then T is decidable.*

EXERCISE 9.3.4 *Let T be axiomatisable. Show that T is decidable if and only if the set of sentences φ such that $T \cup \{\varphi\}$ is consistent is c.e.*

EXERCISE 9.3.5 *Show that if T is decidable, then any finite extension of T is decidable.*

Deduce that any consistent decidable theory T has a complete decidable extension.

There are many model-theoretic techniques for proving completeness of a theory, but unfortunately these do not really belong in a book about computability theory. Examples of complete, and hence decidable, axiomatisable theories include those of densely ordered sets without first or last element, or of algebraically closed fields of given characteristic. Both of these can be proved using a well-known result due to Robert Vaught. I will give the proof because it is both informative and easy!

THEOREM 9.3.6 (Vaught's Completeness Test)
Let T have no finite models, and assume all its models of some infinite cardinality λ are isomorphic. Then T is complete.

PROOF Assume T as in the theorem, but that neither φ nor $\neg\varphi$ are provable in T, for some sentence φ.

Then both $T \cup \{\varphi\}$, $T \cup \{\neg\varphi\}$ are consistent, and have models $\mathfrak{M}_1$, $\mathfrak{M}_2$ — which must be infinite since they are models of T. Using the downward and upward Löwenheim–Skolem Theorems, $\mathfrak{M}_1$, $\mathfrak{M}_2$ can be taken to be of cardinality λ, and hence isomorphic.

But $\mathfrak{M}_1 \models \varphi$ and $\mathfrak{M}_2 \models \neg\varphi$, a contradiction. ▯

For the theory of densely ordered sets without first or last element, one can easily construct an isomorphism between any two countable models. And for algebraically closed fields of given characteristic, we can do it for models of any given uncountable cardinal. But we are already straying too far from our subject. Model theory is a beautiful subject, but best pursued in earnest elsewhere!

I will end this chapter by mentioning one other way of showing decidability. Sometimes one can decide the question of whether $\vdash_T \varphi$, or not, by reducing it to checking whether $\mathfrak{M} \models \varphi$, or not, in a finite number of finite models $\mathfrak{M}$ of T — so long as this essentially covers *all* possible models of T. One can sometimes ease this process by transforming φ into something more approachable — say by using "elimination of quantifiers" to replace φ by a quantifier-free wf with similar model-theoretic properties.

What underlies such an approach is the basic decision procedure for validity of propositional statements. If $\mathcal{A}$ is a formula of propositional calculus, made up from propositional variables and logical connectives in the usual way — then $\mathcal{A}$ is *valid* if it takes truth-value "true" for all possible assignments of true or false to its propositional variables. It is well known that we can computably decide the validity of a given $\mathcal{A}$ by writing down its truth-table showing its truth-value for each such interpretation, using the usual interpretations of the connectives. In effect, we check the validity of $\mathcal{A}$ in a finite number of finite "models" of propositional calculus.

EXERCISE 9.3.7 *Let $SAT(\mathcal{A})$ hold if and only if the propositional formula $\mathcal{A}$ is* satisfiable *— that is, takes truth-value "true" for some assignment of truth-values to its propositional variables.*

Explain why SAT is a computable relation.

EXERCISE 9.3.8 *Let T be a theory whose language contains only finitely many constant, function and predicate symbols. Show that if T has only finite models, then T is decidable.*

EXERCISE 9.3.9 *Show that* **pure monadic predicate calculus** *— that is PC with no function of constant symbols and restricted to predicates of just one argument — is decidable.*

[**Hint:** Consider any interpretation of a wf φ of pure monadic predicate calculus. Show that the members of the domain of the interpretation fall into 2^k equivalence classes, where k is the number of unary predicate symbols appearing in φ. Observe that φ is valid if and only if it is true in every interpretation with $\leq 2^k$ members in its domain.]

On the other hand, just one binary predicate symbol in PC is enough to make it undecidable.

Here are two last exercises for the more intrepid reader.

EXERCISE 9.3.10 *Show that the theory of densely ordered sets with both least and greatest element is complete and hence decidable.*

EXERCISE 9.3.11 *Let the theory SUC of successor have the following axioms involving the successor and zero symbols from PA:*

1. $\forall x_1 \neg (x_1' = \bar{0})$

2. $\forall x_1 (x_1 = \bar{0}) \vee \exists x_2 (x_2' = x_1))$

3. $\forall x_1, x_2 (x_1' = x_2' \rightarrow x_1 = x_2)$

4. $\forall x_1 \neg (x_1^{(n)} = x_1)$ — *for each positive integer n, where* $x_1^{(n)} = x_1^{\overbrace{'' \cdots '}^{n \text{ times}}}$.

Show (a) that SUC has no finite models, (b) that SUC has non-isomorphic countable models, but (c) all models of SUC of cardinality $2^{\aleph_0}$ are isomorphic.

Deduce that SUC is complete.

Part II

Incomputability and
Information Content

Chapter 10

Computing with Oracles

Incomputability emerges at the edge of computability. Its origins are mathematically uncomplicated enough to produce this complex and intimate relationship. And just as the noncomputable universe is woven from the algorithmic fabric of everyday life, so the structures on which we base its analysis are derived from appropriate computable relationships on information content — itself abstracted from the way science describes the material universe.

In this chapter we will describe a model of computationally complex environments due to Alan Turing. This model originated in a difficult and still slightly mysterious paper of Turing's from 1939, shortly before he disappeared into the secret wartime world of Bletchley Park and its clandestine decrypting mission. In it Turing became one of the first of many to try to puzzle out deeper meaning from within Gödel's Theorem.

To many of us, it seems slightly odd, verging on the paradoxical, that given any axiomatisation of arithmetic one can show it incomplete by writing down an undecidable sentence of the theory — a sentence that it is clear to us, but not to the theory, is actually true! How is it that we as mathematicians seem to be able to go beyond the incompleteness encountered in the proof of Gödel's Theorem? For Turing it was a short step to looking for the means for exploring that world of incomputability, so tantalisingly close but just beyond the safety of theories and algorithms. For him — and for us — oracle Turing machines provide the basic exploratory tool.

The aim in this chapter is to define a reducibility between sets which is intuitively *natural*, more *general* than $\leq_m$, and on which we can base a *degree structure* in the same way as we did for $\leq_m$.

10.1   Oracle Turing Machines

Oracle Turing machines contain a new sort of quadruple that enables them to use information from the real world — presented to the machine via an "oracle".

THE INTUITION: *Let $A, B \subseteq \mathbb{N}$. We want B to be computable from A if we can answer "Is $n \in B$?" using an algorithm whose computation from input n uses finitely many pieces of information about membership in A — say answers to "Is $m_0 \in A$?", "Is $m_1 \in A$?", ..., "Is $m_k \in A$?".*

If we wanted to adapt the recursive model to capture this intuition, we could follow Kleene in just adding the characteristic function χ_A of A to the list of initial functions in Definition 2.1.18. We could then define notions of *A-partial recursive* and *A-recursive* just as we previously defined those of partial recursive and recursive.

But we do prefer Turing's machine-based model — it seems to capture the intuition more directly and is the one most often used for measuring the *complexity* of relative computations.

This is the new kind of quadruple Q — called a *query quadruple* — added by Turing in his 1939 paper:

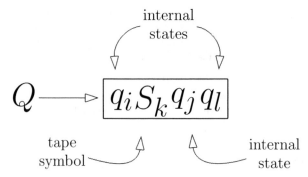

DEFINITION 10.1.1 (Relative Turing Computability)

*(1) An **oracle Turing machine** $\widehat{T}$ is a finite consistent set of quadruples — having both query quadruples and quadruples of the usual kind.*

$\widehat{T}$ *computes in the usual way from input n, except that when it applies $q_i S_k q_j q_l$, it counts up the number of 1's on the tape (say, n), asks the **oracle** $(A,$ say$)$ "Is $n \in A$?" — and if the answer is "**yes**" goes into state q_j, and if "**no**" goes into q_l.*

*(2) A (possibly partial) function ψ is **A-Turing computable** if ψ is computable by an oracle Turing machine with oracle A.*

*(3) Then B is said to be **A-Turing computable** (written $B \leq_T A$) — or **Turing reducible** to A — if χ_B is A-Turing computable.*

REMARK 10.1.2 (1) The program for $\widehat{T}$ is independent of the oracle A — it works for *any* oracle.

(2) We will identify a set A with its characteristic function — in practice that means we often write, for example, $A(n) = 1$ for $n \in A$.

Can we extend $\leq_T$ to relations and functions, while retaining the mathematically neat focus on sets? Yes — intuitively, a function f or a relation R is computably equivalent to its graph. So we just identify $f(\overrightarrow{x})$ or $R(\overrightarrow{x})$ with their coded graphs:

$$\mathrm{Graph}(f) =_{\mathrm{defn}} \{\langle \overrightarrow{x}, y \rangle \mid f(\overrightarrow{x}) = y\}$$
$$\mathrm{Graph}(R) =_{\mathrm{defn}} \{\langle \overrightarrow{x} \rangle \mid R(\overrightarrow{x}) \text{ holds}\}.$$

(3) Turing reducibility is a *generalisation* of many–one reducibility —

$$\boxed{B \leq_m A \implies B \leq_T A}$$

To see this — let $B \leq_T A$ via f and show how to convert the machine T which computes f into an oracle machine $\widehat{T}$ which computes B from A.

First notice that — adapting the solution to Example 2.4.14 — we can assume that from input n, T provides the output as $\varphi_T(n)$ consecutive 1's, with the reading head halting on the leftmost 1 — and in halting state q_M, say, with no q_i in T for any $i > M$. Then $B \leq_T A$ via the oracle machine:

$$T \cup \{ q_M 1 q_{M+1} q_{M+2}, \; q_{M+1} 1 R q_{M+2}, \; q_{M+2} 1 0 q_{M+3}, \; q_{M+3} 0 R q_{M+2} \}.$$

Of course, once we are happy that Turing reducibility fully expresses our intuitive concept of relative computability, we can give a less formal proof — it is clear that if $B \leq_m A$ then we decide if $n \in B$ by asking the oracle the single question: "Is $f(n) \in A$?"

Since $n \in B$ only depends on *positive* answers to questions to the oracle, we call $\leq_m$ — and any other such reducibility — a *positive reducibility*.

(4) Having decided to use 1 to frame questions to the oracle, we cannot simultaneously use it to help the machine store information — so more complex oracle machines do need at least one extra tape symbol apart from 0 and 1. Another way to avoid this conflict would be to use an extra *query tape*. ⬚

To help us familiarise ourselves with the new machines:

EXAMPLE 10.1.3 *Show that $\leq_T$ is reflexive.*

SOLUTION We show $A \;(= \chi_A)$ is A-Turing computable, each $A \subseteq \mathbb{N}$.

A suitable oracle Turing machine is:

$q_0 1 0 q_1$ }   ←   { erase the extra 1 from input n

$q_1 0 q_2 q_3$ }   ←   { ask the oracle "$n \in A$?"

$q_2 0 1 q_3$ }   ← $\begin{cases} \text{if } n \in A, \text{ reinstate the erased 1 for} \\ \text{eventual output, and} \ldots \end{cases}$

$\left. \begin{array}{l} q_3 0 R q_4 \\ q_4 1 0 q_3 \end{array} \right\}$ ← $\begin{cases} \ldots \text{ joining with the case } n \notin A, \text{ move} \\ \text{right in search of remaining 1's to erase} \end{cases}$ $\square$

EXERCISE 10.1.4   *Show that $\leq_T$ is weaker than $\leq_m$ as a relation on $2^{\mathbb{N}}$ — in that there exist $B \leq_T A$ for which $B \not\leq_m A$.*

[**Hint:** Show $A \leq_T \overline{A}$ for all $A \subseteq \mathbb{N}$.]

10.2   Relativising, and Listing the Partial Computable Functionals

We can now go back and replace Turing machines by oracle Turing machines with given oracle in what we did earlier — that is, we can **relativise** the theory of computable functions and c.e. sets to any given oracle A, say. It will be a long time before we come across some theory that actually depends on the individual properties of the oracle A!

The intuitive and technical basis of the Church–Turing Thesis immediately relativises.

> **THE RELATIVISED CHURCH–TURING THESIS**
>
>   *All formalisations of "B computable from A" which are sufficiently reasonable and sufficiently general are equivalent to the intuitive one above — and so can be written $B \leq_T A$.*

So we will call any ψ which is Turing computable from oracle A *either A-computable* (if it is known to be total) *or A-partial computable (A-p.c.)* otherwise.

Using the relativised Church–Turing Thesis we can easily get:

> **EXAMPLE 10.2.1**   *Show that $\leq_T$ is transitive.*

SOLUTION   Say $C \leq_T B$ and $B \leq_T A$.

This means we can computably answer any question "Is $n \in C$?" using answers to finitely many questions "Is $m_0 \in B$?", "Is $m_1 \in B$?", ..., "Is $m_k \in B$?" Since each of these can be computably answered using answers to finitely many questions "Is $p_0 \in A$?", "Is $p_1 \in A$?", ..., "Is $p_l \in A$?" — we can computably answer "Is $n \in C$?" using answers to just finitely many questions about membership in A.

Hence $C \leq_T A$.   □

Corresponding to Definition 4.2.1, we use the same Gödel numbering we used there to get:

DEFINITION 10.2.2   *The eth oracle Turing machine $\widehat{P}_e$ is defined by*

$$\widehat{P}_e = \begin{cases} \widehat{P} & \text{if } gn^{-1}(e) \downarrow = \text{ some oracle Turing program } \widehat{P} \\ \emptyset & \text{otherwise.} \end{cases}$$

REMARK 10.2.3   (1) Notice that we can think of $\widehat{P}_e$ as having arguments n (the input) *and* A (the oracle) — so computing a *functional* — that is a mapping which has arguments which are themselves number theoretic functions or sets.

(2) This explains the Φ in place of the earlier φ in the next definition. Also notice the use of the *sg* function to force these functionals to capture $\leq_T$ as a relation on $2^{\mathbb{N}}$. This is just a convenience, and sometimes we actually need a listing of functionals from functions to functions.   □

DEFINITION 10.2.4 (The eth partial computable functional)

(1) We write Φ for the functional corresponding to the eth oracle Turing machine — where we define:

$$\Phi_e^A(n) = sg\,(\text{the output from } \widehat{P}_e \text{ with input } n \text{ and oracle } A).$$

(2) Then Φ_e^A will denote the — possibly partial — 0–1 valued function computed by $\widehat{P}_e$ with oracle A.

REMARK 10.2.5   Functionals of the form Φ_e are sometimes called *Lachlan functionals*. The list $\{\Phi_i^A\}_{i \in \mathbb{N}}$ contains all sets $X \leq_T A$.   □

EXERCISE 10.2.6 *Given $A \subseteq \mathbb{N}$, show that $\Phi_z^A(x)$ is an A-p.c. function of x and z, such that for each e Φ_e^A is the eth (unary) A-p.c. function.*

That is enough relativisation for the moment. We will return to it with new eyes in Section 10.4.

10.3 Introducing the Turing Universe

Turing reducibility considered over all sets of numbers — that is, as Turing might have viewed it, over the (binary) reals — is what we call the **Turing universe**.

An important first task is to present it in a theoretically approachable form. The following looks much like the earlier Definitions 7.2.1 and 7.2.3.

DEFINITION 10.3.1 *(1) We write $A \equiv_T B$ if $A \leq_T B$ and $B \leq_T A$ — and say A is **Turing equivalent** to B.*

*(2) We define the **Turing degree** — or **degree of unsolvability** — of $A \subseteq \mathbb{N}$ to be*
$$\deg(A) =_{\text{defn}} \{X \subseteq \mathbb{N} \mid X \equiv_T A\}.$$

(3) We write $\mathcal{D}$ for the collection of all such degrees, and define an ordering $\leq$ induced by $\leq_T$ on $\mathcal{D}$ by:
$$\deg(B) \leq \deg(A) \Longleftrightarrow_{\text{defn}} B \leq_T A.$$

The above definition leaves us — or rather you the reader! — with some loose ends to tie up and some basic facts to check.

Corresponding to Lemma 7.2.2:

EXERCISE 10.3.2 *Show that $\mathcal{D}$ is the set $2^{\mathbb{N}}/\equiv_T$ of all equivalence classes of $2^{\mathbb{N}}$ under the equivalence relation $\equiv_T$.*

Just as in Remark 7.2.4 we need to verify:

EXERCISE 10.3.3 *Show that if $A_0 \equiv_T A_1$ and $B_0 \equiv_T B_1$ then $B_0 \leq_T A_0 \Longleftrightarrow B_1 \leq_T A_1$.*
Deduce that $\leq$ is a well-defined ordering on $\mathcal{D}$.

And following Proposition 7.2.5:

EXERCISE 10.3.4   *Show that $\leq$ is a partial ordering on $\mathcal{D}$.*

Elementary set theory gives us some basic properties of $\mathcal{D}$. We reviewed the notions of *countable* and *uncountable* sets in Remark 3.1.16 — remember, X is countable if we can list its elements: $x_0, x_1, \ldots$ Here are two set theoretical facts we need — I will give sketch proofs because they are easy, and you may not have seen them before:

1. | If $X \subseteq$ a countable Y, then X is also countable.

> **PROOF**   Let $Y = \{y_0, y_1, \ldots\}$. Delete from $y_0, y_1, \ldots$ all the y's not in X. The result is a list of the members of X. $\qquad\square$

2. | **CANTOR'S THEOREM:** $2^{\mathbb{N}}$ is not countable

> **PROOF**   To get a contradiction, assume $X_0, X_1, \ldots$ to be a list of all subsets of $\mathbb{N}$. Define
>
> $$S = \{x \in \mathbb{N} \mid x \notin X_x\}.$$
>
> Then (a) $S \subseteq \mathbb{N}$, but
>
> (b) $S \neq X_x$ any $x \in \mathbb{N}$ — since $x \in X_x \iff x \notin S$. $\qquad\square$

In the following theorem notice how we use $\mathbf{a}$, $\mathbf{b}$, etc., to denote degrees, and write $\mathbf{a} = \deg(A)$ — this is just standard practice.

THEOREM 10.3.5
*(i) There is a **least** Turing degree $\mathbf{0} =$ the set of all computable sets.*
(ii) Each Turing degree $\mathbf{a}$ is countably infinite (that is, $|\mathbf{a}| = \aleph_0$).
(iii) The set of degrees $\leq$ a given degree $\mathbf{a}$ is countable — possibly finite (that is, $|\{\mathbf{b} \mid \mathbf{b} \leq \mathbf{a}\}| \leq \aleph_0$).
(iv) $\mathcal{D}$ is uncountable (that is, $|\mathcal{D}| > \aleph_0$).

PROOF   (i) See Exercise 10.3.6.

(ii) Let $\mathbf{a} = \deg(A)$ be some Turing degree. Then

$$\begin{aligned}
\mathbf{a} &= \{X \mid X \equiv_T A\} \\
&\subseteq \{X \mid X \leq_T A\} \\
&\subseteq \{\Phi_i^A \mid \Phi_i^A \text{ is total}\} \\
&\subseteq \{\Phi_i^A \mid i \geq 0\}.
\end{aligned}$$

So $\mathbf{a}$ is a subset of a countable set, and so countable.

To see that $\mathbf{a}$ is also infinite — write

$$A_i = \begin{cases} A \cup \{i\} & \text{if } i \notin A \\ A - \{i\} & \text{if } i \in A. \end{cases}$$

Then for each $i \neq j$ we have $A_i(i) \neq A_j(i) = A(i)$ — giving $A_i \neq A_j$.

But it is easy to check that each $A_i \equiv_T A$. So $\mathbf{a}$ contains $A_0, A_1, \ldots$, all distinct.

(iii) Write $\mathcal{D}(\leq \mathbf{a}) = \{\mathbf{b} \mid \mathbf{b} \leq \mathbf{a}$, and let $A \in \mathbf{a}$.
Then

$$\begin{aligned}
\mathcal{D}(\leq \mathbf{a}) &= \{\deg(X) \mid X \leq_T A\} \\
&= \{\deg(\Phi_i^A) \mid \Phi_i^A \text{ total}\} \\
&\subseteq \{\deg(\Phi_i^A) \mid i \geq 0\}.
\end{aligned}$$

So $\mathcal{D}(\leq \mathbf{a})$ is countable.

(iv) Since every $X \subseteq \mathbb{N}$ appears in some degree — namely $X \in \deg(X)$ — we have

$$\begin{aligned}
2^{\mathbf{N}} &= \bigcup\{\mathbf{a} \mid \mathbf{a} \in \mathcal{D}\} \\
&= \{X \mid X \in \text{ some degree } \mathbf{a}\}.
\end{aligned}$$

To get a contradiction, assume that $\mathcal{D}$ is countable $= \mathbf{a}_0, \mathbf{a}_1, \ldots$, say. Then

$$2^{\mathbf{N}} = \mathbf{a}_0 \cup \mathbf{a}_1 \cup \mathbf{a}_2 \cup \ldots$$

If you already know that a countable union of countable sets is countable, you will already have a contradiction to Cantor's Theorem. But assuming not, let us finish off our proof with what amounts to a proof of this fact.

Since each $\mathbf{a}_i$ is countable by part (ii), we can list its members: $A_{i,0}, A_{i,1}, \ldots$. Then writing $A_{\langle i,j \rangle} = A_{i,j}$, each $i, j \geq 0$, we get

$$2^{\mathbf{N}} = \bigcup\{\mathbf{a} \mid \mathbf{a} \in \mathcal{D}\} = \{A_e \mid e \geq 0\}$$

which is a countable set — contradicting Cantor's Theorem. $\qquad\square$

EXERCISE 10.3.6 *Show (a) that there is a **least** Turing degree $\mathbf{0} =$ the set of all computable sets, but (b) there is no **greatest** member of $\mathcal{D}$.*

REMARK 10.3.7 How do we classify *functions* according to their relative computability? The obvious definition is:

$$f \leq_T \Longleftrightarrow_{\text{defn}} \text{ for all } n \in \mathbb{N} \text{ we can compute } f(n) \text{ using oracle}$$
answers to questions of the form "What is $g(m)$?"

You could formalise this via a simple modification of our oracle Turing machines — or better:

EXERCISE 10.3.8 *Show that $f \leq_T g \Longleftrightarrow \text{Graph}(f) \leq_T \text{Graph}(g)$.*

Then we can get the Turing degrees of functions by defining:

$$\deg(f) =_{\text{defn}} \deg(\text{Graph}(f)).$$

$\Box$

EXERCISE 10.3.9 *Show that every Turing degree is the degree of some function.*

The final exercise in this section shows that $\mathcal{D}$ has a join operation very like that for $\mathcal{D}$.

EXERCISE 10.3.10 *(a) Show that for all $A, B \subseteq \mathbb{N}$:*
  i) $A \leq_T A \oplus B$ and $B \leq_T A \oplus B$, and
  ii) If $A \leq_T C$ and $B \leq_T C$, then $A \oplus B \leq_T C$.
  *(b) If we define the **join** $\mathbf{a} \cup \mathbf{b}$ of Turing degrees $\mathbf{a} = \deg(A)$, $\mathbf{b} = \deg(B)$ by $\mathbf{a} \cup \mathbf{b} = \deg(A \oplus B)$, deduce that $\mathbf{a} \cup \mathbf{b} = \text{lub}\{\mathbf{a}, \mathbf{b}\}$.*

10.4 Enumerating with Oracles, and the Jump Operator

Relativising what we have found out about computably enumerable sets is surprisingly revealing. This is a first indication of the deep and intimate relationship between the information content of degrees and their structural interrelationships.

DEFINITION 10.4.1 *(1) We say B is* **computably enumerable in**
*A — or just A-**c.e.** — if we can computably enumerate the members of B*
with help of an oracle for A.

  (2) We say the degree **b** *is* **computably enumerable in a** *if some $B \in \mathbf{b}$*
is c.e. in $A \in \mathbf{a}$.

  (3) A degree **a** *is* **computably enumerable** *if it contains a c.e. set.*
We write $\mathcal{E}$ for the set of all c.e. Turing degrees.

REMARK 10.4.2 Notice that A is c.e. if and only if it is c.e. in some
computable set. So a degree **a** is c.e. if and only if it is c.e. in **0**. ▯

EXERCISE 10.4.3 *Show that the set $\{X \subseteq \mathbb{N} \mid X$ has c.e. Turing degree$\}$*
is countable.

EXAMPLE 10.4.4 *Show that B is A-c.e. if and only if it is the domain*
of an A-p.c. function.

SOLUTION ($\Rightarrow$) Assume we can A-computably enumerate the elements
of B.

  Define the *semicharacteristic function* S_B of B by:

$$S_B(n) = \begin{cases} 1 & \text{if } n \in B \\ \text{undefined} & \text{if } n \notin B. \end{cases}$$

It is easy to see that S_B is A-partial computable — to compute $S_B(n)$ just
enumerate the elements of B with help from A, setting $S_B(n) = 1$ if we
observe n enumerated into B. And of course $\text{dom}(S_B) = B$.

  ($\Leftarrow$) Let B be the domain of the A-p.c. function ψ.

  Then we can enumerate the elements of B with help from A by successively
setting in motion the computations of $\psi(0), \psi(1), \psi(2), \dots$ — and enumerating
into B any n for which we observe a computation of $\psi(n)$ terminate. ▯

  We can now write $\boxed{W_e^A = \text{dom}\Phi_e^A}$ for each $e \in \mathbb{N}$.

  Then $\{W_e^A\}_{e \in \mathbb{N}}$ is a listing of all A-c.e. sets. It is in fact what we call a
uniformly A-c.e. listing — in that Exercise 10.2.6 tells us we can enumerate
the sets of the list simultaneously, given an oracle for A.

We can also relativise the computable approximations to p.c. functions and c.e. sets we saw earlier.

DEFINITION 10.4.5 *We write*

$$\Phi_{e,s}^A(x) = y \iff_{\text{defn}} x, y, e < s, \text{ and } y \text{ is the output from } \Phi_e^A(x) \text{ in } < s$$
$$\text{steps of the Turing program } \widehat{P}_e,$$

and $W_{e,s}^A = \text{dom}\Phi_{e,s}^A.$

Similarly to Corollary 4.5.3 we can verify some basic properties.

EXERCISE 10.4.6 *Show that:*

(a) $\Phi_{e,s}^A(x) \downarrow$, $\Phi_{e,s}^A(x) = y$ *are A-computable relations of* e, s, x, y.

(b) $\Phi_e^A(x) \downarrow \iff (\exists s)\Phi_{e,s}^A(x) \downarrow$ *and* $\Phi_e^A(x) = y \iff (\exists s)\Phi_{e,s}^A(x) = y$.

We can also relativise the first level of the arithmetical hierarchy.

DEFINITION 10.4.7 *We define the* Σ_1^A *sets by*

$$X \in \Sigma_1^A \iff_{\text{defn}} (\forall x)[x \in X \Leftrightarrow (\exists y)R^A(x, y)], \text{ some } A\text{-computable } R^A.$$

Also $X \in \Pi_1^A$ *if* $\overline{X} \in \Sigma_1^A$ — *and* $\Delta_1^A = \Sigma_1^A \cap \Pi_1^A$.

Of course, as in Exercise 5.2.9 we can show that

$$X \in \Pi_1^A \iff (\forall x)[x \in X \Leftrightarrow (\forall y)R^A(x, y)], \text{ some } A\text{-computable } R^A.$$

Here are relativisations of some other familiar facts:

EXERCISE 10.4.8 *Show that if* $X \subseteq \mathbb{N}$ *is A-computable then* X *is A-c.e.*

EXERCISE 10.4.9 *Show that* X *is A-computable if and only if* X *and* $\overline{X}$ *are A-c.e.* — *if and only if* $X \in \Delta_1^A$.

EXERCISE 10.4.10 *Show that* X *is A-c.e. if and only if* $X \in \Sigma_1^A$.

EXERCISE 10.4.11 *Show that* $X \leq_m A'$ *if and only if* X *is A-c.e.*
[**Hint:** Relativise your solution to Exercise 7.1.6.]

EXERCISE 10.4.12   *Define $K^A = \{x \mid x \in W_x^A\}$. Verify that K^A is A-c.e. but not A-computable.*

It is now time to reveal the power concealed in these inoffensive-looking relativisations. We first define a jump operation on sets by relativising the definition of K_0, and then extend it to a jump on degrees.

DEFINITION 10.4.13   *(1) We define the **jump** A' of a set A to be*

$$A' =_{\text{defn}} \{\langle x, y \rangle \mid x \in W_y^A\} \qquad (= K_0^A).$$

(2) The **(n+1)th jump** *of A is defined to be* $A^{(n+1)} =_{\text{defn}} (A^{(n)})'$.

Then:

THEOREM 10.4.14 (The Jump Theorem)
Let $A, B \subseteq \mathbb{N}$. Then

  (i) A' is A-c.e.
  (ii)   A set B is A-c.e. $\iff B \leq_m A'$ — and hence $A \leq_T A'$.
  (iii) $A' \not\leq_T A$.
  (iv)   If $A \equiv_T B$ then $A' \equiv_T B'$.

PROOF   (i) We have

$$\langle x, y \rangle \in A' \iff x \in W_y^A$$
$$\iff \exists s \underbrace{\left[x \in W_{y,s}^A \right]}_{\text{A-computable}}.$$

So $A' \in \Sigma_1^A$ — and so is A-c.e.

(ii) Use Exercise 10.4.11.
Then $A \leq_m A'$ since A is A-c.e. — and hence $A \leq_T A'$.

(iii) To get a contradiction assume $A' \leq_T A$.
By Exercise 10.4.12, K^A is A-c.e. So by part (ii) of the proof, $K^A \leq_m A'$ — and hence $K^A \leq_T A$ by transitivity of $\leq_T$.
But also by Exercise 10.4.12, K^A is not A-computable — which gives the required contradiction.

(iv) Unlike the other parts, we do not get (iv) by direct relativisation of earlier results. We first need the following:

LEMMA 10.4.15

If $B \leq_T A$ and X is B-c.e., then X is A-c.e.

PROOF   We use the relativised Church–Turing Thesis in the same sort of way we used it to prove transitivity of $\leq_T$.

Say we can enumerate X with help from an oracle for B. Then since we can answer all questions about membership in B with help from an oracle for A, we can enumerate X using such help. ⬜

Then assuming $A \equiv_T B$, we have $A \leq_T B$.

But B' is B-c.e. by part (i) — and so by the lemma, B' is A-c.e. And by part (ii) this means $B' \leq_m A'$.

Similarly we can get $A' \leq_m B'$.

So $A' \equiv_m B'$ — giving $A' \equiv_T B'$. ⬜

The next exercise ties in the jump with our earlier definition in Theorem 7.2.17 of $\mathbf{0}'_m$.

EXERCISE 10.4.16   *Show that $K_0 \equiv_m A'$ for any computable set A — and hence $\mathbf{0}'_m = \deg_m(\emptyset')$.*

[**Hint:** Use part (ii) of the Jump Theorem and Remark 10.4.2.]

REMARK 10.4.17   It is not true generally that if $A \equiv_T B$ then $W_i^A \equiv_T W_i^B$. Part (iv) of the Jump Theorem is very special, and enables us to turn the jump on sets into one on degrees — sometimes called the *Turing jump*. ⬜

DEFINITION 10.4.18   *The **jump** $\mathbf{a}'$ of $\mathbf{a} = \deg(A)$ is given by*

$$\mathbf{a}' =_{\text{defn}} \deg(A').$$

*The $(n+1)^{\text{th}}$ **jump** $\mathbf{a}^{(n+1)}$ of $\mathbf{a}$ is given by*

$$\mathbf{a}^{(n+1)} =_{\text{defn}} (\mathbf{a}^{(n)})' = \deg(A^{(n+1)}).$$

The Jump Theorem now gives:

COROLLARY 10.4.19 (of Definition 10.4.18 and the Jump Theorem)

$$\mathbf{a} < \mathbf{a}' < \mathbf{a}'' < \cdots < \mathbf{a}^{(n)} < \ldots,$$

where $\mathbf{a}^{(n+1)}$ *is c.e. in* $\mathbf{a}^{(n)}$, *each* $n \geq 0$.

PROOF   We just need to observe the following:

- $\mathbf{a}^{(n+1)}$ is c.e. in $\mathbf{a}^{(n)}$ by (i) of the Jump Theorem.

- $\mathbf{a}^{(n+1)} \not\leq \mathbf{a}^{(n)}$ by (iii) of the Jump Theorem.

- $\mathbf{a}^{(n)} \leq \mathbf{a}^{(n+1)}$ from (ii) of the Jump Theorem.   ◻

Taking $\mathbf{a} = \mathbf{0} = \deg(\emptyset)$ we get the ascending sequence $\mathbf{0} < \mathbf{0}' < \mathbf{0}'' < \ldots =$ $\deg(\emptyset) < \deg(\emptyset') < \deg(\emptyset'') < \ldots$ — which shows, for instance, that $\mathcal{D} \neq$ the set $\mathcal{E}$ of all c.e. degrees. We will see later that there even exist degrees $\leq \mathbf{0}'$ which are not c.e. — unlike for the m-degrees $\leq \mathbf{0}'_m$. Another difference — the *only* c.e. degree which consists entirely of c.e. sets is $\mathbf{0}$.

EXERCISE 10.4.20   *Show that if* $\mathbf{b} \leq \mathbf{a}$ *then* $\mathbf{b}^{(n)} \leq \mathbf{a}^{(n)}$ *for all* $n \geq 1$.

EXERCISE 10.4.21   *We say that a degree* $\mathbf{a}$ *is* **arithmetical** *if* $\mathbf{a} \leq \mathbf{0}^{(n)}$ *for some* $n \geq 0$.
    Show that $\mathcal{D} \neq$ *the set* $\mathcal{D}(\text{arith})$ *of all arithmetical degrees.*

Large as the collection $\mathcal{D}(\text{arith})$ of all arithmetical degrees is, we do not have to stop there — we can replace $\mathbb{N}$ with its ordinal ω, and then use transfinite ordinals to index bigger and bigger jumps. Limit ordinals are the tricky levels — and we cannot sensibly take the process beyond ordinals for which we have computable notations — but here is how we do it at the first limit point:

EXERCISE 10.4.22   *Define the* ω-**jump** *by*

$$A^{(\omega)} =_{\text{defn}} \{\langle m, n \rangle \mid m \in A^{(n)}\}, \quad \mathbf{a}^{(\omega)} =_{\text{defn}} \deg(A^{(\omega)}) \text{ with } A \in \mathbf{a}.$$

Show that (i) $B \leq_T A \implies B^{(\omega)} \leq_T A^{(\omega)}$.
(ii) $A^{(n)} \leq_m A^{(\omega)}$ *for all* $n \in \mathbb{N}$, *and hence*
(iii) $\mathbf{a}^{(n)} \leq \mathbf{a}^{(\omega)}$ *for all* $n \in \mathbb{N}$. *And*
(iv) $\mathbf{a}^{(\omega)} \not\leq \mathbf{a}^{(n)}$ *for any* $n \in \mathbb{N}$.

Why would we want to do such a thing, we hear you ask? We will see that Post's Theorem — giving a link between statements in first order arithmetic and the degrees $\mathbf{0}^{(n)}$ — will justify what we have done so far. But it turns out that we need *second order* statements — involving quantification of sets or functions — to say quite basic things. And there are very beautiful links between second order information content and the extended hierarchies in computability theory. But that is another story, and another book.

The following diagram summarises what we know about $\mathcal{D}$ so far — and some things we are still to discover.

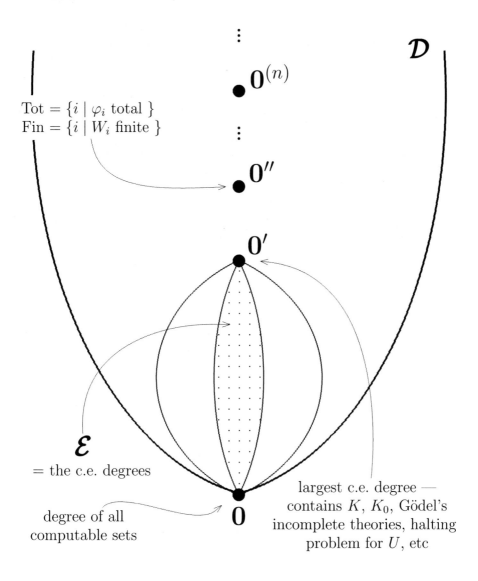

You will see from the above picture that I am claiming some special information content for $\mathbf{0}''$ as well as for $\mathbf{0}$ and $\mathbf{0}'$. In the next section we will focus on Post's Theorem, which reveals special information content at *every* level of the sequence $\mathbf{0} < \mathbf{0}' < \mathbf{0}'' < \dots$. We will also see where the arithmetical degrees get their name.

10.5   The Arithmetical Hierarchy and Post's Theorem

It is time to extend our analysis of the first level of the arithmetical hierarchy to higher levels. We have already done most of the work for this — what we did earlier is the basis for an inductive argument anchored to $\mathbf{0} < \mathbf{0}' < \mathbf{0}'' < \dots$.

We saw that the Σ_1^0 sets were those we could describe in PA — that is, in axiomatic first order arithmetic. The next definition inductively captures all those sets (or more generally *relations*) describable in *true* first order arithmetic — that is, those which are *definable* in the language of PA.

DEFINITION 10.5.1 (The Arithmetical Hierarchy)

(1) $\Sigma_0^0 = \Pi_0^0 = \Delta_0^0 =$ all the computable relations. And for $n \geq 0$:

(2) $\Sigma_{n+1}^0 =$ all relations of the form $(\exists \overrightarrow{y}_l) R(\overrightarrow{x}_k, \overrightarrow{y}_l)$, with $R \in \Pi_n^0$.

(3) $\Pi_{n+1}^0 =$ all relations of the form $(\forall \overrightarrow{y}_l) R(\overrightarrow{x}_k, \overrightarrow{y}_l)$, with $R \in \Sigma_n^0$.

(4) $\Delta_{n+1}^0 = \Sigma_{n+1}^0 \cap \Pi_{n+1}^0$.

R is **arithmetical** *if* $R \in \bigcup_{n \geq 0} (\Sigma_n^0 \cup \Pi_n^0)$.

We get Σ_n^0 sets from this definition since they are really just one-place relations of the form $x \in A$.

EXAMPLE 10.5.2   *Show that* $Tot = \{i \mid \varphi_i \text{ is total}\}$ *is a* Π_2^0 *set — and is also* Δ_3^0.

SOLUTION $Tot \in \Pi_2^0$ follows immediately from:

$$i \in Tot \iff (\forall n)\varphi_i(n) \downarrow$$
$$\iff (\forall n)(\exists s)\varphi_{i,s}(n) \downarrow$$

Also

$$i \in \text{Tot} \iff (\exists m)(\forall n)\varphi_i(n) \downarrow$$
$$\iff (\forall n)(\exists s)(\forall m)\varphi_{i,s}(n) \downarrow .$$

So $\text{Tot} \in \Sigma_3^0 \cap \Pi_3^0$. □

We call the $(\exists m)$ and $(\forall m)$ we added to Tot "dummy" quantifiers. Since we can always do this, we get:

$$\boxed{\Sigma_n^0, \Pi_n^0 \subseteq \Delta_{n+1}^0 \subseteq \Sigma_{n+1}^0, \Pi_{n+1}^0 \text{ etc}}$$

EXERCISE 10.5.3 *Show that Fin $= \{i \mid W_i$ is finite$\}$ is a Σ_2^0 set.*

Notice that many relations describable in PA — like

$$(\exists y)[(\forall z)(x < y < z) \,\&\, (\forall u)(\exists v)(u + v = x + y)]$$

— are not obviously arithmetical. But there are tricks for getting all such relations in the form of a block of quantifiers applied to a computable relation. To write them all down would be tedious, but we will mention those we need as we go along.

EXERCISE 10.5.4 *Write $(\exists y)[(\forall z)(x < y < z) \,\&\, (\forall u)(\exists v)(u + v = x + y)]$ in Σ_3^0 form.*

We can now get the promised links between degrees and their information content by relativising what we did in Chapter 5.

We will also need the *relativised arithmetical hierarchy* Σ_n^A, Π_n^A, Δ_n^A, $n \geq 0$, which we get by relativising Definition 10.5.1 in the same way as we got Definition 10.4.7 of Σ_1^A, etc. — this time we just restart the hierarchy with

(1) $\Sigma_0^A = \Pi_0^A = \Delta_0^A =$ all the A-computable relations.

The only really new ingredient in our proof of Post's Theorem will be a powerful and more quantifier-sensitive relativising principle linking different levels of the arithmetical hierarchy:

LEMMA 10.5.5 (The $\emptyset^{(n)}$-Relativising Principle)
For any $A \subseteq \mathbb{N}$
$$A \text{ is } \Sigma_{n+1}^0 \iff A \text{ is c.e. in } \emptyset^{(n)}.$$

PROOF By induction on n.

As we observed in Remark 10.4.2, we have $A \in \Sigma_1^0 \iff A$ is c.e. in $\emptyset^{(n)}$. For the inductive hypothesis assume that $n > 0$ and for all X:

$$X \in \Sigma_n^0 \iff X \text{ is c.e. in } \emptyset^{(n-1)}.$$

($\Rightarrow$)   Assume $A \in \Sigma_{n+1}^0$ — then there is an $R \in \Pi_n^0$ such that

$$\begin{aligned}
x \in A &\iff (\exists y) R(x, y) \\
&\iff A \in \Sigma_1^R \quad (= \Sigma_1^{\text{Graph}(R)}) \\
&\iff A \in \Sigma_1^{\overline{R}} \quad \text{where } \overline{R} \in \Sigma_n^0.
\end{aligned}$$

But by the inductive hypothesis $\overline{R}$ is c.e. in $\emptyset^{(n-1)}$ — and so by part (ii) of the Jump Theorem

$$\overline{R} \leq_m (\emptyset^{(n-1)})' = \emptyset^{(n)}.$$

So $A \in \Sigma_1^{\emptyset^{(n)}}$ — and so is c.e. in $\emptyset^{(n)}$ by the relativised Normal Form Theorem for c.e. sets.

($\Rightarrow$)   On the other hand — say A is c.e. in $\emptyset^{(n)}$ — say $A = W_i^{\emptyset^{(n)}}$. Since $\emptyset^{(n)}$ is c.e. in $\emptyset^{(n-1)}$, the inductive hypothesis gives $\emptyset^{(n)} \in \Sigma_n^0$. Then

$$\begin{aligned}
x \in A &\iff x \in W_i^{\emptyset^{(n)}} \\
&\iff (\exists s)(\exists y_1, \ldots, y_k, z_1, \ldots, z_l)[\text{We computably get } x \in W_{i,s}^{\emptyset^{(n)}} \\
&\qquad \text{using oracle answers: } \underbrace{y_1, \ldots, y_k \in \emptyset^{(n)}}_{\Sigma_n^0 \text{ relation}} \,\&\, \underbrace{z_1, \ldots, z_l \in \overline{\emptyset^{(n)}}}_{\Pi_n^0 \text{ relation}}].
\end{aligned}$$

A straightforward quantifier manipulation now shows $A \in \Sigma_{n+1}^0$.   ☐

This is now the main ingredient in Post's Theorem, and the key to its other parts. We need one more useful term — a generalisation of the earlier Definition 7.3.4 of *m-complete*:

DEFINITION 10.5.6   *A is Σ_n^0-complete if $A \in \Sigma_n^0$ and $X \leq_m A$ for every Σ_n^0.*

    Π_n^0-complete and Δ_n^0-complete are similarly defined.

EXERCISE 10.5.7   *Show that $\emptyset'$ is Σ_1^0-complete.*

THEOREM 10.5.8 (Post's Theorem)

Let $A \subseteq \mathbb{N}$ and $n \geq 0$. Then:

(a) $\emptyset^{(n+1)}$ *is Σ_{n+1}^0-complete.*

(b) $A \in \Sigma_{n+1}^0 \iff A$ *is c.e. in $\emptyset^{(n)}$.*

(c) $A \in \Delta_{n+1}^0 \iff A \leq_T \emptyset^{(n)}$.

PROOF (a) For each $A \subseteq \mathbb{N}$ we have

$A \in \Sigma_{n+1}^0 \iff A$ is $\emptyset^{(n)}$-c.e. (by Lemma 10.5.5)

$\iff A \leq_m (\emptyset^{(n)})' = \emptyset^{(n+1)}$ (by part (ii) of the Jump Theorem).

In particular $\emptyset^{(n+1)} \in \Sigma_{n+1}^0$, and so by the above is Σ_{n+1}^0-complete.

(b) This is just Lemma 10.5.5.

(c) We have the following equivalences

$A \in \Delta_{n+1}^0 \iff A \in \Sigma_{n+1}^0 \cap \Pi_{n+1}^0$

$\iff A$ and $\overline{A} \in \Sigma_{n+1}^0$

$\iff A$ and $\overline{A}$ c.e. in $\emptyset^{(n)}$ (by part (b))

$\iff A \leq_T \emptyset^{(n)}$ (Complementation Theorem relativised). $\quad\square$

We call the $(\exists m)$ and $(\forall m)$ we added to Tot "dummy" quantifiers. Since we can always do this, we get:

$$\Sigma_n^0, \Pi_n^0 \subseteq \Delta_{n+1}^0 \subseteq \Sigma_{n+1}^0, \Pi_{n+1}^0, \text{ etc.}$$

Post's Theorem for $n = 1$ is specially useful — particularly part (c) which tells us something new:

$$A \in \Delta_2^0 \iff A \leq_T \emptyset'$$

This is often applied in a more user-friendly form. To describe this we need:

DEFINITION 10.5.9 *We say a computable sequence $\{A^s\}_{s \geq 0}$ of finite sets is a Δ_2^0-approximating sequence for A if*

$$(\forall x)\mathrm{Lim}_s \, A^s(x) \text{ exists } = A(x).$$

And we call A^s the approximation to A at stage s.

This is also a good time to bring in some notation and terminology related to approximations.

NOTATION AND TERMINOLOGY

(1) If $\mathcal{A}$ is an expression involving approximations, we write $\mathcal{A}[s]$ for $\mathcal{A}$ evaluated at stage s — so for example $\Phi_i^A[s] = \Phi_{i,s}^{A^s}$.

(2) If X is a set, we write $X{\restriction}n$ for (the characteristic function of) X restricted to arguments $< n$.

(3) Let $\mathcal{A}^A$ be an expression whose computation involves a finite set of queries "$n_0 \in A$?", ... , "$n_k \in A$?" to an oracle for A.
We call $max\{n_i \mid 0 \le i \le k\}$ the **use** *of (the computation of) $\mathcal{A}^A$.*

(4) We write $\varphi_i^A(x)$ for the use of $\Phi_i^A(x)$. We assume $\varphi_{i,s}^A(x) < s$.

Here is the promised user-friendly description of the sets Turing reducible to $\emptyset'$. And it is used again and again.

PROPOSITION 10.5.10 (The Limit Lemma)
$A \le_T \emptyset'$ *if and only if A has a Δ_2^0-approximating sequence.*

PROOF ($\Rightarrow$) If $A \le_T \emptyset'$ then $A = \Phi_i^K$ for some i. Define

$$A^s = \Phi_i^K[s], \quad \text{each } s \ge 0.$$

Then A^s is finite — since $x \in A^s \Rightarrow x < s$ — and $A^s = \Phi_i^K[s]$ is computable from s.

Finally — given x — choose s^* and $z >$ the use of $\Phi_i^K(x)$, so that for all $s \ge s^*$ we have $\Phi_{i,s}^{K{\restriction}z}(x) \downarrow= A(x)$ and $K^s{\restriction}z = K{\restriction}z$.

Then $\text{Lim}_s A^s(x)$ exists $= \Phi_i^K(x)[s^*] = A(x)$.

(We call $\{A^s\}_{s \ge 0}$ here a *standard* Δ_2^0-approximating sequence for A.)

($\Leftarrow$) On the other hand — if A has Δ_2^0-approximating sequence $\{A^s\}_{s \ge 0}$ we immediately get:

$$x \in A \iff (\exists t)(\forall s > t)\, x \in A^s$$
$$\iff (\forall t)(\exists s > t)\, x \in A^s,$$

giving $A \in \Delta_2^0$. $\qquad\qquad$ □

What else does Post's Theorem tell us about the arithmetical hierarchy?

Many hierarchies *collapse* after some point — in the sense that all sufficiently high levels of the hierarchy are identical. Not so the arithmetical hierarchy.

EXERCISE 10.5.11 *Show that for all $n \geq 0$ we have:*

$$\Sigma_n^0, \Pi_n^0 \subsetneq \Delta_{n+1}^0 \subsetneq \Sigma_{n+1}^0, \Pi_{n+1}^0.$$

Note: We often leave off the superscript 0 — writing Σ_1 for Σ_1^0, etc. Doing this we can picture the arithmetical hierarchy:

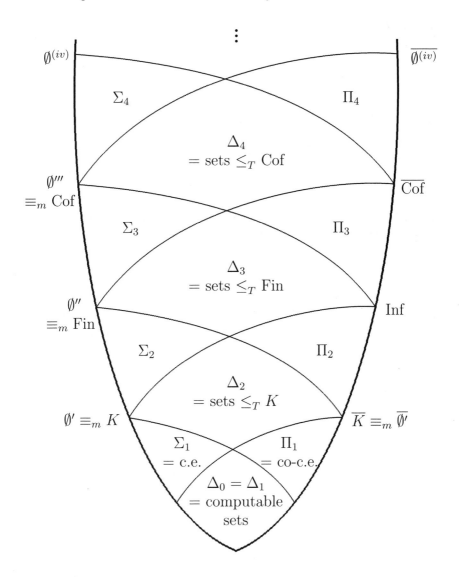

There are a few loose ends we can tie up before moving on — the first being:

EXERCISE 10.5.12   *Show that* $Cof \in \Sigma_3$.

Here is a rather harder quantifier analysis, which can be used to provide an example of a Σ_4-complete set.

EXERCISE 10.5.13   *Given a degree* $\mathbf{a}$, *let* $G(\mathbf{a}) = \{e \mid W_e \in \mathbf{a}\}$. *Show that if* $A \in \mathbf{a}$, *we have* $G(\mathbf{a}) \in \Sigma_3^A$.

EXERCISE 10.5.14   *Show that for all* $m, n \geq 0$ *we have*

$$\Sigma_n^{\emptyset^{(m)}} = \Sigma_{n+m}.$$

Deduce that $G(\mathbf{0}') \in \Sigma_4$.

The fact that these examples are actually complete in their classes involves quite technical proofs. We have already seen that K is Σ_1-complete. I will just give you one of the simplest of the remaining unproved cases.

Let us first get a bit of the Limit Lemma for Σ_2 sets.

EXERCISE 10.5.15   *We say a computable sequence* $\{A^s\}_{s \geq 0}$ *of finite sets is a* Σ_2^0**-approximating sequence** *for* A *if for all* x

$$x \in A \iff (\exists s^*)(\forall s \geq s^*)x \in A^s.$$

Show that every $A \in \Sigma_2$ *has a* Σ_2-*approximating sequence.*

[**Hint:** Use Post's Theorem to get $A = W_e^K$, some e — and then define $A^s = \{x \mid x \in W_e^K[s] \cap W_e^K[s+1]$ and $W_e^K(x)[s], W_e^K(x)[s+1]$ have the same use$\}$.]

EXERCISE 10.5.16   *Show that if the* Σ_2-*approximating sequence in Exercise 10.5.15 is defined as in the hint, then it has infinitely many* **thin** *stages — that is, stages* s *at which* $A^s \subseteq A$.

EXAMPLE 10.5.17   *Show that Fin is* Σ_2-*complete.*

SOLUTION   Let A be Σ_2 — and by Exercises 10.5.15 and 10.5.16 let $\{A^s\}_{s \geq 0}$ be a Σ_2-approximating sequence for A with infinitely many thin stages. Then

$$x \in A \iff (\exists s^*)(\forall s \geq s^*)\, x \in A^s$$
$$\iff \{s \mid x \notin A^s\} \text{ is finite.}$$

Choosing f computable such that for all x $W_{f(x)} = \{s \mid x \notin A^s\}$, we find $A \leq_m$ Fin via f.

So — remembering Exercise 10.5.3 — we have Fin is Σ_2-complete. $\quad\Box$

EXERCISE 10.5.18   *Show that Inf and Tot are Π_2-complete.*

Here are two more analogues of results proved earlier for Σ_1 sets.

EXERCISE 10.5.19   *Show that if $B \leq_m A \in \Sigma_n$, then $B \in \Sigma_n$.*

EXERCISE 10.5.20   *(i) Show that if $A, B \in \Sigma_n$, then so are $A \cup B$ and $A \cap B$.*
(ii) Show that if $A, B \in \Pi_n$, then so are $A \cup B$ and $A \cap B$.

All the above results for the arithmetical hierarchy can, of course, be relativised. We can take this without further comment. Do you have any lingering doubts? Then just check through the details of the relativised version of Post's Theorem:

EXERCISE 10.5.21   *Let $A, B \subseteq \mathbb{N}$ and $n \geq 0$. Then:*
(a) $A^{(n+1)}$ is Σ_{n+1}^A-complete.
(b) $B \in \Sigma_{n+1}^A \iff B$ is c.e. in $A^{(n)}$.
(c) $B \in \Delta_{n+1}^A \iff B \leq_T A^{(n)}$.

REMARK 10.5.22   You might remember how in Exercise 7.1.7 we showed that K_0 is actually Σ_1-complete with respect to 1-reducibility — not just m-reducibility. Well, for what it is worth, most of the other completeness results above also go through automatically with $\leq_1$ in place of $\leq_m$. $\quad\Box$

10.6   The Structure of the Turing Universe

Degree structures and hierarchies are two complementary ways of looking more closely at the universe of incomputable — or computable — objects.

We are all familiar with the hierarchical structure of science itself — within the life sciences, say, the fragmented focus on the quantum level, on atoms, on molecules, on cells, on multicellular organisms, on social structures ... . And how within this descriptive framework the dynamic relationships at each level have to be investigated within the local constraints operating at each level.

Keeping to our mathematical perspective — let us now see how different levels of the arithmetical hierarchy reveal an analogous dynamic infrastructure — its analysis based on a detailed examination of algorithmic relationships.

To start with we will follow Stephen Kleene and Emil Post's seminal paper on Turing degrees from 1954. The development of a more contemporary presentation and range of techniques will be put off until later.

Notation for Strings

We are about to get our hands dirty. We will reveal degree theoretic structure by working with all sorts of approximations to reals with special properties. To do this we will need to take hold of the basic building blocks.

DEFINITION 10.6.1   *1) A* **string***, σ or τ etc., is a finite sequence of 0's and 1's,   e.g., 00111010 is a string.   Equivalently, a string σ or τ is an initial segment of a characteristic function.*

2) We write $\sigma(x) = 0$ or 1 according as the $x + 1^{\text{th}}$ member — if it exists — of σ is 0 or 1. E.g., if $\sigma = 0100$, we get $\sigma(1) = 1$ and $\sigma(2) = 0$.

We also define the **length** *of σ to be $|\sigma| =_{\text{defn}}$ the number of entries in σ.   E.g., $|0100| = 4$.*

And the **concatenation** *$\sigma^\frown \tau$ of σ and τ consists of σ followed by τ. E.g., $0100^\frown 10 = 010010$.*

We write $\sigma \subseteq \tau$ — τ is an **extension** *of σ — if for all $x < |\sigma|$ we have $\sigma(x) \downarrow \Rightarrow \tau(x) \downarrow = \sigma(x)$.   E.g., $01 \subseteq 0100$.*

We write $\sigma \subset A$ if σ is a **beginning** *— i.e., an initial segment — of χ_A. E.g., $1010 \subset$ Ev.*

σ, τ are **compatible** *if $\sigma \subseteq \tau$ or $\tau \subseteq \sigma$ — otherwise we write $\sigma \mid \tau$. E.g., $011 \mid 0100$. If $\sigma \mid \tau$, we write $y(\sigma, \tau) = \mu y [\sigma(y) \neq \tau(y)]$.*

3) We write $\emptyset =$ the empty string and $S =$ the set of all strings.

4) The **lexicographical ordering** *of S is given by*

$$\sigma \leq \tau \iff_{\text{defn}} \sigma \mid \tau \text{ and } \sigma(y(\sigma, \tau)) < \tau(y(\sigma, \tau)), \text{ or } \sigma \subseteq \tau.$$

REMARK 10.6.2   (1) You should think of strings as beginnings of characteristic functions — which is precisely the way we will use them. Then a string σ can be used as an oracle insofar as its finite information content is enough to answer queries to it. So we can write $\Phi_e^\sigma(x) = y$ whenever there is a computation of $\widehat{P_e}$ from input x with output y during which all oracle queries are answered by σ, and concern numbers $< |\sigma|$.

(2) Actually, the sort of strings we have just defined are *binary* strings. But we sometimes need more general strings which are a sequence of *numbers* — that is, beginnings σ of functions $f : \mathbb{N} \to \mathbb{N}$, not just of characteristic functions. There is some notation which encompasses both sorts of string. We write $2^{<\omega}$ for the binary strings and $\omega^{<\omega}$ for the more general kind. This is closely related to the notation 2^ω for the set of all characteristic functions and to the way we write ω^ω for the set of all functions $\mathbb{N} \to \mathbb{N}$, of course.

In Section 13.4 such strings will be organised into *trees* of sets or functions satisfying special requirements. You will see rather different trees — trees of *strategies* — in Chapter 12. ⬜

What we will show first is a very basic fact.

THEOREM 10.6.3 (Kleene–Post, 1954)
$\mathcal{D}$ *is not linearly ordered — that is, there exist Turing degrees* **a** *and* **b** *such that* **a** $\not\leq$ **b** *and* **b** $\not\leq$ **a**.

PROOF We need to construct two sets A and B such that $A \not\leq_T B$ and $B \not\leq_T A$.

To do this, we will satisfy for all $i \geq 0$ the *requirements*:

$$\mathcal{R}_{2i} : \qquad A \neq \Phi_i^B$$
$$\mathcal{R}_{2i+1} : \qquad B \neq \Phi_i^A.$$

And to do this, we construct sequences

$$\sigma_0 \subset \sigma_1 \subset \sigma_2 \subset \dots$$
$$\tau_0 \subset \tau_1 \subset \tau_2 \subset \dots$$

of strings — and take

$$A \, (= \text{the characteristic function of } A) \; = \; \bigcup_{s \geq 0} \sigma_s$$
$$B \, (= \text{the characteristic function of } B) \; = \; \bigcup_{s \geq 0} \tau_s.$$

The idea is that at stage $e+1$ we build σ_{e+1}, τ_{e+1} in such a way as to ensure that $A \supset \sigma_{e+1}$ and $B \supset \tau_{e+1}$ do satisfy the eth requirement in the above list.

The Construction

Stage 0.

Define $\sigma_0 = \tau_0 = \emptyset$.

Stage $e + 1 = 2i + 1$.

We deal with $\mathcal{R}_{2i}$.

Assume $\sigma_0 \subset \sigma_1 \subset \ldots \subset \sigma_e$ and $\tau_0 \subset \tau_1 \subset \ldots \subset \tau_e$ are already defined.
Let $x_e = |\sigma_e|$ $(=$ the least x such that $\sigma(x) \uparrow)$.
Look for a string $\tau \supset \tau_e$ such that $\Phi_i^\tau(x_e) \downarrow$.

CASE I. τ *exists.*

Then computably choose such a τ and let $\Phi_i^\tau(x_e) \downarrow = k$ — and define

$$\sigma_{e+1} = \sigma_e{}^\frown (1 - k)$$
$$\tau_{e+1} = \tau \supset \tau_e.$$

Outcome: If $A, B \supset \sigma_{e+1}, \tau_{e+1}$ then

$$\Phi_i^B(x_e) \downarrow = \Phi_i^\tau(x_e) = k \neq \tau_{e+1}(x_e) = A(x_e).$$

So $\mathcal{R}_e$ is satisfied.

CASE II. *No such τ exists.*
Define

$$\sigma_{e+1} = \sigma_e{}^\frown 0$$
$$\tau_{e+1} = \tau_e{}^\frown 0.$$

Outcome: If $B \supset \tau_e$ then $\Phi_i^B(x_e) \uparrow$. Otherwise let w be greater than $|\tau_e|$ and the use of $\Phi_i^B(x_e)$ — then $\tau = B{\restriction}w$ would stop us being in Case II.
So $\mathcal{R}_e$ is again satisfied.

Stage $e + 1 = 2i + 2$ is similar — with $\tau, \sigma, 2i + 1$ in place of $\sigma, \tau, 2i + 1$.
And so $\mathcal{R}_e$ is satisfied by the construction for all $e \geq 0$ — and the theorem follows.   ▢

A closer look at the above proof gives something stronger — some fine structure at the level of the Δ_2 sets.

> **COROLLARY 10.6.4**
> *The local structure $\mathcal{D}(\leq \mathbf{0}')$ is not linearly ordered.*

PROOF We need to verify that we can carry out the above construction of A and B using an oracle from $\mathbf{0}'$.

First notice — At stage $e+1$ we can effectively define σ_{e+1}, τ_{e+1} *as long as we can tell whether we are in Case I or Case II.*

But Case I applies at stage $e+1$ if and only if

$$(\exists t, \tau)[\tau \supset \tau_e \, \& \, \Phi_{i,t}^{\tau}(x_e) \downarrow]. \tag{10.1}$$

Define

$$V = \{\langle \pi, \pi' \rangle \mid (\exists t, \tau)[\tau \supset \pi \, \& \, \Phi_{i,t}^{\tau}(|\pi'|) \downarrow]. \tag{10.2}$$

Then

(1) V is c.e., and so is $\leq_m K_0 \in \mathbf{0}'$, and

(2) Case I applies at stage $e+1$ if and only if $\langle \tau_e, \sigma_e \rangle \in V$.

So we can carry out the construction of A, B computably in $V \leq_T K_0 \in \mathbf{0}'$.
So — as required — $\deg(A), \deg(B) \leq \mathbf{0}'$. □

REMARK 10.6.5 You might be wondering what the point of using Equation (10.2) was? Why did the conclusion $\deg(A), \deg(B) \leq \mathbf{0}'$ not follow straight from the fact that Equation (10.1) is in Σ_1 form?

Well, there is a subtle point to notice here. Equation (10.1) is what applies at just one stage $e+1$ — it is quite possible to put together infinitely many such questions and get something very far from being Σ_1. So what we did in Equation (10.2) was showed how all the stages could be made to depend on *just one* Σ_1 set. What made this possible was a *uniformity* in the way Equation (10.1) applied at each stage.

In future we may omit such extra details, and just say that the construction depends on a *uniformly* Σ_1 list of oracle questions. □

EXERCISE 10.6.6 *Show that there exists an infinite sequence $\mathbf{a}_0, \mathbf{a}_1, \ldots$ of degrees $\leq \mathbf{0}'$ such that for each $i \neq j$ $\mathbf{a}_i | \mathbf{a}_j$.*

More generally:

EXERCISE 10.6.7 *We say $\mathcal{A} \subset \mathcal{D}$ is* **independent** *if $\mathbf{a} \nleq \mathbf{a}_1 \cup \cdots \cup \mathbf{a}_n$ for each finite $\{\mathbf{a}_1, \ldots, \mathbf{a}_n\} \subseteq \mathcal{A}$, and each $\mathbf{a} \in \mathcal{A} - \{\mathbf{a}_1, \ldots, \mathbf{a}_n\}$.*

Show that there exists an infinite independent set $\{a_1, a_2, \dots\}$ *below* $0'$.

[**Hint:** Build sets $A_0, A_1, \dots$ satisfying all requirements of the form $A_k \neq \Phi_i^{A_{k_1} \oplus \cdots \oplus A_{k_n}}$ with $k \neq$ any of the k_i's.]

REMARK 10.6.8 Here are some comments for the more abstract-minded reader. You can happily ignore them if you want.

The statements we proved in the last two exercises were statements about $\mathcal{D}$. If we allowed ourselves just the usual first order language to make such statements — variables for degrees, finitary quantification, logical connectives, and a special symbol $\leq$ for the ordering on $\mathcal{D}$, we would have the *first order language* of $\mathcal{D}$. The set $\mathrm{Th}(\mathcal{D})$ of all true statements about $\mathcal{D}$ in that language is called the *first order* — or *elementary* — *theory* of $\mathcal{D}$.

We would like to answer: Just how *complicated* is the structure $\mathcal{D}$?

Well, if $\mathrm{Th}(\mathcal{D})$ is undecidable — just like the true first order theory of $\mathbb{N}$ turned out to be — $\mathcal{D}$ is certainly complicated. But we cannot prove that yet.

We can refine this measure of complicatedness, though. Just as we classified statements in arithmetic using the arithmetical hierarchy, we can rate sentences of $\mathrm{Th}(\mathcal{D})$ according to the complexity of quantification used. So the set of true sentences about $\mathcal{D}$ using just existential quantifiers is called the Σ_1 — or *existential* — *theory* of $\mathcal{D}$.

The statement we proved in Exercise 10.6.7 actually used infinitary quantification. But we can now use it to show that the Σ_1 theory of $\mathcal{D}$ is *decidable*.

Here is how, in a nutshell.

An existential sentence φ of $\mathrm{Th}(\mathcal{D})$ can only say something of the form: "There are degrees $x_1, \dots, x_n$ in a certain partial order in relation to each other" — for example, φ could say $(\exists x_1, x_2)[x_1 \leq x_2 \,\&\, x_2 \not\leq x_1]$. But — if we take our independent set $\{a_1, a_2, \dots\}$, we can form finite joins of the a_i's to get a finite set of degrees in any given partial ordering asked for by φ — so long as the ordering asked for is consistent. For example, for the φ just given we could take $x_1 = a_1$ and $x_2 = a_1 \cup a_2$. Then the formation of the join makes $x_1 \leq x_2$, and the independence ensures $x_2 \not\leq x_1$. We say that the given partial ordering can be *embedded* in $\mathcal{D}$.

This means all we need to do to decide if a sentence φ of the Σ_1 theory of $\mathcal{D}$ is true is to check if the ordering it asks for is realisable *anywhere*. If it is, then φ is true in $\mathcal{D}$.

This result is not fantastically exciting in itself. But we have initiated a powerful approach to gauging the complexity of the noncomputable universe. Now let us return to the main development. ▯

We saw in Exercise 10.4.20 that the Turing jump is order preserving. It immediately follows that since $0 \leq$ every $a \in \mathcal{D}$, we must have every jump $a' \geq 0'$. Can we pin down exactly *which* degrees $\geq 0'$ are jumps — or, in standard terminology, are *Turing complete*? Here is a deservedly celebrated theorem from 1957 which gives the best possible answer. It is sometimes called the *Friedberg Completeness Theorem*, of course, but "complete" is a much overworked term in mathematics!

THEOREM 10.6.9 (The Friedberg Jump Inversion Theorem)
If $\mathbf{c} \geq \mathbf{0}'$, then there exists an $\mathbf{a} \in \mathcal{D}$ such that $\mathbf{c} = \mathbf{a}'$.

PROOF I actually like Friedberg's original proof, but the one I will give here — due to Hartley Rogers — is clever and concise, and most people seem to prefer it.

Let $\mathbf{c} = \deg(C) \geq \mathbf{0}'$.

We aim to construct $A = \bigcup_{n \geq 0} \sigma_n$ such that $A \oplus \emptyset' \equiv_T A' \equiv_T C$.

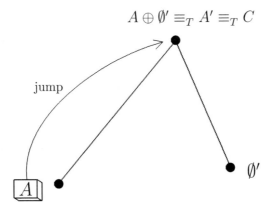

$$A \oplus \emptyset' \equiv_T A' \equiv_T C$$

The strategy will be:

(1) To make each of $A \oplus \emptyset'$, A' and C retrievable from the construction of A, and

(2) To make the construction retrievable from each of $A \oplus \emptyset'$, A' and C.

The main thing needed for (1) is to make the construction tell us if $i \in A'$ each i, done at stage $2i + 1$ — and to tell us the value of $C(i)$ each i, done at stage $2i + 2$.

Achieving (2) just depends on our taking care not to make the construction too complicated — which mainly shows up in the way we decide whether or not $i \in A'$ at stage $2i + 1$.

Remember from Definition 10.4.13 that $X'[s] = \{\langle i, j \rangle \mid i \in W_{j,s}^X\} = K_0^A[s]$.

The Construction

Stage 0.

   Define $\sigma_0 = \emptyset$.

Stage $n + 1 = 2i + 1$.

   Ask: Is there a string $\sigma \supset \sigma_n$ such that $i \in K_0^\sigma[\,|\sigma|\,]$?

   CASE I: If σ exists, take the least such σ and define $\sigma_{n+1} = \sigma$.

   CASE II: Otherwise, define $\sigma_{n+1} = \sigma_n$.

Stage $n + 1 = 2i + 2$.

   Define $\sigma_{n+1} = \sigma_n ^\frown C(i)$.

LEMMA 10.6.10

   The construction can be carried out computably in each of C, $A \oplus \emptyset'$ and A' — that is $\{\sigma_n\}_{n \geq 0} \leq_T C$, $A \oplus \emptyset'$ and A'.

PROOF (of lemma)   We can carry out stage $2i + 1$ computably, once we know whether Case I or Case II applies. But this depends on the Σ_1 question: $(\exists? \sigma \supset \sigma_n)(i \in K_0^\sigma[\,|\sigma|\,])$. And as for Corollary 10.6.4, this is uniformly computable from $\emptyset'$ — which is $\leq_T C$, $A \oplus \emptyset'$ and A'.

   Stage $2i + 2$ is computable from C, obviously. It is also computable from $A \leq_T A \oplus \emptyset'$, since $\sigma_{2i+2} = \sigma_{2i+1} ^\frown A(\,|\sigma_{2i+1}|\,)$ — and hence also from A', since $A \leq_T A'$.   ☐

LEMMA 10.6.11

   C, $A \oplus \emptyset'$ and A' are computable from the construction — that is C, $A \oplus \emptyset'$ and $A' \leq_T \{\sigma_n\}_{n \geq 0}$.

PROOF (of lemma)   Assume we are given $\{\sigma_n\}_{n \geq 0}$. Then:

   (1) Since $C(i) = \sigma_{2i+2}(\,|\sigma_{2i+1}|\,)$, we get C — and hence also $\emptyset'$ — from $\{\sigma_n\}_{n \geq 0}$.

   (2) We obviously get A computably from $\{\sigma_n\}_{n \geq 0}$.
   So $A \oplus \emptyset' \leq_T \{\sigma_n\}_{n \geq 0}$.

(3) Finally:

$$i \in A' \iff i \in K_0^A \implies (\exists s \geq 0, \sigma \supset \sigma_{2i})(i \in K_0^\sigma[s]$$
$$\implies (\exists \sigma \supset \sigma_{2i})(i \in K_0^\sigma[\,|\sigma|\,]) \implies \sigma_{2i+1} \supsetneq \sigma_{2i}.$$

Conversely:

$$\sigma_{2i+1} \supset \sigma_{2i} \implies i \in K_0^{\sigma_{2i+1}}[\,|\sigma_{2i+1}|\,] \implies i \in K_0^A.$$

So $A' \leq_T \{\sigma_n\}_{n \geq 0}$.

Putting Lemmas 10.6.10 and 10.6.11 together, we get the result.

In fact, the proof gives a bit more.

COROLLARY 10.6.12
If $\mathbf{c} \geq \mathbf{0}'$, *there is a degree* $\mathbf{a}$ *such that* $\mathbf{a}' = \mathbf{a} \cup \mathbf{0}' = \mathbf{c}$.

What this says is that the degree $\mathbf{a}$ we constructed has jump as small as it can be in relation to $\mathbf{a}$ — since by the Jump Theorem 10.4.14 we always have $\mathbf{a}' \geq \mathbf{a}$ and $\geq \mathbf{0}'$. Such degrees are unusual. Later on we will call them *generalised low* degrees.

EXERCISE 10.6.13   *Show that there is a degree* $\mathbf{b}$ *such that* $\mathbf{b}' = \mathbf{0}''$ *but* $\mathbf{b} \not\leq \mathbf{0}'$ *and* $\mathbf{0}' \not\leq \mathbf{b}$.

You will have to wait until Part III for some real richness of degree structure. Before then I should mention that there are many other reducibilities with important modelling roles.

Turing reducibility is the dominant classical reducibility, and many of the other reducibilities are refinements of that — we have already seen the classical *strong reducibility* $\leq_m$.

DEFINITION 10.6.14   (1) *If* $\Phi_i^A(x) \downarrow$ *for all* $i, x \geq 0$, $A \subseteq \mathbb{N}$, *we say* Φ_i *is a* **computable functional**.

(2) *If further* $m(i,x) = \max\{\,\varphi_i^A(x) \mid A \in 2^\omega\,\}$ *is a computable function of* i, x, *we say* Φ_i *is a* **truth-table reduction**.

(3) *We say* B *is* **truth-table reducible** *to* A — *written* $B \leq_{tt} A$ — *if* $B = \Phi^A$ *for some truth-table reduction* Φ.

Intuitively: Given $m(i,x)$ we can make up a table whose first column is all possible beginnings $A \upharpoonright m(i,x)$ of A of length $m(i,x)$, and whose second column is the corresponding value, 0 or 1, of $\Phi_i^A(x) = \Phi_i^{A \upharpoonright m(i,x)}(x)$.

EXERCISE 10.6.15 *Show that* $B \leq_m A \implies B \leq_{tt} A \implies B \leq_T A$. *Give an example of sets* A, B *such that* $B \leq_{tt} A$ *but* $B \not\leq_m A$.

EXERCISE 10.6.16 *Show that* $\leq_{tt}$ *is reflexive and transitive.*

This means that we can define $\equiv_{tt}$ in the usual way and get a degree structure $\boldsymbol{D}_{tt}$ with a partial ordering $\leq$ induced by $\leq_{tt}$.

There are other such strong reducibilities. There is *bounded truth-table* — or *btt* — reducibility $\leq_{btt}$, which we get by restricting ourselves to truth-table reductions Φ which come with some fixed n bounding the number of oracle questions it can ask. There is *weak truth-table* — or *wtt* — reducibility $\leq_{wtt}$, got by allowing Turing reductions Φ which come with some computable f, with the use $\varphi^A(x)$ bounded by $f(x)$ for all A. Just as the use functions of wtt-reductions are better behaved than those of Turing reductions in general, the degree structure $\boldsymbol{D}_{wtt}$ has some nicer properties than does $\boldsymbol{D}$.

Other mathematically interesting reducibilities can be got be relativising and extending the arithmetical hierarchy — one gets notions of *arithmetical in* and *hyperarithmetic in*, with their corresponding degree structures. You can take this approach further, getting reducibilities derived from hierarchies based on quantification over sets or functions, or even more general quantification.

At the other end of the spectrum — if you are interested in more obviously "real world" computability — you can bound the time or space allowed for the carrying out of Turing computations. For computations we can actually carry out, we usually need bounds which are some sort of polynomial function of the input.

But there is a very important classical model of relative computability which is *different* from Turing reducibility and cannot be viewed as a variation on what we have already done. We will look at that in the next chapter.

EXERCISE 10.6.17 *We say that* B *is* **arithmetical in** A — *written* $B \leq_{\text{arith}} A$ — *if* $B \in \Sigma_n^A$ *for some* $n \geq 0$.
Say $deg(B) = \mathbf{b}$, $deg(A) = \mathbf{a}$. *Show that:*

(i) $B \leq_{\text{arith}} A \iff \mathbf{b} \leq \mathbf{a}^{(n)}$ *for some* $n \geq 0$.

(ii) $\leq_{\text{arith}}$ *is reflexive and transitive.*

(iii) $\leq_{\text{arith}}$ *is a strictly weaker reducibility than Turing reducibility — in that* $B \leq_T A \Rightarrow B \leq_{\text{arith}} A$, *but there exist* A, B *such that* $B \leq_{\text{arith}} A$ *and* $B \not\leq_T A$.

EXERCISE 10.6.18 *(i) Writing $A \equiv_{\text{arith}} B$ for "$A \leq_{\text{arith}} B$ and $B \leq_{\text{arith}} A$", show that $\equiv_{\text{arith}}$ is an equivalence relation on $2^{\mathbb{N}}$.*

(ii) Writing $\mathcal{D}_{\text{arith}}$ for the equivalence classes under $\equiv_{\text{arith}}$, define

$$\deg_{\text{arith}}(B) \leq \deg_{\text{arith}}(B) \iff B \leq_{\text{arith}} A.$$

Show that $\leq$ is a partial ordering on $\mathcal{D}_{\text{arith}}$.

EXERCISE 10.6.19 *Show that:*

(i) There is a least member $\mathbf{0}_{\text{arith}}$ of $\mathcal{D}_{\text{arith}}$ consisting of the set of all arithmetical sets.

(ii) Each arithmetical degree $\mathbf{a}_{\text{arith}}$ is countably infinite.

(iii) The set of arithmetical degrees $\leq$ a given degree $\mathbf{a}_{\text{arith}}$ is countable.

(iv) $\mathcal{D}_{\text{arith}}$ is uncountable.

Chapter 11

Nondeterminism, Enumerations and Polynomial Bounds

You are having trouble solving a mathematics problem and enter your department looking for help. You knock on Professor A's office door. There is no reply. You wait. Unfortunately Professor A has had an accident while working in his garden, and will never again be coming to work. You wait forever. Your mathematics problem remains unsolved ...

A realistic scenario? Of course not — nobody behaves that stupidly — there is something wrong here with the oracle Turing machine model. In real life, having got no help from Professor A, you move on to Professor B's office, and if she is not there — you try Professor C. Professor C, always helpful and knowledgeable, gives you just the information you need, and you happily complete your project.

In this chapter we learn how to model computability relative to sources of auxiliary information that are not always responsive. The model we come up with will involve *nondeterministic* computations — mirroring the guessing at Professors A, B and C. There will also be important applications to situations in which the oracles do always answer— or where there is even no oracle used — but where there are time constraints which make us guess at optimal routes to successfully terminating computations. The extent to which this guessing actually helps in this context is a famous — and as yet unsolved — problem.

11.1 Oracles versus Enumerations of Data

What makes computation relative to partial information *different* is that *we cannot computably tell* whether our oracle is going to answer or not. In our new model, there is not just one computation dependent on guaranteed oracle responses to our queries. It is more like real life where our search for useful knowledge is only partially rewarded and new information emanates from our environment in a fairly surprising and unpredictable way.

But this is a slightly different way of looking at things. It seems now we are getting *two* new models. One is where we focus on what we *do* — how we *guess* and pursue different alternative computations. This leads to *nondeterministic*

Turing machines.

The other viewpoint emphasises how knowledge is *delivered* to us by our source of outside information — in a sense *enumerated*, and according to a timetable not under our own control. This leads to a notion of *enumeration reducibility*. But do not worry — both turn out to be equivalent.

Let me pin down the models more precisely, before I say any more.

APPROACH I — Nondeterministic (oracle) Turing machines

DEFINITION 11.1.1 *T is a* **nondeterministic** *(oracle) Turing machine if it is an (oracle) Turing machine* **with no consistency condition** *on the quadruples.*

So, a computation of such a nondeterministic T follows the same conventions as previously — except that when confronted with $q_i S_k A q_j$, $q_i S_k A' q_l$ the machine makes an arbitrary choice of which quadruple to obey — that is, *guesses*.

How do we get any sense out of such a machine? The next definition makes all clear.

DEFINITION 11.1.2 *Let f and g be (possibly partial) functions $\mathbb{N} \to \mathbb{N}$. We say f is* **nondeterministically computable** *by T [from g] if for each $n \in \mathbb{N}$*

(a) $f(n) \downarrow \iff$ there is a terminating computation of T from input n [using oracle g], and

(b) every terminating computation of T from input n [with oracle g] gives output $f(n)$.

We write $f \leq_{NT} g$ if f is nondeterministically computable by some nondeterministic T from g.

REMARK 11.1.3 (1) The intuition behind this definition is that if we allow ourselves to occasionally guess the next step of our computation, we might find we can compute something we could not previously — either because computing without guesses did not terminate — or if we are concerned about time bounds — took too long.

But if we allow guesses, we do need to know that if a computation does terminate, the output is guaranteed to be correct — hence condition (b) of the definition.

(2) Is this intuition borne out? For the classical case — where we do not care about how much time our computations take — the answer is known. Here is our analysis:

Assume we are given an oracle for g. We can lay out *all possible* computations from a given input relative to g on a *computation tree*. Each time we guess, the tree branches into alternative computations — it is a bit like how alternative histories arise in quantum theory!

If there are no guesses, the tree has just one completely determined g-computable branch. And we can even arrive at this situation if the tree itself is g-computable — that is, each computational step in the tree can be completed immediately without any uncertainty about whether an oracle for g answers. We just systematically travel round the computation tree in a completely determined g-computable way, looking for a terminating branch from which to read off our output. Any such branch gives the correct answer. If there is a terminating computation, we do eventually find it. The search is not very efficient maybe, but here we do not care.

When do we get this g-computable computation tree? We certainly do if there are *no* questions to the oracle! So:

f is nondeterministically computable $\Longleftrightarrow$ f is computable.

And the same happens if we do have an oracle, but it always replies.

If g is total, we have $f \leq_{NT} g \Longleftrightarrow f \leq_T g$.

So in these cases we get nothing new by guessing. The situation is very different if there are oracles for *partial* functions involved, as we shall see later. And if we start caring about efficiency of computations, we get the famous unsolved problem we alluded to earlier. $\Box$

APPROACH II — Enumeration reducibility ($\leq_e$)

Intuitively:

We want $B \leq_e A$ to mean we can computably enumerate the members of B from an enumeration of the members of A — where this enumeration of B does not depend on the *order* in which A is enumerated.

We have in mind the real world where we make scientific calculations according to the available data, but where the eventual answers we get do not depend on the order of discovery of these data.

Here is a picture:

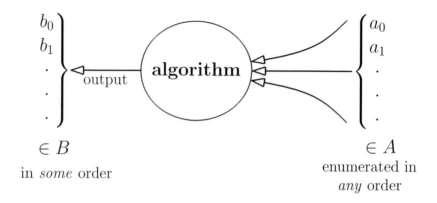

By now you should be happy to work with the intuitive notion — but it is always important to have a precise formal definition in reserve. One was first provided by Friedberg and Rogers back in 1959.

For this we will need a computable coding $\langle n, D \rangle$ of pairs n, D with $n \in \mathbb{N}$ and D a finite subset of $\mathbb{N}$ — and we can get this, for instance, by computably listing all finite sets $D_0, D_1, \ldots$ using some coding/Gödel numbering, and then setting $\langle n, D_i \rangle = \langle n, i \rangle$.

Then:

DEFINITION 11.1.4   *(1) An* **enumeration operator** *(or* **e-operator***)* Ψ *is a c.e. set — where for any* $A \subseteq \mathbb{N}$

$$n \in \Psi_i^A \iff_{\text{defn}} (\exists \text{ a finite } D \subseteq A)[\langle n, D \rangle \in \Psi].$$

So $\Psi^A = \{n \mid \langle n, D \rangle \in \Psi \text{ for some finite } D \subseteq A\}$.

(2) We say B is **enumeration reducible** *(or* **e-reducible***) to A — written $B \leq_e A$ — if $B = \Psi^A$ for some enumeration operator Ψ.*

If we rewrite $W_i = \Psi_i$ we immediately get a corresponding listing $\{\Psi_i\}_{i \in \mathbb{N}}$ of the e-operators — along with computable approximations $\{\Psi_{i,s}\}_{i,s \in \mathbb{N}}$ with all the usual properties.

Notice that we have started using index i in place of e for obvious reasons. And renaming W_i avoids our confusing Ψ_i^A with W_i^A — closely linked but *not* the same object. We do have:

EXAMPLE 11.1.5   *Show that if $B \leq_e A$ then B is A-c.e.*

SOLUTION   Say $B = \Psi^A$ — then

$$n \in B \iff n \in \Psi^A$$
$$\iff (\exists \text{ a finite } D \subseteq A)[\langle n, D \rangle \in \Psi]$$
$$\iff (\exists s, i)[D_i \subseteq A \,\&\, \langle n, D_i \rangle \in \Psi_s].$$

So $B \in \Sigma_1^A$ by the relativised Post's Theorem.   ☐

But:

EXERCISE 11.1.6   *Show that if $X \leq_e A$ c.e. then X is c.e.*
Deduce that although $\overline{K}$ is c.e. in K, we do not have $\overline{K} \leq_e K$.

EXERCISE 11.1.7   *Show that X is c.e. if and only if $X \leq_e K$.*
Deduce that if we define $U_i = \Psi_i^K$, then $\{U_i\}_{i \geq 0}$ is a standard listing of all c.e. sets.

For the exact relationship between enumeration reducibility and "computably enumerable in", you will have to wait until later — see Selman's Theorem at the end of this section. Just notice for now that $\leq_e$ is another *positive* reducibility — remember from Remark 10.1.2 that $\leq_m$ was our first one — so the reduction $X \leq_e K$ cannot depend on information about $\overline{K}$.

What is important here is how we reconcile Approaches I and II to extending Turing reducibility. At the moment the two approaches seem to be talking about different things. Let us change that. It is easy to see that enumeration reducibility gives us a natural notion of relative computability between partial functions. And it is this notion that turns out to be just the same as $\leq_{NT}$.

THEOREM 11.1.8
Let f, g be partial functions. Then

$$f \leq_{NT} g \iff \text{Graph}(f) \leq_e \text{Graph}(g).$$

PROOF (sketch)   ($\Rightarrow$) Let T be a nondeterministic Turing machine computing f from g. Here is an algorithm for enumerating an e-operator Ψ:

(1) Given any finite set D, let oracle D be the finite function g coded by D — that is, such that $D = \mathrm{Graph}(g)$.

(2) For each $\langle m, n \rangle, D$ start computing the computation tree for input m to T with oracle D.

(3) Whenever one observes a terminating branch with output n, say, enumerate $\langle \langle m, n \rangle, D \rangle$ into Ψ.

EXERCISE 11.1.9 *Verify that Ψ is an e-operator such that*

$$\mathrm{Graph}(f) = \Psi^{\mathrm{Graph}(g)}.$$

($\Leftarrow$) Let Ψ be an e-operator such that $\mathrm{Graph}(f) = \Psi^{\mathrm{Graph}(g)}$.

Let T be the nondeterministic Turing machine which implements the following computation trees:

(1) Assume, given input m, oracle g.

(2) There is an infinite branch along which we enumerate the members $\langle \langle m, n \rangle, D \rangle$ of Ψ — say using a computable h with range Ψ, and enumerating $h(0), h(1), \dots$ (At every point on this branch we guess that we need to continue searching for a $\langle \langle m, n \rangle, D \rangle \in \Psi$ such that oracle $D \subseteq g$.)

(3) The finite branches of the tree emanate from nodes at which we observe a new $\langle \langle m, n \rangle, D \rangle \in \Psi$ enumerated. We guess that oracle $D \subseteq g$ and implement along this branch the necessary queries to oracle g. And,

(4) Reward confirmation of oracle $D \subseteq g$ with a terminating branch with output n — and punish each branch evidencing $D \not\subseteq g$ with a Turing subroutine preventing the branch from terminating.

I will not ask you to fill in the details of the Turing program for T! ⬚

EXERCISE 11.1.10 *Let A be any incomputable set, with semicharacteristic function S_A (defined as in Example 10.4.4). Show that*

$$\chi_A \leq_{NT} S_{A \oplus \overline{A}} \qquad but \qquad \chi_A \not\leq_T S_{A \oplus \overline{A}}.$$

[**Hint:** Show that $\mathrm{Graph}(\chi_A) \leq_e \mathrm{Graph}(S_{A \oplus \overline{A}})$ via the e-operator

$$\{ \langle \langle n, 1 \rangle, \{ \langle 2n, 1 \rangle \} \rangle \mid n \in \mathbb{N} \} \cup \{ \langle \langle n, 0 \rangle, \{ \langle 2n + 1, 1 \rangle \} \rangle \mid n \in \mathbb{N} \}$$

— but on the other hand $\chi_A \leq_T S_{A \oplus \overline{A}}$ would imply that $\chi_A \leq_T$ the constant function 1.]

The beauty of Theorem 11.1.8 is that it shows that e-reducibility is not only a very natural way of comparing the enumerability of sets — it also provides us with a very convincing way of comparing the computability of partial functions. Even better — it does not disagree with Turing reducibility where deterministic Turing machines worked fine:

COROLLARY 11.1.11

If f and g are total,

$$\text{Graph}(f) \leq_e \text{Graph}(g) \iff f \leq_T g.$$

PROOF   By Theorem 11.1.8 we have $f \leq_{NT} g \iff \text{Graph}(f) \leq_e \text{Graph}(g)$ — and in (2) of Remarks 11.1.3 we noted that if g is total, then $f \leq_{NT} g \iff f \leq_T g$. $\qquad\qquad\Box$

I will leave you to prove a couple of basic properties of e-reducibility, which should not be unexpected.

EXERCISE 11.1.12   *Show that $\leq_e$ is (a) reflexive and (b) transitive.*

It would be nice to leave you to prove the next fact too — which we promised earlier — but you will probably need some help:

THEOREM 11.1.13 (Selman's Theorem, 1971)

For any $A, B \subseteq \mathbb{N}$

$$A \leq_e B \iff \forall X \, [\, B \text{ c.e. in } X \Rightarrow A \text{ c.e. in } X \,].$$

PROOF (sketch)   The left-to-right implication I *will* leave to you.

Conversely, assume that $A \not\leq_e B$. We will construct a $C = \cup_{s \geq 0} C_s$ such that B is c.e. in C but A is not c.e. in C.

We satisfy "B c.e. in C" by imposing an overall requirement

$$\exists \langle x, y \rangle \in C \iff x \in B \tag{11.1}$$

for each $x \geq 0$. Call a finite $D \supseteq C_s$ *admissible* if it satisfies Equation (11.1) with D in place of C, but with the right-to-left half of Equation (11.1) restricted to $x \leq s$ (so that the admissible D's can be enumerated from an enumeration of B and a *finite* amount of information about $\overline{B}$).

We satisfy $A \neq W_s^C$ (at stage $s+1$) by looking for some admissible $D \supseteq C_s$ with $x \in W_s^D - A$. If D exists, choose $C_{s+1} = D$ giving $A \neq W_s^C$. Otherwise, either $x \in A - W_s^D$ for some x, all admissible D (so $A \neq W_s^C$ again), or

$$\forall x \, (x \in A \iff \exists \text{ an admissible } D \text{ such that } x \in W_s^D),$$

giving $A \leq_e B$, a contradiction. $\qquad\qquad\Box$

Not satisfied with discovering such a beautiful theorem, Selman went on to prove in his 1971 paper that $\leq_e$ has an even more special role — it is actually a *maximal* transitive subrelation of "c.e. in". That is, not only do we have $\leq_e$ transitive and $B \leq_e A \Rightarrow B$ is c.e. in A, but for every transitive reducibility $\leq_r$ between sets, either

(a) $(\exists A, B)[\, B \leq_r A$ but B not c.e. in $A\,]$, or

(b) $(\exists A, B)[\, B \leq_e A$ but $B \not\leq_r A\,]$.

Here is one last fact for future use; relative to a total function f, there is no difference between being c.e. in f and being e-reducible to its graph:

EXERCISE 11.1.14 *Show that $A \subset \mathbb{N}$ is c.e. in f total if and only if $A \leq_e \mathrm{Graph}(f)$.*

[**Hint:** We have an enumeration of $\mathrm{Graph}(f) \iff$ we have an oracle for f.]

11.2   Enumeration Reducibility and the Scott Model for Lambda Calculus

I am not sure what is more surprising — the way in which enumeration reducibility crops up in all sorts of unexpected places. Or how e-reducibility is not even mentioned in almost every basic text on computability, logic or theoretical computer science.

Enumeration operators arose in a particularly elegant and unexpected way back in 1975, when Dana Scott used them for his countable version of the graph model for the lambda calculus. *Lambda calculus* — and lambda computability — was developed by Church and his student Stephen Kleene from around 1934 onwards and gives another important model of computability. In more recent years it has provided the basis for numerous functional programming languages (Lisp, Scheme, ML, Haskell, etc.), and as such is more important to computer science than to the basic theory of computability — which partly explains why it was omitted from my select list of such models in Chapter 2. But now is the time to say just a little about how it works — and the role of enumeration operators. For more detail a standard reference is H.P. Barendregt's *The Lambda Calculus*. For the lambda calculus and its relationship to programming see Ravi Sethi's *Programming Languages*.

The lambda calculus puts functions centre-stage — it formalises the ways in which functions can be formed, combined and used for computation. It does this via rules for manipulating λ-*terms* — the "wfs" of lambda calculus.

DEFINITION 11.2.1   *A* **λ-term** *is inductively defined to be either:*

(1) A **variable** *V — think: "V is a function variable" — or*

(2) An **abstraction** *λV.E where V is a variable and E is a λ-term — think: "λV.E is the function E of variable V " — or*

(3) An **application** *$(E_1 E_2)$ where E_1, E_2 are λ-terms — think: "$(E_1 E_2)$ is the value of function E_1 applied to argument E_2."*

We can assume V chosen from some given infinite list, but we will not be too pedantic about it.

These rules are so simple they inevitably allow expressions which will look strange, conditioned as we are to function notation such as $f(x)$, with f and x being very different kinds of objects. For instance, $λV.(VV)$ is a perfectly acceptable λ-term, even though in real "life" we are not used to expressions like $f(f)$! But to make our λ-terms more readable, we will allow ourselves to use f's and x's — but always remember, this is the *untyped* lambda calculus — here such qualitative distinctions between a function and its argument do not exist. We also allow abbreviations such as $λx.Ex$ for $λx.(Ex)$, (fgh) for $((fg)h)$ (left associativity of application) or $λxy.E$ for $(λx.(λy.E))$ (multiple arguments).

EXERCISE 11.2.2   *Find the λ-term informally represented by $λxyz.x(yz)E$.*

On a technical note — notice that $λV$ in $λV.E$ "takes charge of" — or *binds* — the variable V in the same way that quantifiers do in logic. An occurrence of a variable V only becomes *bound* when in the *scope E* of a *binding operator* $λV$ — otherwise it is *free*. For example, f occurs both bound and free in $(fλf.fx)$:

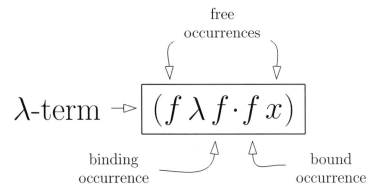

If E has no free variables, we say E is a *closed* term.

EXERCISE 11.2.3   *Let Free(E) = the set of variables occurring free in E. Verify that:*

(a) Free(x) = {x},

(b) Free($E_1 E_2$) = Free(E_1)$\cup$Free(E_2),

(c) Free($\lambda x.E$) = Free(E) − {x}.

A useful piece of notation is $[E'/V]E$ — which denotes the result of replacing each free occurrence of V in E with E'.

Let us play with the intended meaning of a particular λ-term before setting out the formal rules for manipulating λ-terms in general.

EXAMPLE 11.2.4   *Interpret the λ-term $(\lambda f.\lambda x.f(fx))(\lambda y.y)x$.*

SOLUTION   Here is how we mentally interpret our given λ-term:

$$
\begin{aligned}
(\lambda f.\lambda x.f(fx))(\lambda y.y)x &= ([(\lambda y.y)/f]\lambda x.f(fx))x \\
&= (\lambda x.(\lambda y.y)((\lambda y.y)x)))x \\
&= (\lambda x.(\lambda y.y)(x))x \\
&= (\lambda x.x)x \\
&= x
\end{aligned}
$$

So $(\lambda f.\lambda x.f(fx))(\lambda y.y)x$ turns out to be just a complicated way of describing the variable x! $\Box$

Anyway — we see that our intended interpretation tells us to include in the lambda calculus some simple rules such as are used in the above example for rewriting or *reducing* λ-terms — and these provide the basis for computation within this framework. A computation will be a series of reductions of a λ-term, the aim being to "simplify" the term to a *normal form* — that is, one which can be no further "simplified".

I just want to give you the rules now — they really are very simple. But we first need a second look at substitutions — like $[(\lambda y.y)/f]\lambda x.f(fx)$ in Example 11.2.4. Unlike that one, $[(\lambda y.xy)/f]\lambda x.f(fx)$ would *not* be valid — the free variable x in $\lambda y.xy$ gets *captured* by the binding operator λx in this substitution.

So we define:

CAPTURE FREE SUBSTITUTION

If the free variables of E' have no bound occurrences in E — then $[E'/V]E$ is a **capture free substitution** and = the result of replacing each free occurrence of V in E with E'.

EXERCISE 11.2.5 *Say which of the following substitutions are capture free:* (a) $[y/x]((\lambda z.zx)(\lambda x.x))$, (b) $[\lambda x.xy/x]((\lambda y.xy)(yz))$ — *and evaluate any which are.*

We can now state the rules for rewriting terms. I will continue to write "=" between terms reduced to each other, instead of introducing some (more formally correct) equivalence relation.

REDUCTION RULES

• **α-reduction**: $\lambda x.E = \lambda z.[z/x]E$ if z not free in E — allows renaming of bound variables.

• **β-reduction**: $(\lambda x.E_1)E_2 = [E_2/x]E_1$ if this substitution is capture free — reduces the complexity by implementing an application.

• **η-reduction**: $(\lambda x.E)x = E$ if x not free in E — allows removal of redundant abstractions.

Our rogue substitution $[(\lambda y.xy)/f]\lambda x.f(fx)$ is now no obstacle:

EXAMPLE 11.2.6 *Reduce $(\lambda f.\lambda x.f(fx))(\lambda y.xy)$ to $\lambda z.x(xz)$.*

SOLUTION   We proceed as follows:

$$
\begin{aligned}
(\lambda f.\lambda x.f(fx))(\lambda y.xy) &= (\lambda f.\lambda z.f(fz))(\lambda y.xy) \quad (\alpha\text{-reduction}) \\
&= [(\lambda y.xy)/f]\lambda z.f(fz) \quad (\beta\text{-reduction}) \\
&= \lambda z.(\lambda y.xy)((\lambda y.xy)z) \\
&= \lambda z.(\lambda y.xy)([z/y]xy) \quad (\beta\text{-reduction}) \\
&= \lambda z.(\lambda y.xy)(xz) \\
&= \lambda z.[xz/y]xy \quad (\beta\text{-reduction}) \\
&= \lambda z.x(xz)
\end{aligned}
$$

So we achieved the reduction by using α-reduction to change the bound variable x to z, and then using β-reduction with the capture free substitution $[(\lambda y.xy)/f]\lambda z.f(fz)$. $\Box$

EXERCISE 11.2.7   *Justify the reduction of $(\lambda f.\lambda x.f(fx))(\lambda y.y)x$ to x in Example 11.2.4 by identifying which reduction rule is used at each step of the reduction.*

Notice that the reduction in Example 11.2.4 is to a term x which cannot be further β-reduced — or η-reduced for that matter.

DEFINITION 11.2.8   *If a term E cannot be further β- or η-reduced, we say it is in* **normal form**.

Equivalently, E is in normal form when it contains no *redex* — that is, no subterm of the form $((\lambda.xE_1)E_2)$.

EXERCISE 11.2.9   *Show that $(\lambda x((\lambda z.zx)(\lambda x.x)))y$ has normal form y.*

Remember — computability in the lambda calculus will be reduction to normal form. So you will not be surprised to find that normal forms do not always exist.

EXAMPLE 11.2.10   *Show that $(\lambda x.xx)(\lambda x.xx)$ has no normal form.*

SOLUTION   Consider the following sequence of β-reductions:

$$
\begin{aligned}
(\lambda x.xx)(\lambda x.xx) &= [(\lambda x.xx)/x](xx) \quad (\beta\text{-reduction}) \\
&= (\lambda x.xx)(\lambda x.xx) \\
&= (\lambda x.xx)(\lambda x.xx) \quad (\beta\text{-reduction}) \\
&= (\lambda x.xx)(\lambda x.xx) \quad (\beta\text{-reduction}) \\
&= \quad \dots
\end{aligned}
$$

There is clearly no alternative sequence of reductions that avoids this non-terminating sequence. $\Box$

EXERCISE 11.2.11   *Say whether or not there exists a normal form for the λ-term $(\lambda x.xxx)(\lambda x.xxx)$.*

What is worse — as well as normal forms not always existing, there is no way of telling computably whether a given term has a normal form! If you think of the normal form as the "value" to which the term evaluates, this is where the Halting Problem rears its head again in a superficially different guise.

If there were just one procedure for reducing a given term, things might be a little simpler. But there may be a whole tree of possible reduction sequences for a given term — some of the branches of which may terminate in a normal form, and others of which may be infinite — like that for Example 11.2.10. Luckily there is a theorem — called the *Church–Rosser Property* — which tells us that if there is a normal form for a term, then *every* reduction sequence gives the *same* normal form.

EXERCISE 11.2.12 *Does $(\lambda x.y)((\lambda x.xx)(\lambda x.xx))$ have a normal form?*

In fact λ-computability turns out to be equivalent to Turing computability. To dig Turing computability out of the lambda calculus you do have to find counterparts to the numbers amongst the λ-terms. There are various ways of doing that — here is a variation on Church's original definition:

DEFINITION 11.2.13 *Define the* **Church numerals** *to be:*

$$0 = \lambda f.\lambda x.x$$
$$1 = \lambda f.\lambda x.(fx)$$
$$2 = \lambda f.\lambda x.(f(fx))$$
$$\dots$$

$$n = \lambda f.\lambda x.\overbrace{(f(f\dots(f\,x)\dots))}^{n\ \text{times}} = \lambda f.\lambda x.(f^n x), \quad \text{etc.}$$

Intuitively, n in lambda calculus is the function which given a function argument f returns the nth iterate of f.

How do we get functions like $+$ and $\times$ in lambda calculus?

EXAMPLE 11.2.14 *Let* **add** *be the* λ-*term* $\lambda mn.(\lambda fx.((mf)(nfx)))$. *Show that* (**add** $2\,3$) $= 5$.

SOLUTION   Here is the reduction:

$$(\mathbf{add}\,2\,3) = (\lambda mn.(\lambda fx.((mf)(nfx)))\,2\,3)$$
$$= \lambda gy.((2g)(3gy))$$
$$= \lambda gy.(((\lambda fx.(f^2x))g)((\lambda fx.(f^3x))gy))$$
$$= \lambda gy.((\lambda x.(g^2x)(\lambda x.(g^3x))y)$$
$$= \lambda gy.((\lambda x.(g^2x)g^3y)$$
$$= \lambda gy.(g^2g^3y)$$
$$= \lambda fx.(f^5x)$$
$$= 5 \qquad\qquad \square$$

Obviously the reduction works just the same way if we replace $2, 3$ by any other pair of numbers p, q. So **add** represents $+$ in a very natural way. Notice how we represent n-ary functions as a sequence of one-place functions — a method associated with the logician Haskell Curry, and hence called *Currying*.

EXERCISE 11.2.15   Let **mult** be the λ-term $\lambda mn.(\lambda fx.((m\,(nf))x))$. Given $p, q \in \mathbb{N}$, show that $(\mathbf{mult}\,p\,q) = p \times q$.

It now makes sense to define generally:

DEFINITION 11.2.16   A k-ary function f is λ-**computable** if there exists a λ-term E such that for all $m_1, \ldots, m_k \in \mathbb{N}$

$$f(m_1, \ldots, m_k) = n \iff Em_1 \ldots m_k \text{ has normal form } n.$$

REMARK 11.2.17   You might be thinking at this point "Isn't this definition of computability all a bit contrived, not much like how we actually compute?"

Not at all! Firstly — notice that numbers arise from our experience of counting in different contexts. We get n through our involvement with the iteration of a process n times. So the association of n with $\lambda f.\lambda x.(f^n x)$ is entirely natural.

And when we compute a value of a function f, what we actually do is take *some* description of $f(n)$ and reduce this to its simplest form, namely, the numerical value of $f(n)$. So again — λ-computability is a very accurate description of the real world.

Of course nobody uses the pure lambda calculus for programming. But it does provide a basis for a whole family of modern-day functional programming languages. $\square$

Incidentally, there is another striking unsolvable problem concerning the lambda calculus. Two λ-terms are *equal* if they have the same normal forms. Then we cannot computably decide whether two terms are equal or not.

Here is the theorem that tells us, via the Church–Turing Thesis, that we do have another model of computability.

THEOREM 11.2.18

A function is lambda computable if and only if it is recursive.

There is no need for us to prove this result.

The Scott model for lambda calculus

Anyway, the lambda calculus is as useful in its way as are axiomatic theories — so we do need to verify its soundness by exhibiting mathematically simple models. One problem in finding a model for the (untyped) lambda calculus is that of giving an interpretation of the term (VV). It is not usually possible for a function to be applied to itself as we think of a function as being an object of higher *type* than its arguments — where for example numbers are of type 0, functions from numbers to numbers are of type 1, etc. This is where e-operators come in. For e-operators there is no difficulty. An e-operator can be applied to a c.e. set, and the e-operator itself is essentially a c.e. set.

We will briefly outline the details of how Scott formulated his countable model. Our interpretation is the collection of all c.e. sets. So unlike other models, this one will be *countable*.

To interpret a λ-term with free variables we will need a mapping η from the variables to the c.e. sets, which we will call an *assignment*. An assignment η will describe an environment within which we carry out in a natural way what the term tells us to do. We will write $[\![E]\!]_\eta$ for the interpretation of E as a c.e. set, relative to η. Then we can think of the term as describing a "recipe" for deriving $[\![E]\!]_\eta$, with an occurrence of application telling us to unpack a c.e. set as an e-operator and apply it to another c.e. set, and an abstraction telling us to pack up an e-operator as a c.e. set.

This means we will interpret application as an evaluation of an e-reduction: If E_1, E_2 are λ-terms, then $(E_1 E_2)$ is interpreted as an unpacking of the e-operator coded up by $[\![E_1]\!]_\eta$, which is then applied to $[\![E_2]\!]_\eta$. That is:

$$[\![(E_1 E_2)]\!]_\eta = \{m \mid (\exists D)(\langle m, D\rangle \in [\![E_1]\!]_\eta \,\&\, D \subset [\![E_2]\!]_\eta)\}.$$

And abstraction describes how we pack up an e-operator, whose recipe is given by a λ-term E: $\lambda V.E$ becomes

$$[\![(\lambda V.E)]\!]_\eta = \{\langle m, D\rangle \mid m \in [\![E]\!]_{\eta(D/V)}\}$$

where $[\![E]\!]_{\eta(D/V)}$ denotes the interpretation of the term E according to the assignment η, except that every free occurrence of V is interpreted as D instead of $\eta(V)$.

As you would expect, multiple abstraction entails e-operators of more than one argument. For example $\lambda V_1 V_2.E$ becomes

$$[\![\lambda V_1 V_2.E]\!]_\eta = \{\langle m, D, D'\rangle \mid m \in [\![E]\!]_{\eta(D/V_1)(D'/V_2)}\},$$

where the corresponding e-operator Θ is given by

$$m \in \Theta^{A,B} \iff \exists \langle m, D, D'\rangle \in [\![\lambda V_1 V_2.E]\!]_\eta \, (D \subseteq A \,\&\, D' \subseteq B).$$

Here is something a little more formal:

DEFINITION 11.2.19 *(1) If A is a c.e. set then Ψ_A is the enumeration operator defined by it, namely*

$$x \in \Psi_A^B \Leftrightarrow (\exists D \text{ finite})(\langle x, D\rangle \in A \,\&\, D \subseteq B).$$

(2) If Θ is an enumeration operator then $\mathrm{Pack}(\Theta)$ is a well-defined c.e. set defining it, namely

$$\langle x, D\rangle \in \mathrm{Pack}(\Theta) \Leftrightarrow x \in \Theta^D.$$

EXERCISE 11.2.20 *If Ψ is an e-operator, show that $\Psi_{\mathrm{Pack}(\Psi)} = \Psi$.*

Then:

DEFINITION 11.2.21 (Plotkin 1972, Scott 1975) *Let η be an assignment of c.e. sets to the variables of lambda calculus. With every λ-term E we inductively associate a c.e. set $[\![E]\!]_\eta$:*
  (1) $[\![x]\!]_\eta = \eta(x)$,
  (2) $[\![(E_1 E_2)]\!]_\eta = \Psi_{[\![E_1]\!]_\eta}([\![E_2]\!]_\eta)$,
  (3) $[\![\lambda x.E]\!]_\eta = \mathrm{Pack}(\lambda X.[\![E]\!]_{\eta(X/x)})$,
where $\lambda X.[\![E]\!]_{\eta(X/x)}$ is the function $W \mapsto [\![E]\!]_{\eta(W/x)}$.

REMARK 11.2.22 It is easy to check inductively that this definition gives a derivation of $[\![E]\!]_\eta$ which uniformly uses only *positive* information about the sets assigned by η to the free variables of E.

This is trivial for $[\![x]\!]_\eta = \eta(x)$ in (1).

It follows immediately for (2) from:

EXERCISE 11.2.23 *Show that Ψ_X^Y is an enumeration operator of both arguments X and Y.*

So if $[\![E_1]\!]_\eta$ and $[\![E_2]\!]_\eta$ are positively derived from the sets assigned by η to their free variables, then so is

$$[\![(E_1 E_2)]\!]_\eta = \Psi_{[\![E_1]\!]_\eta}([\![E_2]\!]_\eta).$$

For (3) we just notice that if $[\![E]\!]_{\eta(W/x)}$ is derived uniformly from positive information about $\eta(x_1), \ldots, \eta(x_k), W$, then

$$[\![\lambda x.E]\!]_\eta = \mathrm{Pack}(\lambda X.[\![E]\!]_{\eta(X/x)})$$

is derived uniformly from positive information about $\eta(x_1), \ldots, \eta(x_k)$. ⬚

It is now straightforward to check that Definition 11.2.21 produces what it is intended to:

THEOREM 11.2.24
For any term E and assignment η, $[\![E]\!]_\eta$ is well defined from η as a c.e. set. In particular, if E is closed then $[\![E]\!]_\eta$ is a c.e. set $[\![E]\!]$ not depending on η.

PROOF Let E have free variables $x_1, \ldots, x_k$.

Then by Remark 11.2.22 we have $[\![E]\!]_\eta \leq_e \eta(x_1), \ldots, \eta(x_k)$ c.e. And then just as for Exercise 11.1.6, we get $[\![E]\!]_\eta$ c.e. ⬚

And we can easily verify now that our interpretation does provide a model of lambda calculus:

THEOREM 11.2.25
If the λ-term E_1 is reducible to the term E_2, then $[\![E_1]\!]_\eta = [\![E_2]\!]_\eta$ for any given assignment η.

PROOF The main thing is to prove the interpretation models β-reduction. That is, if

$$(\lambda x.E_1)E_2 = [E_2/x]E_1$$

is an instance of β-reduction, and η is a given assignment, then

$$[\![(\lambda x.E_1)E_2]\!]_\eta = [\![\,[E_2/x]E_1]\!]_\eta.$$

But:

$$
\begin{aligned}
[\![(\lambda x.E_1)E_2]\!]_\eta &= \Psi_{[\![\lambda x.E_1]\!]_\eta}([\![E_2]\!]_\eta) && \text{by (2) of Definition 11.2.21} \\
&= \Psi_{\mathrm{Pack}(\lambda X.[\![E_1]\!]_{\eta(X/x)})}([\![E_2]\!]_\eta) && \text{by definition of } [\![\lambda x.E]\!]_\eta \\
&= (\lambda X.[\![E_1]\!]_{\eta(X/x)})([\![E_2]\!]_\eta) && \text{by Exercise 11.2.20} \\
&= [\![E_1]\!]_{\eta([\![E_2]\!]_\eta/x)} && \text{by definition of } \lambda X.[\![E]\!]_{\eta(X/x)} \\
&= [\![\,[E_2/x]E_1]\!]_\eta && \text{by induction on } E_1.
\end{aligned}
$$

Then, by induction on the number of reduction steps, the interpretation also preserves any reduction of a term E_1 to a term E_2. ▯

So of course, if E_1, E_2 are also closed, then we can write $[\![E_1]\!] = [\![E_2]\!]$.

We finish with an interesting byproduct of this modelling role of the e-operators. We will derive a fixed point theorem for e-operators from one for λ-terms.

EXERCISE 11.2.26 *We define the* **fixed point combinator** *Y to be* $Y =_{\mathrm{defn}} \lambda f.((\lambda x.f(x\,x))\,(\lambda x.f(x\,x)))$.
Show that any λ-term F has fixed point YF:

$$YF = F(YF).$$

EXERCISE 11.2.27 *Let Ψ be any e-operator. Show that Ψ can be expressed as $\Psi = \Psi_{[\![x]\!]_\eta}$ for suitable choice of assignment η.*
Hence, deduce that every e-operator Ψ has a c.e. fixed point W:

$$\Psi^W = W.$$

[**Hint:** Use the previous exercise and your choice of x and η to show that

$$\Psi([\![Yx]\!]_\eta) = [\![Yx]\!]_\eta. \,]$$

EXERCISE 11.2.28 *Show that for any assignment η*

$$[\![Yx]\!]_{\eta(\emptyset/x)} = \emptyset.$$

What does this tell you about fixed points for the e-operator $\Psi = \emptyset$?

REMARK 11.2.29 Fixed points play an important role in translating recursions into lambda calculus. In practice this means we can use Y to express functions like $+$ and $\times$ as fixed points of terms got straight from their recursive definitions. ▯

11.3 The Enumeration Degrees and the Natural Embedding of the Turing Degrees

Since — by Exercise 11.1.12 — $\leq_e$ is reflexive and transitive, we can use it to define a degree structure. In fact, there are *two* such degree structures in the literature, one on sets, the other on partial functions.

DEFINITION 11.3.1 *Let* $A, B \subseteq \mathbb{N}$.

(1) Define $A \equiv_e B \iff_{\text{defn}} A \leq_e B \,\&\, B \leq_e A$.

The **enumeration degree** *— or* **e-degree**, *written* $\deg_e(A)$ *— of* A *is*

$$\deg_e(A) =_{\text{defn}} \{X \mid X \equiv_e A\}.$$

We define $\deg_e(A) \leq \deg_e(B) \iff_{\text{defn}} A \leq_e B$.

We write $\boldsymbol{\mathcal{D}}_e =_{\text{defn}}$ *the set of all e-degrees with the ordering* $\leq$.

(2) The **partial degree** *of a partial function* f *is*

$$\deg(f) =_{\text{defn}} \{g \mid \operatorname{Graph}(f) \equiv_e \operatorname{Graph}(g)\}$$
$$= \{g \mid f \equiv_{NT} g\} \quad \text{(by Theorem 11.1.8)}.$$

We write $\boldsymbol{\mathcal{P}} =$ *the set of all partial degrees, with ordering* $\leq$ *defined by* $\deg(f) \leq \deg(g) \iff_{\text{defn}} \operatorname{Graph}(f) \leq_e \operatorname{Graph}(g) \iff f \leq_{NT} g$.

(3) We say that an e-degree $\mathbf{a}_e$ *is* **total** *if there is a total function* f *with* $\operatorname{Graph}(f) \in \mathbf{a}_e$.

We write **TOT** $=$ *the set of total e-degrees.*

EXERCISE 11.3.2 *Show that* $\leq$ *on* $\boldsymbol{\mathcal{D}}_e$ *— and hence on* $\boldsymbol{\mathcal{P}}$ *also — is a well-defined partial ordering.*

REMARK 11.3.3 For total functions, the definition of $\deg(f)$ agrees with that via $\equiv_T$ — since for f, g total Corollary 11.1.11 gives

$$\boxed{f \equiv_T g \iff \operatorname{Graph}(f) \equiv_e \operatorname{Graph}(g)}$$

⬚

The following theorem now confirms the naturalness of all these degree structures. It says that the partial degrees and the e-degrees are just the same structures, as are the Turing degrees and the total e-degrees — and that the Turing degrees form a *substructure* of $\boldsymbol{\mathcal{D}}_e$.

THEOREM 11.3.4
$\mathcal{D} \cong \mathbf{TOT} \subseteq \mathcal{D}_e \cong \mathcal{P}$.

PROOF   We get $\mathbf{TOT} \subseteq \mathcal{D}_e$ immediately from Definition 11.3.1, of course.

(1)   Prove $\mathcal{D}_e \cong \mathcal{P}$

Define $\theta : \mathcal{P} \to \mathcal{D}_e$ by $\theta(\deg(f)) = \deg_e(\mathrm{Graph}(f))$. Then

$$\deg(f) \le \deg(g) \iff \mathrm{Graph}(f) \le_e \mathrm{Graph}(g) \quad (\text{definition of} \le \text{on } \mathcal{P})$$
$$\iff \deg_e(\mathrm{Graph}(f)) \le \deg_e(\mathrm{Graph}(g))$$
$$\iff \theta(\deg(f)) \le \theta(\deg(f)), \tag{11.2}$$

so that θ is *order preserving*.

And from the equivalence (11.2), with Exercise 11.3.2, we get

$$\deg(f) = \deg(g) \iff \theta(\deg(f)) = \theta(\deg(f)).$$

So θ is *well defined* — that is, does not depend on the choice of $f \in \deg(f)$ — and is *one–one*.

To see that θ is *onto* $\mathcal{D}_e$, we just need to show every $\mathbf{a}_e$ contains a set of the form $\mathrm{Graph}(f)$, for some function f. But it is easy to see that for every $A \subset \mathbb{N}$ we have $A \equiv_e \mathrm{Graph}(S_A) = \{\langle x, 1 \rangle \mid x \in A\}$, where S_A is the semicharacteristic function of A.

We call θ the **natural isomorphism** between $\mathcal{D}_e$ and $\mathcal{P}$.

---

(2)   Prove $\mathcal{D} \cong \mathbf{TOT}$

We just take $\iota : \mathcal{D} \to \mathbf{TOT}$ to be θ restricted to the set of all *total* functions f. That is: $\iota(\deg(f)) = \deg_e(\mathrm{Graph}(f))$, for all total functions f.

We call ι the **natural embedding** of $\mathcal{D}$ into $\mathcal{D}_e$.

If f is total, $\iota(\deg(f)) = \deg_e(\mathrm{Graph}(f)) \in \mathbf{TOT}$, so ι is *well defined*. Also ι is *onto* straight from the definition of ι and $\mathbf{TOT}$. And we get ι *one–one* and *order preserving* directly from the same properties for θ. $\square$

The picture we are starting to build up of the noncomputable universe will be a complex one, but one which keeps on revealing underlying form and coherence in unexpected and captivating ways.

We have extended "computability in the real world" — that is, computability relative to context — in different ways: on the one hand introducing e-reducibility as a natural alternative to the oracle model, and on the other allowing nondeterministic computation. And now we see the structures corresponding to these approaches converging in a very satisfying and informative way:

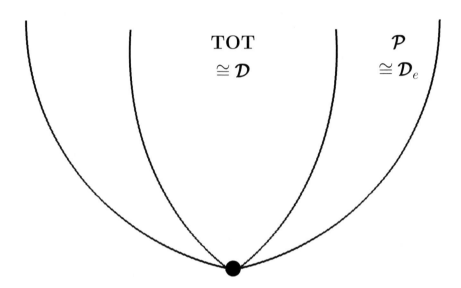

Having extended $\mathcal{D}$, our new structures still have many of the basic properties we found earlier. Obviously part (iv) of Theorem 10.3.5 and $\mathcal{D} \cong \mathbf{TOT} \subseteq \mathcal{D}_e$ tell us that $\mathcal{D}_e$ is uncountable. Also:

EXERCISE 11.3.5 *Let* $\mathbf{a} \in \mathcal{D}_e$. *Show that*
(i) $\mathbf{a}$ *is countably infinite, and*
(ii) $\mathcal{D}_e(\leq \mathbf{a})$ *is countable.*

EXERCISE 11.3.6 *Show that* $\mathcal{D}_e$ *has a least degree* $\mathbf{0}_e$ *consisting of all c.e. sets.*

EXERCISE 11.3.7 *Show that* $\mathcal{P}$ *has a least degree* $\mathbf{0}$ *consisting of all p.c. functions.*

EXERCISE 11.3.8   Let $\deg_e(A) \cup \deg_e(B) =_{\text{defn}} \deg_e(A \oplus B)$.

Show that $\deg_e(A) \cup \deg_e(B) = \text{lub}\{\deg_e(A), \deg_e(B)\}$.

REMARK 11.3.9   Some people find it confusing that the Turing degrees are embedded as a substructure of the enumeration degrees, whereas the Turing degree of a set A may be a very different thing from the enumeration degree of A. This is because when we write $\deg_e(A)$ or $B \leq_e A$, we really are computing relative to the information contained in the *set* A, but $\deg(A)$ or $B \leq_T A$ is based on computation relative to the *characteristic function* of A.

Sometimes though the natural embedding ι of $\mathcal{D}$ in $\mathcal{D}_e$ does put $\deg(A)$ and $\deg_e(A)$ in exact correspondence. Exercise 11.3.13 below shows that the natural embedding of the set $\mathcal{E}$ of c.e. Turing degrees is precisely the set of all Π_1 e-degrees — where we follow the usual convention of saying a degree has property P if it contains a set with property P. ☐

Since $\mathcal{P}$ and $\mathcal{D}_e$ are really the same structure, let us simplify our notation by writing "$f \leq_e g$" for "$\text{Graph}(f) \leq_e \text{Graph}(g)$". Then the definition of the natural embedding also simplifies to

$$\iota(\deg(f)) = \deg_e(f)$$

REMARK 11.3.10   There are lots of other situations where it is simpler to suppress the coding, and compute relative to sets of objects which are not numbers.

For instance it is easy to code *strings* — so why not allow ourselves to compute relative to sets of strings? ☐

The next exercise shows that the usual join on the Turing degrees agrees with the join defined on the e-degrees in Exercise 11.3.8.

EXERCISE 11.3.11   *Show that if $a, b \in \mathcal{D}$, then $\iota(a \cup b) = \iota(a) \cup \iota(b)$.*

[**Hint:** Show that $\chi_{A \oplus B} \equiv_e \chi_A \oplus \chi_B$.]

EXAMPLE 11.3.12   *Show that for each set $A \subseteq \mathbb{N}$ we have*

$$\chi_A \equiv_e A \oplus \overline{A} \equiv_e A',$$

so that A' is always of total e-degree.

SOLUTION We get $\chi_A \equiv_e A \oplus \overline{A}$ easily, since we can enumerate the graph of χ_A from an enumeration of $A \oplus \overline{A}$, and vice versa.

For the second e-equivalence, $A, \overline{A} \leq_m A'$ gives $A, \overline{A} \leq_e A'$, so that we have $A \oplus \overline{A} \leq_e A'$.

On the other hand we have A' c.e. in A, and so is $\leq_e \chi_A \equiv_e A \oplus \overline{A}$ by Exercise 11.1.14.
◻

EXERCISE 11.3.13 *Show that*

(i) If $\overline{A} \leq_e A$, we have $\chi_A \equiv_e A \oplus \overline{A} \equiv_e A$, so that $\deg_e(A)$ is total.

(ii) If A is c.e., then $\iota(\deg(\overline{A})) = \deg_e(\overline{A})$.

(iii) Defining $\mathbf{0}'_e = \deg_e(\overline{K})$, deduce that $\iota(\mathbf{0}') = \mathbf{0}'_e > \mathbf{0}_e$.

EXERCISE 11.3.14 *Let f be a total function.*

Show that $\overline{f} \leq_e f$ and, hence, that $\chi_f \equiv_e f$.

Deduce that a degree $\mathbf{a} \in \mathcal{D}_e$ is total if and only if it contains a characteristic function.

Other naturally arising sets turn out to have total e-degree:

EXERCISE 11.3.15 *Let $D = A - B$ be a d.c.e. set (d.c.e. was defined in Exercise 5.2.19) where A, B have standard c.e. approximating sequences $\{A^s\}_{s \geq 0}, \{B^s\}_{s \geq 0}$.*

Show that $D \equiv_e \overline{C} \equiv_e \chi_C$, where C is the c.e. set given by

$$C = \{\langle s, x \rangle \mid x \in A^s \,\&\, x \in B^t, \text{ some } t \geq s\}.$$

Deduce that each d.c.e. e-degree is Π_1 and hence total.

Notice that we also have $C \leq_T D$ in the above exercise, but it can be shown that C is not always $\equiv_T D$.

The above exercise might lead you to expect *all* n-c.e. sets to have total e-degree. However, there turn out to be 3-c.e. e-degrees which are far from being total.

At this point it would be nice to get some order into the flow of miscellaneous information about the new degree structures. What are the main questions raised by what we have done so far?

GENERAL QUESTIONS

(1) In Exercise 11.3.13 we defined $\mathbf{0}'_e = \deg_e(\overline{K}) = \iota(\mathbf{0}') > \mathbf{0}_e$. Can we generalise this to get a **jump operator** on the e-degrees which agrees with the natural embedding of the Turing jump?

(2) We also found that familiar information content — such as a co-c.e. or d.c.e. set — made an e-degree total. Can we find a nice information theoretic characterisation of the total e-degrees?

(3) What about structure? Is $\mathcal{D}_e = \mathbf{TOT}$ or $\cong \mathcal{D}$? How do we extend the methods of Section 10.6 to $\mathcal{D}_e$?

(4) What more can we say about the important local structure $\mathcal{D}_e(\leq \mathbf{0}'_e)$? How does it relate to the arithmetical hierarchy?

Well, this is a very big topic. But we can get some interesting and attractive results which will not get us into too much deep water.

The easiest question to answer is the one on the jump. We cannot take the e-jump to be the e-degree of A', since we already saw from Example 11.3.12 and Exercise 11.3.13 that we can have $A \equiv_e A'$ — this happens for instance if $A \in \Pi_1$. A'' does not suffer from this problem. But since A'' uses negative information about A, it is not *e-invariant* — that is, we can have $A \equiv_e B$ but $A'' \not\equiv_e B''$. So $\deg_e(A'')$ would not be well defined as a jump on e-degrees. Here is a good definition:

DEFINITION 11.3.16   Let $K_A = \{x \mid x \in \Psi_x^A\}$.

Then the **e-jump** of a set A is $J_e^A =_{\text{defn}} A \oplus \overline{K}_A$. And the **jump** of an e-degree $\mathbf{a} = \deg_e(A)$ is defined to be $\mathbf{a}' = \deg_e(A \oplus \overline{K}_A) = \deg_e(J_e^A)$.

We iterate the jump in the usual way to obtain the nth jump $\mathbf{a}^{(n)}$ of $\mathbf{a}$.

EXERCISE 11.3.17   Show that if $A \leq_e B$, then $J_e^A \leq_e J_e^B$.

Deduce that (i) The e-jump is well defined — that is, if $A \equiv_e B$ then $J_e^A \equiv_e J_e^B$.

(ii) The e-jump is order-preserving — that is, if $\mathbf{a} \leq \mathbf{b}$ in the e-degrees, then $\mathbf{a}' \leq \mathbf{b}'$.

Also, the e-jump does jump:

EXERCISE 11.3.18   Given an e-degree $\mathbf{a}$, show that $\mathbf{a} < \mathbf{a}'$.

[**Hint:** To show that $\overline{K}_A \not\leq_e A$, assume that $\overline{K}_A = \Psi_i^A$ for some $i \in \mathbb{N}$, and get a contradiction.]

Here is the really essential property of the e-jump:

PROPOSITION 11.3.19

The e-jump agrees with the natural embedding of the Turing jump — that is, for each $A \subseteq \mathbb{N}$ we have $\iota(\deg(A')) = \deg_e(J_e(\chi_A))$.

PROOF   We need to show that $J_e(\chi_A) \equiv_e \chi_{A'}$.

(1) $\boxed{\text{Prove } J_e(\chi_A) \leq_e \chi_{A'}}$

Since $A, \overline{A}, K_{\chi_A}$ are c.e. in A, we have $A, \overline{A}, K_{\chi_A} \leq_m A'$. So

$$\chi_A \leq_e A' \leq_e \chi_{A'}.$$

Also $\overline{K}_{\chi_A} \leq_m \overline{A'}$ gives $\overline{K}_{\chi_A} \leq_e \overline{A'} \leq_e \chi_{A'}$.
So $J_e(\chi_A) = \chi_A \oplus \overline{K}_{\chi_A} \leq_e \chi_{A'}$.

(2) $\boxed{\text{Prove } \chi_{A'} \leq_e J_e(\chi_A)}$

Since A' is c.e. in A we have $A' \leq_e \chi_A$, by Exercise 11.1.14.
To get $\overline{A'} \leq_e \overline{K}_{\chi_A}$ we first need:

LEMMA 11.3.20
K_{χ_A} *is* Σ_1^A-*complete*.

PROOF   If we can show that K_{χ_A} is creative relative to A, then the result follows from the relativised form of Theorem 11.1.14.

The argument turns out to be not much different from that for proving K creative in the proof of Theorem 6.1.3.

It is straightforward to write K_{χ_A} in Σ_1^A form.

Taking our cue from the proof of Theorem 6.1.3, we take the A-creative function for K_{χ_A} to be the identity function $f : x \mapsto x$.

Let $\{U_i^A\}_{i \geq 0}$ be the standard listing of the A-c.e. sets we get from relativising Exercise 11.1.7.

Assume $U_i^A \subseteq \overline{K}_{\chi_A}$. Then $i \notin U_i^A$, since otherwise we would have $i \in K_{\chi_A} \cap U_i^A$, a contradiction. So $f(i) = i \in \overline{K}_{\chi_A} - U_i^A$, as required.   ◻

From the lemma we get $A' \leq_m K_{\chi_A}$, so $\overline{A'} \leq_m \overline{K}_{\chi_A}$, giving $\overline{A'} \leq_e \overline{K}_{\chi_A}$, as required.

We now put this together with $A' \leq_e \chi_A$ to get

$$\chi_{A'} \leq_e \overline{K}_{\chi_A} \leq_e J_e(\chi_A),$$

and the theorem is proved.   □

Most importantly — we now know that $\mathbf{0} < \mathbf{0}' < \mathbf{0}'' < \cdots < \mathbf{0}^{(n)} < \ldots$ naturally embeds to $\mathbf{0}_e < \mathbf{0}'_e < \mathbf{0}''_e < \cdots < \mathbf{0}_e^{(n)} < \ldots$. It also follows that the total e-degrees below $\mathbf{0}'_e$ are all Δ_2.

EXERCISE 11.3.21   *Show that $\mathcal{D}_e(\leq \mathbf{0}'_e)$ is not linearly ordered.*

You may have noticed we defined $\mathbf{0}'_e = \deg_e(\overline{K})$ in Exercise 11.3.13, and then — taking $\mathbf{0}_e = \deg_e(\chi_\emptyset)$, say — get $\mathbf{0}'_e = \deg_e(J(\chi_\emptyset))$ from the general Definition 11.3.16 of the e-jump. You might like to check that these definitions are just the same:

EXERCISE 11.3.22   *Show that $\overline{K} \equiv_m \overline{K}_{\chi_\emptyset}$.*
   Deduce that $\overline{K} \equiv_e J_e(\chi_\emptyset)$.

There are various kinds of information content which turn out to characterise the total e-degrees. The nicest — and the most recent — uses the notion of an *introreducible* set:

DEFINITION 11.3.23   *An infinite set $A \subseteq \mathbb{N}$ is **introreducible** if A is computable from every infinite subset of A.*

EXERCISE 11.3.24   *Given W c.e., let*

$$A = \{\langle x, s \rangle \mid W^s \restriction x = W \restriction x\}.$$

Show that A is an introreducible d.c.e. set having the same Turing degree as W. Show also that $\overline{W} \equiv_e A$.

THEOREM 11.3.25 (Carl Jockusch)
   TOT $=$ *the set of e-degrees of introreducible sets.*

PROOF (1) | Prove **TOT** $\subseteq$ the set of e-degrees of introreducible sets |

Let $f : \mathbb{N} \to \mathbb{N}$ be total.
To find an introreducible set of the same e-degree as the graph of f, define

$$A = \{\sigma \in \omega^{<\omega} \mid \sigma \subset f\} = \text{ the set of beginnings of } f.$$

Keeping in mind Remark 11.3.10 — we first notice that $f \equiv_e A$. This is because an enumeration of the beginnings of f enables us to enumerate the graph of f, and vice-versa.

We also see that A is introreducible. If B is an infinite subset of A and $\sigma \in \omega^{<\omega}$, we can compute, using B, some $\tau \in B$ with $|\tau| \geq |\sigma|$. Then $\sigma \in A$ or not according as $\sigma \subseteq \tau$ or not.

(2) | Prove the set of e-degrees of introreducible sets $\subseteq$ **TOT** |

Say A is introreducible. We show that $A \equiv_e \chi_A$, giving $\deg_e(A) \in$ **TOT**.

Say A is c.e. in a given X. Then — relativising Exercise 5.1.10 — one can get an infinite subset C of A which is Turing computable from X. But then $A \leq_T C \leq_T X$. But this means $\chi_A \leq_T X$, and so χ_A is c. e. in X.

But then by Selman's Theorem 11.1.13, and the fact that X could be *any* set, one gets $\chi_A \leq_e A$. Since trivially $A \leq_e \chi_A$, we get $A \equiv_e \chi_A$, as required.

☐

EXERCISE 11.3.26 *Show that every Turing degree is the degree of an introreducible set.*

It is now time to look at our last two general questions.

11.4 The Structure of $\mathcal{D}_e$ and the Arithmetical Hierarchy

We can find out a lot about the structure of $\mathcal{D}_e$ from the natural embedding of $\mathcal{D}$ in $\mathcal{D}_e$. For instance, Corollary 10.6.4 immediately tells us that

| The structure $\mathcal{D}_e(\leq 0'_e)$ — and hence $\mathcal{D}_e$ — is not linearly ordered. |

But we must first deal with an important and overdue structural question:

| Is $\mathcal{D}_e =$ **TOT** $\cong \mathcal{D}$? |

In 1955 the Russian mathematician Yuri Medvedev described a special kind of e-degree which is very far from being total:

DEFINITION 11.4.1 *We say an e-degree* $\mathbf{a} > \mathbf{0}_e$ *is* **quasi-minimal** *if* $\mathbf{0}_e$ *is the only total e-degree* $\leq \mathbf{a}$.

Medvedev went on to prove:

THEOREM 11.4.2
There does exist a quasi-minimal e-degree.

PROOF   We construct $A \subseteq \mathbb{N}$ satisfying all the requirements:

$$\mathcal{R}_{2i} : \quad A \neq W_i$$
$$\mathcal{R}_{2i+1} : \quad \text{If } \Psi_i^A \text{ is the graph of a total function, then } \Psi_i^A \text{ is c.e.}$$

We build A via strings $\sigma_0 \subset \sigma_1 \subset \ldots \sigma_n \subset \ldots \subset \chi_A$.
The idea is to satisfy requirement $\mathcal{R}_n$ by a clever choice of σ_{n+1}.

The strategy for satisfying $\mathcal{R}_{2i}$:

Choose $\sigma_{2i+1} \supset \sigma_{2i}$ to make $A(|\sigma_{2i}|) = \sigma_{2i+1}(|\sigma_{2i}|) \neq W_i(|\sigma_{2i}|)$.

The strategy for satisfying $\mathcal{R}_{2i+1}$:

Try to choose $\sigma_{2i+1} \subset A$ to make some $\langle x, y \rangle, \langle x, z \rangle \in \Psi_i^A$ with $y \neq z$ — that is, so that Ψ_i^A is not *single valued* — with the outcome that Ψ_i^A *cannot* be the graph of a function.
   If this primary strategy fails, we will be able to argue that if Ψ_i^A is the graph of a total function, then it must be c.e.

The Construction

Stage 0.

Define $\sigma_0 = \emptyset$.
   At stage $n + 1$ assume σ_n already constructed.

Stage $n + 1 = 2i + 1$.

Define $\sigma_{n+1} = \sigma_n {}^\frown (1 - W_i(|\sigma_n|))$.

Outcome: Since $A \supset \sigma_{n+1}$, we have $A(|\sigma_n|) = 1 - W_i(|\sigma_n|) \neq W_i(|\sigma_n|)$. So $\mathcal{R}_{2i}$ is satisfied.

Stage $n + 1 = 2i + 2$.

Let $\sigma^+ = \{x \mid \sigma(x) \downarrow = 1\}$.
Look for a $\sigma \supset \sigma_n$ such that for some $x, y, z \in \mathbb{N}$ with $y \neq z$ we have $\langle x, y \rangle, \langle x, z \rangle \in \Psi_i^{\sigma^+}$.
CASE I. If σ exists, define such a $\sigma = \sigma_{n+1}$.

Outcome: Since $A \supset \sigma_{n+1}$, we have Ψ_i^A not single valued. So $\mathcal{R}_{2i+1}$ is satisfied.

CASE II. Otherwise, define $\sigma_{n+1} = \sigma_n \frown 0$.

Outcome: We show — if Ψ_i^A is the graph of a total function, then it is c.e. In which case $\mathcal{R}_{2i+1}$ is again satisfied.

PROOF (of outcome) Assume Ψ_i^A the graph of a total function. We show:

$$\langle x, y \rangle \in \Psi_i^A \iff (\exists \sigma \supset \sigma_n) \left[\langle x, y \rangle \in \Psi_i^{\sigma^+} \right]$$
$$\iff (\exists \sigma \supset \sigma_n \,\&\, s \geq 0) \left[\langle x, y \rangle \in \Psi_{i,s}^{\sigma^+} \right],$$

so that Ψ_i^A is Σ_1 and so c.e.

($\Rightarrow$) If $\langle x, y \rangle \in \Psi_i^A$ then $\langle x, y \rangle \in \Psi_i^{\sigma^+}$, some $\sigma \subset A$.

($\Leftarrow$) Conversely, say $\langle x, y \rangle \in \Psi_i^{\sigma^+}$, some $\sigma \supset \sigma_n$. And let z be such that $\langle x, z \rangle \in \Psi_i^A$, where $\langle x, z \rangle \in \Psi_i^{\tau^+}$ with $\tau \subset A$.
Choose $\pi \supset \sigma_n$ with $\pi^+ \supset \sigma^+ \cup \tau^+$.
Then $\langle x, y \rangle, \langle x, z \rangle \in \Psi_i^{\sigma^+ \cup \tau^+} \subseteq \Psi_i^{\pi^+}$. So $y = z$, since Case II applies. ☐

This outcome completes the proof of the theorem. ☐

We can of course get a local version of this theorem by inspecting where A lies in the arithmetical hierarchy. But how exactly *do* we relate the arithmetical hierarchy to $\mathbf{0}_e, \mathbf{0}'_e, \mathbf{0}''_e, \ldots, \mathbf{0}_e^{(n)}, \ldots$? Is there an e-version of Post's Theorem? Here it is:

THEOREM 11.4.3
For each $A \subseteq \mathbb{N}, n \geq 0$ we have $\deg_e(A) \leq \mathbf{0}_e^{(n)} \iff A \in \Sigma_{n+1}$.

PROOF   The most important case is $n = 1$, telling us that:

$$\mathcal{D}_e(\leq \mathbf{0}'_e) = \text{the set of all } \Sigma_2 \text{ e-degrees}$$

I will just prove this special case — leaving you to fill in some messy details of the full inductive step.

($\Rightarrow$) Say $\deg_e(A) \leq \mathbf{0}'_e$, so that $A \leq_e \overline{K}$. Let $A = \Psi_i^{\overline{K}}$. Then

$$x \in A \iff x \in \Psi_i^{\overline{K}}$$
$$\iff (\exists s^*)(\forall s > s^*)\,[\,x \in \Psi_i^{\overline{K}}[s]\,] \in \Sigma_2.$$

($\Leftarrow$) If $A \in \Sigma_2$, then it is c.e. in K — and so, by the usual argument, is c.e. in, and hence $\leq_e$, $K \oplus \overline{K} \in \mathbf{0}'_e$.   □

You should now be able to prove:

EXERCISE 11.4.4   *Show that there is a non-total e-degree $< \mathbf{0}'_e$.*

[**Hint:** Verify that there is an oracle from $\mathbf{0}'$ which enables you to carry out the construction of the quasi-minimal A in Theorem 11.4.2.]

11.5   The Medvedev Lattice

The enumeration degrees in their turn can be viewed as part of an even more extensive structure.

Medvedev's idea back in 1955 was that sometimes one is interested in *more* than reducing some particular problem to some other one. Sometimes it made more sense to see how a whole *class* of related problems might be reduced to any representative of another class. His idea has become quite useful in recent years, even to people who are interested in very basic issues concerning relative computability.

DEFINITION 11.5.1   *(1) Call a class $\mathcal{A} \subseteq \omega^\omega$ a* **mass problem**.

(2) We say that $\mathcal{A}$ is **reducible to** *a mass problem $\mathcal{B}$ if $\mathcal{A} \supseteq \Phi(\mathcal{B})$ for some partial computable functional $\Phi : \omega^\omega \to \omega^\omega$ with $\mathcal{B} \subseteq \text{dom}\Phi$.*

$\mathcal{A}$ is **Medvedev equivalent to** *$\mathcal{B}$ if $\mathcal{A}$, $\mathcal{B}$ are reducible to each other.*

(3) The **degree of difficulty** *$[\mathcal{A}]$ of $\mathcal{A} \subseteq \omega^\omega$ is the set of all $\mathcal{B}$ equivalent to $\mathcal{A}$. We write $[\mathcal{A}] \leq [\mathcal{B}]$ if $\mathcal{A}$ is reducible to $\mathcal{B}$.*

The resulting structure

$$\mathfrak{M} = \langle\, \text{degrees of difficulty}, \leq \,\rangle$$

is called the **Medvedev lattice**.

I will leave you to verify the basic facts underlying this definition:

EXERCISE 11.5.2   *Show that Medvedev equivalence is an equivalence relation on mass problems.*

EXERCISE 11.5.3   *Show that the ordering $\leq$ on the degrees of difficulty is a well-defined partial ordering.*

The next two exercises show that $\mathfrak{M}$ lives up to its name — it is indeed a lattice.

EXERCISE 11.5.4   *Show that $\mathfrak{M}$ has a lub operation defined by*

$$[\mathcal{A}] \vee [\mathcal{B}] = [\mathcal{A} \oplus \mathcal{B}]$$

where $\mathcal{A} \oplus \mathcal{B} = \{\, f \oplus g \mid f \in \mathcal{A} \ \& \ g \in \mathcal{B} \,\}$.

EXERCISE 11.5.5   *Show that $\mathfrak{M}$ has a glb operation defined by*

$$[\mathcal{A}] \wedge [\mathcal{B}] = [\mathcal{A} \cup \mathcal{B}].$$

A set $A \subseteq \mathcal{N}$ gives rise to two important examples of mass problems:

(1) $\mathbf{S}_A = \{\chi_A\}$, or *problem of solvability* of A, and

(2) $\mathbf{E}_A = \{f \in \omega^\omega \mid \text{range}(f) = A\}$, or *problem of enumerability* of A.

These give rise, respectively, to the substructures of the *degrees of solvability* and the *degrees of enumerability*.

Some more basic, easy-to-verify facts include:

EXERCISE 11.5.6   *Show that $\mathfrak{M}$ has minimum element $[\omega^\omega]$, and maximum element $[\emptyset]$.*

EXERCISE 11.5.7   *Show that for any $A, B \subseteq \mathbb{N}$, $\deg(A) \leq \deg(B)$ if and only if $[\mathbf{S}_A] \leq [\mathbf{S}_B]$ — and hence $\boldsymbol{D}$ can be embedded in $\mathfrak{M}$ as the degrees of solvability.*

EXERCISE 11.5.8   *Show that for any $A, B \subseteq \mathbb{N}$, $\deg_e(A) \leq \deg_e(B)$ if and only if $[\mathbf{E}_A] \leq [\mathbf{E}_B]$ — and so $\boldsymbol{D}_e$ can be embedded in $\mathfrak{M}$ as the degrees of enumerability.*

One can also show that the embedding of $\boldsymbol{D}$ into $\mathfrak{M}$ induced by this embedding of $\boldsymbol{D}_e$ coincides with the embedding of $\boldsymbol{D}$ into $\mathfrak{M}$ as the degrees of solvability. For a much fuller description of the known properties of the Medvedev lattice one should see the 1996 survey paper of Sorbi.

All this can be represented in the picture:

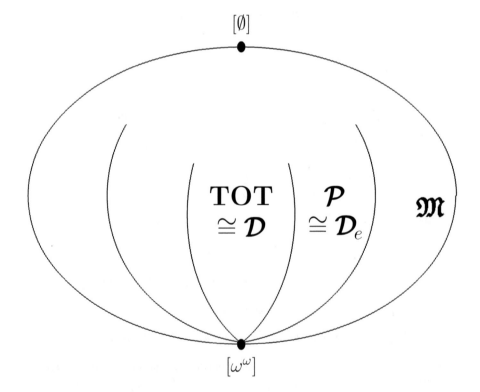

A difficult and interesting question is:

> To what extent can we *describe* $\boldsymbol{D}$ within $\boldsymbol{D}_e$ or $\mathfrak{M}$, or $\boldsymbol{D}_e$ within $\mathfrak{M}$?

There are different ways of making *describe* precise. Here is the usual way:

> **DEFINITION 11.5.9** *A relation R on a structure $\mathbf{M}$ is* **definable** *in $\mathbf{M}$ if we can write down a wf in the first order language for $\mathbf{M}$ which is satisfied by exactly the same set of objects from $\mathbf{M}$ as R is.*

Here are two simple examples:

- **0** is definable in $\boldsymbol{D}$ by the wf:

$$\varphi(x) \iff {}_{\text{defn}} (\forall y)\,[\,x \leq y\,].$$

- The join $\mathbf{x} = \mathbf{y} \cup \mathbf{z}$ is definable in $\boldsymbol{D}$ by the wf:

$$\varphi(x, y, z) \iff {}_{\text{defn}} y \leq x \,\&\, z \leq x \,\&\, (\forall w)\,[\,(y \leq w \,\&\, z \leq w) \Rightarrow x \leq w\,].$$

We will say more about definability in general in Section 12.1. Most of the questions about definability of $\boldsymbol{D}$ or $\boldsymbol{D}_e$ are very difficult. Surprisingly, a simple and elegant definition of $\boldsymbol{D}$ — in the shape of the degrees of solvability — in $\mathfrak{M}$ was found by Dyment in 1976. They are the degrees of difficulty satisfying in $\mathfrak{M}$:

$$\varphi(x) \iff {}_{\text{defn}} (\exists y)[x < y \,\&\, (\forall z)(x < z \implies y \leq z)].$$

In other words — for those of you familiar with the terminology — the degrees of solvability are characterised as those degrees $[\mathcal{A}]$ for which $\mathfrak{M}(> [\mathcal{A}])$ forms a *principal filter* in $\mathfrak{M}$.

11.6   Polynomial Bounds and P =? NP

The sort of computability we have studied so far excludes a whole class of functions from being in any practical sense computable. It indicates principles governing all physical processes. It is the basis for grand mathematical structures within which we are all constrained.

It also provides the basis for measuring what we *can* compute *in practice*, and that is what we briefly look at now. What we must face up to is that what we have come to call computable functions are in the main quite beyond the capacity not only of the computer on your desk, but beyond *any* computer based on Turing principles, in the foreseeable lifetime of our Universe.

In this short section we point the way towards the sort of practical refinements of the pure theory that computer scientists have come up with. Once again nondeterminism appears. If you have practical bounds on computing resources, improvements in the *efficiency* of programs may make incomputable functions computable. The conjecture is that nondeterminism may do this. This is a huge area of research with problems all its own. I can do little more than whet your appetite for further investigations.

Complexity theory is concerned with getting more sensitive and practically relevant measures of how computable a function is, or how solvable a particular problem may be. It has developed a theory of such measures and told us something of their characteristics, a landmark here being the early work of Manuel Blum. It has come up with particular measures, even hierarchies, such as those based on polynomial bounds on computing time. And it has successfully located all sorts of interesting and practically relevant problems in mathematics, logic and computer science within this framework.

Here is a cook's tour — in more ways than one you will discover! — of a small part of this fascinating subject.

Measuring complexity of computation

There are lots of ways of gauging the complexity of a partial computable function φ_i. (You can easily adapt what follows to n-ary functions if you want.) Here is what Manuel Blum thought any *complexity measure* should be like back in 1967. A few minutes thought, and I think you will agree.

DEFINITION 11.6.1   Let $C_i(x)$ be a *2-place function* $\mathbb{N}^2 \to \mathbb{N}$.

We say C is a **complexity measure** *if it satisfies:*

(1) $C_i(x) \downarrow$ *if and only if* $\varphi_i(x) \downarrow$.

(2) $C_i(x) \leq y$ *is a computable relation of i, x and y — where $C_i(x) \leq y$ is false if $C_i(x) \uparrow$.*

Conditions (1)–(2) on C are called the **Blum axioms***.*

Here are some examples of useful complexity measures:

EXAMPLE 11.6.2 *Let the **time measure** t_i be defined by*

$$t_i(x) = \text{ the least } s \text{ such that } \varphi_{i,s}(x) \downarrow.$$

Show that $t_i(x)$ is a complexity measure.

SOLUTION $t_i(x)$ obviously satisfies axiom (1). For axiom (2), $t_i(x) \leq y$ holds if and only if $\varphi_{i,y}(x) \downarrow$, which is a computable relation. □

The time measure is a very natural one, and convenient at a practical level. Here is another useful measure, got using the Turing model again:

EXAMPLE 11.6.3 *Take the **space measure** to be*

$$M_i(x) = \begin{cases} \text{the maximum number of cells visited} \\ \text{by the reading head in computing } \varphi_i(x) & \text{if } \varphi_i(x) \downarrow \\ \uparrow & \text{otherwise.} \end{cases}$$

Show that $M_i(x)$ is a complexity measure.

SOLUTION Again $M_i(x)$ obviously satisfies axiom (1). But what about axiom (2)?

Clearly we can decide $M_i(x) \leq y$ if we follow through the computation of $\varphi_i(x)$ and find that either:

(a) the reading head visits more than y cells — giving $M_i(x) \leq y$ false, or

(b) the computation halts with $\leq y$ cells visited — so $M_i(x) \leq y$ is true.

How do we avoid an infinitary decision process if the computation never stops, but the head only ever visits $n \leq y$ cells?

We can describe the contents of these n cells at stage s of the computation via a string of length n. If we insert into the string the current internal state, positioning it to the left of the symbol from the cell currently observed by the reading head, we get an *instantaneous description* of the computation at stage s. Then notice:

In this case there are at most finitely many instantaneous descriptions possible during the computation of $\varphi_i(x)$.

So this gives a third finitary outcome to the decision process:

(c) the computation eventually has a repeated instantaneous description —

in which case the computation enters an infinitary loop, so that $M_i(x) \leq y$ is false.

Since cases (a) – (c) cover all possibilities, axiom (2) is satisfied. So $M_i(x)$ is a complexity measure. $\quad\Box$

EXERCISE 11.6.4   *Is $C_i(x) = \varphi_i(x)$ a complexity measure?*

[**Hint:** Consider $\varphi_i =$ the semicharacteristic function of K, and investigate the computability of $C_i(x) \leq 1$.]

A nice thing about the Blum axioms is how they allow complexity measures based on different models of computability.

EXERCISE 11.6.5   *Let P_i be a URM program which computes φ_i. Let*

$$
C_i(x) = \begin{cases} \text{the number times we jump in} \\ \text{using } P_i \text{ to compute } \varphi_i(x) & \text{if } \varphi_i(x) \downarrow \\ \uparrow & \text{otherwise.} \end{cases}
$$

Show that $C_i(x)$ a complexity measure.

[**Hint:** To verify axiom (2), use the fact that a computation of P_i from any input x can perform at most $\ell(P_i)$ consecutive steps without jumping.]

But the simplicity of the axioms means we allow complexity measures which are broadly correct, but which can give bizarre estimates of complexity in selected cases.:

EXERCISE 11.6.6   *Let A be a finite subset of Tot. Define*

$$
C_i'(x) = \begin{cases} C_i(x) & \text{if } i \notin A \\ 0 & \text{if } i \in A, \end{cases}
$$

where C is a complexity measure.

Is C' a complexity measure?

EXERCISE 11.6.7   *If C is a complexity measure, show that for every i $C_i(x)$ is partial computable.*

Starting out from the Blum axioms, one can find out all sorts of interesting things relevant to the search for efficient programs. The most famous such result is Blum's *Speedup Theorem* which says there is a computable f with no "optimal" program for computing it. In fact:

> **THEOREM 11.6.8 (The Blum Speedup Theorem)**
> *Let g be computable, and let C be a complexity measure. Then there is a computable f such that given $\varphi_i = f$ there is a j with $\varphi_j = f$ and*
>
> $$g(C_j(x)) < C_i(x) \quad \text{for all } x \geq \text{ some } n_0 \in \mathbb{N}.$$

Of course, the n_0 in the theorem could be very large, which cuts down the practical significance of such a result!

Anyway, it is not clear how such results advance our investigation of functions which are computable *in practice*. Computability is our focus. Arguing in such broad terms does not help us frame our new refined notion of "computable". If we look closely enough, different complexity measures give varying results. So we will pass on without further delay.

Polynomial Time Computability

Our idea is to come up with a class of programs which allows all those we commonly run on computers, and which excludes those for which the amount of computing time or storage space needed balloons hopelessly on larger inputs. In doing this we adopt a model of computability which is relatively "honest" about the work involved in a computation — Turing machines — and evaluate the size of inputs honestly. For this it is convenient to formulate a number x as a string of 1's as usual, and write $|x|$ for the *length* of x in the usual way.

Actually most programs used in practice satisfy $t_i(x) \leq k|x|^2$ for some given $k > 0$. While for a lot of mathematically acceptable algorithms we only have $t_i(x) \leq k2^{|x|}$, completely hopeless on a computer. So a reasonably safe definition of f being "practically computable" — or *tractable* — might be:

> **DEFINITION 11.6.9** $f : \mathbb{N} \to \mathbb{N}$ is **polynomial time computable** *if $f = \varphi_i$ for some i, and there is a polynomial $p(n)$ such that $t_i(x) \leq p(|x|)$, each $x \geq 0$.*

Here is some useful notation:

> **DEFINITION 11.6.10** *Given functions $f, g : \mathbb{N} \to \mathbb{N}$, we write $f(n) = O(g(n))$ if $f(n) \leq kg(n)$ for some $k > 0$ and all $n > $ some $n_0 \in \mathbb{N}$.*

Intuitively, $f(n) = O(g(n))$ means that g *grows faster than* f. If $s \leq t$, then n^t grows faster than n^s. So given a polynomial $p(n) = a_o + a_1 n + \cdots + a_r n^r$, we have $p(n) = O(n^r)$.

EXERCISE 11.6.11   *Show that f is polynomial time computable if and only if for some i, $f = \varphi_i$ and there is some $r \geq 0$ such that $t_i(x) = O(|x|^r)$.*

To bring algorithms with inputs other than numbers under the polynomial time umbrella, we do not code everything as numbers as in the classical case, since the codings themselves have a nontrivial complexity. Instead we allow alphabets Σ other than just 1. Then inputs will be chosen from the set $\Sigma^* = \Sigma^{<\omega}$ of finite strings of symbols — or *words* — chosen from Σ.

DEFINITION 11.6.12   *(1) Let $\Sigma = S_1, \ldots, S_m$ be a Turing machine alphabet. We call a subset L of Σ^* a **language** on alphabet Σ.*

*(2) We say a Turing machine T **accepts** a word $x \in \Sigma^*$ if T halts on input x. T **accepts** the language L consisting of all words T accepts — and further, T **accepts** L **in polynomial time** if there is a polynomial $p(n)$ such that whenever T accepts x, it halts on input x in $\leq p(|x|)$ steps.*

*(3) The language L is **polynomial time decidable** if there is a Turing machine that accepts L in polynomial time.*

(4) Assume given Σ. We write $\mathbf{P} =$ the set of all polynomial time decidable languages.

Sets such as $\mathbf{P}$ are known as *complexity classes*. There are very many of them, based on different ways of bounding the various computing resources, such as time or storage space. Pinning down their exact inter-relationships is a major task of *complexity theory*.

EXAMPLE 11.6.13   *Show that if f and g are polynomial time computable, then so is the composition $f \circ g$ of f and g.*

SOLUTION   Let T, T' compute f, g with corresponding polynomial time bounds $p(n) = O(n^r)$ and $q(n) = O(n^s)$.

Consider input x. To compute $f(g(x))$ we can input x to T' to get $g(x)$ in $\leq q(|x|)$ steps. Notice that the output here cannot be greater than $|x| + q(|x|)$, since T' cannot print more symbols than steps in the computation.

We can then input $g(x)$ to T, and get $f(g(x))$ in $\leq p(|g(x)|)$ steps, that is, in $\leq p(|x| + q(|x|))$ steps.

But this tells us — recalling Exercise 2.4.13 — that we can build a Turing machine $\widehat{T}$ which computes $f(g(x))$ with polynomial bound $= O(n^{rs})$. ⬚

EXERCISE 11.6.14   *Show that if $L, M \in \mathbf{P}$, then so are $\Sigma^* - L$, $L \cup M$, and $L \cap M$.*

REMARK 11.6.15   Clearly, the notion of *polynomial time decidable* is intended to be the polynomial time counterpart of a computable set. It might seem a little strange to base the definition on *acceptability*, when in the classical context acceptability leads to the notion of c.e. rather than computable. But forget that analogy — say L is polynomial time decidable by T with polynomial time bound $p(m)$. Then if T does not accept $x \in \Sigma^*$, we can decide this in $p(|x|)$ computational steps of T with input x (see Exercise 11.6.14 below). The point is that one does not get a polynomial bounded counterpart to the unsolvability of the halting problem. In fact it is very difficult to show particular solvable problems do not belong in $\mathbf{P}$.

There are other important differences between the classical and the polynomial bounded theories. Remember how in Remark 11.1.3 we described how to extract from the tree of computations, arising from a given input x accepted by a nondeterministic T, an equivalent deterministic computation? With polynomial time bounds that argument fails. ⬚

If we replace the deterministic machines in Definition 11.6.12 with nondeterministic ones we get:

DEFINITION 11.6.16   *Let L be a language on alphabet Σ.*

*(1) We say a nondeterministic Turing machine T **accepts** a word $x \in \Sigma^*$ if there is a terminating computation of T from input x.*

*As before, T **accepts** the language L consisting of all words T accepts — and T **accepts** L **in polynomial time** if there is a polynomial $p(n)$ such that whenever T accepts x, it has a terminating computation on input x which takes $\leq p(|x|)$ steps.*

*(2) The language L is **nondeterministically polynomial time decidable** if there is a nondeterministic Turing machine that accepts L in polynomial time.*

(3) We write $\mathbf{NP} =$ the set of all nondeterministically polynomial time decidable languages.

EXERCISE 11.6.17 *Show that (i)* **P** $\subseteq$ **NP***, and (ii) if* $L \in$ **NP***, then* L *is computable.*

If you now think of **NP** as the polynomial time bounded counterpart of "c.e.", then you will expect to be able to prove:

EXERCISE 11.6.18 *If* $L, M \in$ **NP***, then so are* $L \cup M$ *and* $L \cap M$.

This is the basic unanswered question we have been referring to:

OPEN QUESTION: Does **P** = **NP**?

The question is basic in the sense that **P** $\neq$ **NP** would mean we can expect to compute *more* in practice by using guessing strategies. While on the other hand, **P** = **NP** would point to a whole range of **NP** problems with as yet undiscovered polynomial time bounded solutions. So either outcome would dramatically re-focus the attention of computer scientists. For the moment no one is quite sure how to go about solving this problem.

We will briefly look at two related ways of getting a handle on the problem. One is to look at polynomial time bounded reducibilities and hope something comes out of the corresponding structures. The other is to build up a range of candidates whose exact status in relation to **P** would settle the whole problem.

The way the two approaches come together is very analogous to the role of $\leq_m$ and K in Chapter 7. We showed K to be m-complete. We found all sorts of naturally arising c.e. sets to be m-equivalent to K. We will aim to produce a polynomial time bounded version of this scenario, with **P** replacing the computable sets and **NP** the c.e. ones. There will, of course, be one vital ingredient missing from the picture — an analogue of Theorem 5.3.1 showing K c.e. but not computable!

I will first give a polynomial time bounded version of many–one reducibility, also known as *Karp reducibility*.

DEFINITION 11.6.19 *A language* M *is* **polynomial time reducible** *to* L — *written* $M \leq_p L$ *or* $M \leq_m^P L$ — *if there is a polynomial time computable function* f *such that*

$$x \in M \iff f(x) \in L.$$

It is easy to see that $\leq_p$ is transitive:

EXERCISE 11.6.20 *Show that if $N \leq_p M$ and $M \leq_p L$, then $N \leq_p L$.*

Also, you can get an analogue of Theorem 7.1.3:

EXERCISE 11.6.21 *Show (i) if $M \leq_p L \in$ **P**, then $M \in$ **P**, and (ii) if $M \leq_p L \in$ **NP**, then $M \in$ **NP**.*

It is now natural to define:

DEFINITION 11.6.22 *(1) We say L is **NP**-hard if $M \leq_p L$ for every $M \in$ **NP**.*

*(2) An **NP**-hard L is **NP**-complete if $L \in$ **NP**.*

EXERCISE 11.6.23 *Show that if there exists an **NP**-complete $L \in$ **P**, then we have* **P** $=$ **NP**.

And that is almost the end of the story (so far). Except for some remarkable results of Stephen Cook from 1971, giving us a whole range of **NP**-complete problems, many of them quite natural ones from combinatorics and graph theory. We may not have a polynomial time bounded analogue K yet, but we do have a good candidate. Remember — Exercise 9.3.7 — that SAT is the set of propositional formulas which are *satisfiable*. And that SAT is computable. It turns out you can prove a lot more:

THEOREM 11.6.24 (Cook's Theorem)
SAT is **NP**-complete.

You should go to a computer science text for a full proof. I would recommend *Computability, Complexity, and Languages* by Martin Davis (yes, the same one who helped solve Hilbert's Tenth Problem), Ron Sigal and Elaine Weyuker.

Anyway, armed with this theorem, you can now use $\leq_p$ to reduce SAT to all sorts of other problems, which then turn out to be **NP**-hard or **NP**-complete.

EXERCISE 11.6.25 *Show that if M is **NP**-complete and $M \leq_p L$, then*

*L is **NP**-hard. And hence, if in addition $L \in$ **NP**, then L is **NP**-complete.*

The most famous of these "other problems" which turn out to be **NP**-complete is, of course, the *Travelling Salesman Problem* (TSP) — which involves finding a Hamilton circuit of minimal weight in a given complete weighted graph. You can get this result from Richard Karp's 1972 proof that (HC) — the problem of finding a Hamilton circuit for any given graph G — is **NP**-complete.

EXERCISE 11.6.26   *Show that HC$\leq_p$ TSP.*

REMARK 11.6.27   For most theoretical computer scientists, **P** $=?$ **NP** becomes less and less a computability-theoretic problem, and more and more to do with the details of very specific combinatorial problems.

It is not that one cannot develop polynomial analogues of much of the infrastructure of the Turing universe — see below. There is a perfectly good polynomial time degree structure derived from $\leq_p$, and a polynomial time counterpart of the Turing universe. There is a good polynomial time version of the arithmetical hierarchy — which collapses if **P** $=$ **NP**.

What has really driven researchers in a very different direction is a little paper of Baker, Gill and Solovay from 1975. They note that elementary diagonalising techniques from classical computability theory — as in the proof of the Kleene–Post Theorem 10.6.3 — *relativise*. They then produce suitable oracles A and B, such that **P** $=?$ **NP** relativised to A in a natural way is *true* — but **P** $=?$ **NP** relativised to B is *false*! Well, there are lots of ways in which relativisation breaks down in computability theory. But the computability appearing in the proofs of Baker, Gill and Solovay is very simple. Their work certainly adds to the puzzle.   ⬚

Polynomial Bounded Computability Theory

Most people believe **P** $\neq$ **NP**. If they are right, and however it is proved, all sorts of problems regarding resource bounded computability will become both more approachable and more relevant. I will finish with a glimpse of some of the concepts and basic results of theory as it stands now. Firstly, let us get a polynomial versions of (deterministic) Turing reducibility, also known as *Cook reducibility*. I will revert to languages being sets of numbers using alphabet 1.

DEFINITION 11.6.28 *A set B is* **polynomial time computable** *in A — written $B \leq_T^P A$ — if it is accepted by a Turing machine with oracle A in polynomial time.*

If $B \leq_T^P A$ and $A \leq_T^P B$ we write $B \equiv_T^P A$

EXERCISE 11.6.29 *Show that $\leq_T^P$ is reflexive and transitive.*

EXERCISE 11.6.30 *Show that if $B \leq_T^P A \in \mathbf{P}$, then $B \in \mathbf{P}$.*

EXERCISE 11.6.31 *Show that $\equiv_T^P$ is an equivalence relation.*

Show that the resulting degree structure — called the **PTIME degrees** *— has a partial ordering $\leq$ induced by $\leq_T^P$ and a least element $\mathbf{0} = \mathbf{P}$.*

Notice that it really only makes sense to consider the PTIME degrees of computable sets.

Richard Ladner — one of a number of computability theorists turned successful computer scientist — shows that Kleene–Post type constructions can, with a little ingenuity, be polynomial bounded. His seminal result proves the existence of a *minimal pair* of PTIME degrees — that is, degrees $\mathbf{a}, \mathbf{b} > \mathbf{0}$ such that $\mathbf{0}$ is the only PTIME degree $\leq \mathbf{a}$ and $\mathbf{b}$:

THEOREM 11.6.32 (Ladner, 1975)
There exist computable sets $A, B \notin \mathbf{P}$, such that if $D \leq_T^P A$ and $D \leq_T^P B$, then $D \in \mathbf{P}$.

The proof is basically a Kleene–Post type argument, but with some subtle "spreading out" of the work to be done in computing a $D \leq_T^P A$ and $\leq_T^P B$, so as to get it computable in polynomial time. I would very much recommend Ladner's paper *On the Structure of Polynomial Time Reducibility* as a clear introduction to such methods.

You can define a polynomial time bounded version of nondeterministic Turing reducibility in the obvious way:

DEFINITION 11.6.33   *A set B is* **nondeterministically polynomial time computable** *in A — written* $B \leq^P_{NT} A$ *— if it is accepted by a nondeterministic Turing machine with oracle A in polynomial time.*

But there is no corresponding degree structure.

You can probably see why $\leq^P_{NT}$ is not transitive in general. The problem is that if $C \leq^P_{NT} B$ via oracle Turing machine T, and $B \leq^P_{NT} A$ via T', then to decide if $x \in C$ using oracle A, you need to run T with input x — and when you need an oracle answer to "$y \in B$?", say, you need to run T' with input y. But if $y \notin B$, you will need to examine *all* the computation branches for T' with input y to verify this. And even though the lengths of the individual branches are polynomial bounded, the total number of computational steps checked is not.

Of course $\leq^P_{NT}$ is still an important and interesting reducibility. We cannot reject reality because its mathematics is messier than we would like.

On the other hand, $\leq^P_{NT}$ is closely related to polynomial time bounded versions of e-reducibility, which do have nice properties such as giving degree structures with least degree = **NP**. For instance, Selman's *polynomial enumeration reducibility* $\leq_{pe}$ is a maximal transitive subreducibility of $\leq^P_{NT}$ — just like $\leq_e$ in relation to "c.e. in".

But I am already telling you more than you need to know! It is time to terminate this part of the book.

Part III

More Advanced Topics

Chapter 12

Post's Problem: Immunity and Priority

Many naturally occurring incomputable sets turn out to be computably enumerable. We will now look at how people have approached the question of:

> **Just how rich is the Turing structure of the computably enumerable sets?**

We will see that there are very different ways of looking at this question, each with its own strengths and technical beauties. An intuitively satisfying approach — first tried by Post — is to look for links between natural information content and relations on $\mathcal{D}$. Another is to delve into the intricacies of $\mathcal{D}$ by directly constructing interesting features of the Turing universe. It is the relationship between these approaches which seems to have a special potential for modelling aspects of the material Universe. I should mention that we are entering an area in which the techniques are quite hard to handle even at the classical level — and it is not surprising that their wider potential is largely unrealised.

12.1 Information Content and Structure

So the search is for richness of information corresponding to local degree theoretic structure. Ideally we would like something corresponding to the arithmetical hierarchy below $\emptyset'$. A forlorn hope? Well, not completely. We can use jump inversion to bring aspects of that beautifully natural hierarchy down to the local level. The resulting *high/low hierarchy* — simultaneously appearing in different papers of Bob Soare and myself around October of 1972 — provides an invaluable frame of reference at the local level. But it is hard to characterise in terms of *natural* information content, or to describe in the local structure of $\mathcal{D}$.

DEFINITION 12.1.1 (The High/Low Hierarchy)

(1) The **high/low hierarchy** *is defined by*

$$\mathbf{High}_n = \{\mathbf{a} \le \mathbf{0}' \mid \mathbf{a}^{(n)} = \mathbf{0}^{(n+1)}\}, \quad \mathbf{Low}_n = \{\mathbf{a} \le \mathbf{0}' \mid \mathbf{a}^{(n)} = \mathbf{0}^{(n)}\},$$

for each $n \ge 1$.

(2) If $\deg(A) \in \mathbf{High}_n$ we say A and $\deg(A)$ are **high$_n$**. *We similarly define the* **low$_n$** *sets and degrees.*

(3) For $n = 1$ we often drop the subscript — A and $\deg(A)$ are **low** *if $A' \in \mathbf{0}'$, and* **high** *if $A' \in \mathbf{0}''$.*

Intuitively — a is high$_n$ or low$_n$ according as $\mathbf{a}^{(n)}$ takes its greatest or least possible value.

I have widened the focus to $\mathcal{D}(\le \mathbf{0}')$, since we can learn a lot about $\mathcal{E}$ by putting it in its local context. If the high/low hierarchy is thought of as defining a *horizontal* stratification of $\mathcal{D}(\le \mathbf{0}')$, there is another very important hierarchy whose effect is *vertical*. The **n-c.e. hierarchy** — independently devised by Putnam and Gold around 1965 — inductively builds on the way we got the d.c.e. sets from the c.e. sets in Exercise 5.2.19.

DEFINITION 12.1.2 (The n-C.E. Hierarchy)

(1) Let $A \in \Delta_2^0$ have standard approximating sequence $\{A^s\}_{s \ge 0}$.

We say A is **n-c.e.** *if $A^s(x)$ changes value at most n times, each $x \in \mathbb{N}$ — that is, if for all $x \in \mathbb{N}$*

$$\mid \{s \mid A^s(x) \ne A^{s+1}(x)\} \mid \; \le n.$$

We say A is **ω-c.e.** *if there is a computable bound on the number of such changes — that is, if there is some computable f such that for all $x \in \mathbb{N}$*

$$\mid \{s \mid A^s(x) \ne A^{s+1}(x)\} \mid \; \le f(x).$$

(2) A degree $\mathbf{a}$ is **d.c.e.** *if it contains a d.c.e. set. And for each $n \ge 2$ we write $\mathbf{D}_n$ for the nth level of the corresponding* **n-c.e. hierarchy** *of n-c.e. degrees.*

It turns out that the n-c.e. sets are exactly those btt-reducible to K. Another way of looking at the n-c.e. hierarchy of sets is that it classifies the set of all finite Boolean combinations — that is, combinations formed using unions and intersections — of computably enumerable sets and their complements.

EXERCISE 12.1.3 *(i) Show that A is 1-c.e. if and only if A is c.e., and 2-c.e. if and only if A is d.c.e.*

(ii) Show that if A is n-c.e., then A can be expressed as a Boolean combination of n c.e. sets and their complements.

[**Hint:** Show $A = (\ldots((A_1 \cap \overline{A}_2) \cup A_3) \cap \ldots)$ where

$$x \in A_n \iff_{\text{defn}} |\,\{s \mid A^s(x) \neq A^{s+1}(x)\}\,| = n. \ ]$$

This is the local framework we get from these two basic hierarchies:

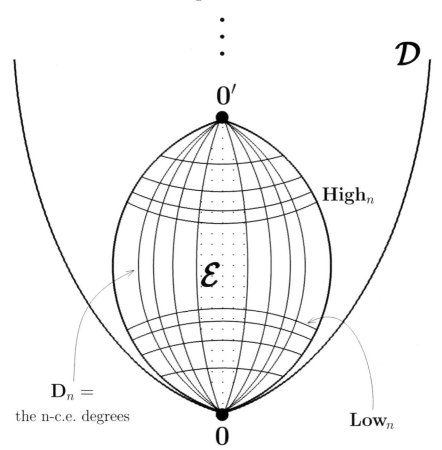

Please do not be misled by my diagram into thinking that either of the hierarchies eventually includes everything below $\mathbf{0}'$. Far from it. This can only be achieved by extending the levels of the hierarchies into the transfinite, with not very satisfactory results.

There are other hierarchies based on genericity, randomness, etc, but they soon run out of local relevance. I will mention one other one here due to Jockusch and Shore, which does extend well beyond the local context. It slips in for its close relationship to the n-c.e. hierarchy.

DEFINITION 12.1.4 (The n-CEA Hierarchy)

(1) We say A is **CEA** *B if A is c.e. in B and $A \geq_T B$.*

(2) Inductively define $\emptyset$ to be **0-CEA***, and then A to be $(n+1)$-**CEA** if A is CEA some n-CEA B.*

(3) The **n-CEA hierarchy** *of degrees is the one got from this n-CEA hierarchy of sets.*

EXERCISE 12.1.5 (Lachlan)   *Show that every d.c.e. set A is also 2-CEA.*
[**Hint:** Show that A is CEA the c.e. set $\{\langle x, s \rangle \mid x \in A^s - A\}$.]

EXERCISE 12.1.6   *Show that every n-c.e. set A is also n-CEA.*

You cannot get a converse to this last exercise, of course — I will leave you to think up some obvious counterexamples.

REMARK 12.1.7   We now have some very fine hierarchies, but as yet know nothing much beyond their definitions. The information content we have for the high/low hierarchy is horrible, buried in iterations of jumps and their inversions. As for Turing structure, all our hierarchies might as well be the dark side of the moon before the space age. It was Emil Post back in the 1940s — before anyone had discovered the local hierarchies — who pointed the way. Let me try and say something about why Post's work is still so important to us, and how his approach relates to basic science. And why certain difficult technical problems in computability theory promise to have far-reaching implications. If you are impatient, you can go straight to the next section.

The diagram below is an attempt to show how computability relates to the primary ingredients of science. These are firstly *observation* — that is, our experience of interacting with the Universe. And then mathematical *descriptions*, or information content, pinning down plausible relationships on the Universe in a widely communicable form. Computability is intrinsic to both, and at the same time stands outside, its theory a sort of meta-science.

Process, causality, algorithmic content — all basic aspects – perhaps *the most* basic aspect of the real world of observation. And it is computability theory — suitably fleshing out and qualifying the Church–Turing Thesis — which mathematically models this. But this is not the only such modelling process. Science routinely builds much more specific mathematical models of natural phenomena, codifying all sorts of observed data into general laws. What is different about computability is that it also has something to say about this extraction of information content as an aspect of the real world. It has the potential to explain how this *information* content — natural laws — relates to the basic *algorithmic* content of the Universe.

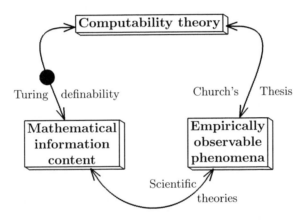

The technical expression of this relationship is the notion of *Turing definability*. It is basic to understanding how the beautiful descriptions science gives us of the real world actually derive their material substance.

Definability in a structure — remember Definition 11.5.9 — is a key mathematical concept, and not widely understood. Let us say a little more about it. It is easy to give an intuitive idea of what definability is and how it relates to another useful notion, that of *invariance*. This is not necessarily because the notions are very simple ones (they are not), but because they do correspond to phenomena in the real world which we already, at some level, are very familiar with.

As one observes a rushing stream, one is aware that the dynamics of the individual units of flow are well understood. But the relationship between this and the continually evolving forms manifest in the streams surface is not just too *complex* to analyse — it seems to depend on globally emerging relationships not derivable from the local analysis. The form of the changing surface of the stream appears to constrain the movements of the molecules of water, while at the same time being traceable back to those same movements. The mathematical counterpart is the relationship between familiar operations and relations on structures, and globally arising new properties based on those locally encountered ones. The emergence of form from chaos, of global relations within turbulent environments, is a particularly vivid metaphor for the assertion of definability, or invariance. Let us take a simple mathematical example from arithmetic.

Given the usual operation $+$ of addition on the set $\mathbb{Z}$ of integers, it is easy to see that the set Ev of even integers is describable from $+$ within $\mathbb{Z}$ via the formula

$$x \in Ev \iff (\exists y)(y + y = x).$$

So all we mean by a relation being *definable* from some other relations and/or functions on a given domain is that it can be *described* in terms of those relations and/or functions in some agreed standard language. Of course, there are languages of varying power we can decide on. In the above example, we have used very basic first order language, with finitary quantification over individual elements — we say that Ev is *first order definable* from $+$ over $\mathbb{Z}$. What has happened is that we started off with just an arithmetical operation on $\mathbb{Z}$ but have found it distinguishes certain subsets of $\mathbb{Z}$ from all its other subsets. Intuitively, we first focused on a dynamic flow within the structure given locally by applications of the form $n + m$ to arbitrary integers m, n. But then, standing back from the structure, we observed something global — $\mathbb{Z}$ seemed to fall into two distinct parts, with flow relative to even integers constrained entirely within Ev, and flow from outside Ev being directed into Ev — with Ev being a maximal such subset of $\mathbb{Z}$. From within the structure, $+$ is observable and can be algorithmically captured. Further than that, we are dealing with "laws" which cannot be related to the local without some higher analysis. This feature of the integers is not, of course, a deep one, but it does act as a basic metaphor for other ways in which more or less unexpected global characteristics of structures emerge quite deterministically from local infrastructure.   ⬜

For the following definition, the *first order language* for $\mathcal{D}$ is one with just the basic variables, brackets, quantifiers and logical connectives, and one 2-place symbol for the ordering $\leq$.

DEFINITION 12.1.8   *(1) Let $R(\mathbf{x}_1, \ldots, \mathbf{x}_k)$ be a relation on $\mathcal{D}$.*

*We say that R is **Turing definable** — or **definable in $\mathcal{D}$** — if there is some first order formula $\varphi(x_1, \ldots, x_k)$ in the language for $\mathcal{D}$ such that for all $\mathbf{a}_1, \ldots, \mathbf{a}_k \in \mathcal{D}$ $R(\mathbf{a}_1, \ldots, \mathbf{a}_k)$ holds if and only if $\varphi(\mathbf{a}_1, \ldots, \mathbf{a}_k)$ holds in $\mathcal{D}$.*

*(2) If $\mathcal{A}$ is some family of sets, we say $\mathcal{A}$ is **Turing definable** if the set of all Turing degrees of members of $\mathcal{A}$ is Turing definable.*

EXAMPLE 12.1.9 *Show that* **0** *is Turing definable.*

SOLUTION   **0** is definable in $\mathcal{D}$ via the formula

$$\varphi(\mathbf{x}) \iff {}_{\text{defn}} (\forall \mathbf{y})\, [\, \mathbf{x} \leq \mathbf{y}\,]. \qquad \qquad []$$

One can, of course, talk about definability in other structures. For instance, $\mathbf{0}_e$ is definable in $\mathcal{D}_e$, and $\mathbf{0}'$ is definable in $\mathcal{D}(\leq \mathbf{0}')$ and in $\mathcal{E}$. We have already looked at definability in the Medvedev lattice.

EXERCISE 12.1.10 *Show that the join is Turing definable — that is, show that the relation* $\mathbf{x} \cup \mathbf{y} = \mathbf{z}$ *is definable in* $\mathcal{D}$.

The notion of *invariance* gives a useful, if slightly more abstract, way of looking at definability. Being able to uniquely *describe* a feature of a structure is a measure of its uniqueness. But some feature of a structure may be quite unique, without one being able to describe that uniqueness in everyday language. Mathematically, we use the notion of *automorphism* to capture the idea of a *reorganisation* of a structure which does not change any of its properties. A feature of that structure is *invariant* if it is left fixed by any automorphism of the structure. Obviously if one can uniquely describe such a feature, it must be invariant, but not necessarily conversely.

DEFINITION 12.1.11   *(1) Let* $R(\mathbf{x}_1, \ldots, \mathbf{x}_k)$ *be a relation on* $\mathcal{D}$.

Then R is **Turing invariant** *if for every automorphism* $\psi : \mathcal{D} \to \mathcal{D}$ *and every* $\mathbf{a}_1, \ldots, \mathbf{a}_k \in \mathcal{D}$ *we have* $R(\mathbf{a}_1, \ldots, \mathbf{a}_k)$ *holds if and only if we have* $R(\psi(\mathbf{a}_1), \ldots, \psi(\mathbf{a}_k))$ *holds in* $\mathcal{D}$.

(2) If $\mathcal{A}$ *is some family of sets, we say* $\mathcal{A}$ *is* **Turing invariant** *if the set of all Turing degrees of members of* $\mathcal{A}$ *is Turing invariant.*

Only in recent years have we become aware that much of the past fifty years' research into computability has actually been about Turing definability and invariance. Current research focuses on getting optimal Turing definitions of the various levels of the high/low and n-c.e. hierarchies we have been looking at.

For more background information the joint article by myself and George Odifreddi is an approachable source. For more technical material try volume II of Odifreddi's mammoth *Classical Recursion Theory*.

The next section brings us back down to earth again.

12.2   Immunity Properties

You have already seen an example of how information content and the structure of the incomputable relate. Simplicity of a set guarantees it m-degree strictly between $\mathbf{0}_m$ and $\mathbf{0}'_m$ (Corollary 7.3.3).

Looking at this more closely, this was achieved by asking the complement of a c.e. set to be *immune*. Immunity is a sort of *sparseness* of a set. And not *sparseness* in the sense that there are "not many" prime numbers — here we mean *sparse* inasmuch as it is hard to computably locate infinitely many members of the set. The primes are positively omnipresent in this sense!

DEFINITION 12.2.1 $A \subseteq \mathbb{N}$ is **immune** *if and only if*
   (i)   *A is infinite, and*
   (ii) $(\forall e)[W_e$ *infinite* $\Rightarrow W_e \not\subseteq S]$.

EXERCISE 12.2.2   *Show that if A is introreducible and not computable, then A is immune.*

   Hence show that every Turing degree $> \mathbf{0}$ contains an immune set, and so the set of degrees of immune sets is Turing definable.

[**Hint:** Take A as in part (1) of the proof of Theorem 11.3.25.]

Emboldened by his discovery of the role of simplicity in the m-degrees, Post hoped he might find some stronger immunity property that played the same role for the Turing degrees. Although he failed to do this, the idea gave rise to what became known as:

POST'S PROBLEM

   Find some natural property $\mathcal{P}$ of computably enumerable sets such that if A satisfies $\mathcal{P}$ then $\emptyset <_T A <_T \emptyset'$.

I have deliberately stated this problem as vaguely as Post originally seems to have conceived it. We want the property to be "natural", since otherwise we could just take $\mathcal{P}$ to be "A is c.e. and $\emptyset <_T A <_T \emptyset'$ ", which would tell us nothing! The more literal minded have asked that $\mathcal{P}$ should be describable in the lattice $\mathcal{E}$, just as simplicity is, although it is not clear Post was so restrictive. He himself considered properties not obviously *lattice theoretic* — which has led to another whole industry, searching for lattice theoretic definitions in unlikely places.

DEFINITION 12.2.3 *A property* $\mathcal{P}$ *of sets (c.e. or otherwise) is* **lattice theoretic** *if it is definable in the first order language of the set* $\mathcal{E}$ *of c.e. sets with set inclusion* $\subseteq$.

The set $\boldsymbol{A}$ *of degrees of a lattice theoretic class of c.e. sets is also said to be* **lattice theoretic**.

EXAMPLE 12.2.4 *Let* $Comp(X)$ *hold if and only if* X *is computable. Show that Comp is lattice theoretic.*

SOLUTION The definability of Comp in $\mathcal{E}$ follows from the equivalence

$$\mathrm{Comp}(X) \iff (\exists Y)\,[\,X \cup Y = \mathbb{N} \ \& \ X \cap Y = \emptyset\,]$$

in $\mathcal{E}$. □

EXERCISE 12.2.5 *Let* $Fin(X)$ *hold if and only if* X *is finite. Show that Fin is lattice theoretic.*

It is easy to see why the lattice theoretic property of being simple does not ensure incompleteness for Turing computability.

EXAMPLE 12.2.6 *Show that every c.e. degree* $\mathbf{a} > \mathbf{0}$ *contains a simple set.*

SOLUTION Given $A \in \mathbf{a}$ c.e. but not computable, let $a_0, a_1, \ldots$ be a computable enumeration of its members without repetitions — remember Exercise 5.1.11. Define the *deficiency set* for this enumeration to be

$$D = \{s \mid (\exists t > s)\,[\,a_t < a_s\,]\,\}.$$

We show that the set $\overline{D}$ of *nondeficiency stages* is a simple set of degree $\mathbf{a}$.

D is obviously Σ_1 and so c.e. Since A is infinite, so is $\overline{D}$.

We also have $s \in \overline{D} \iff A \restriction a_s \subseteq \{a_0, \ldots, a_s\}$. It immediately follows that $D \leq_T A$.

But if we have $A \restriction a_s \subseteq \{a_0, \ldots, a_s\}$, we can compute $A(x)$ for every $x \leq a_s$. So given any infinite subset of $\overline{D}$, we can compute infinitely many beginnings of A.

Since $\overline{D}$ itself is infinite, this immediately gives $A \leq_T D$. So $A \leq_T D$.

But on the other hand, if W_i is an infinite subset of $\overline{D}$, we can *computably* enumerate infinitely many beginnings of A — contradicting A not being computable. So D is simple.   □

EXERCISE 12.2.7 *Show that the nondeficiency set $\overline{D}$ in the above proof is introreducible.*

Immunity properties such as those sought by Post are roughly speaking those *hereditary under inclusion*.

DEFINITION 12.2.8 *A property $\mathcal{P}$ of infinite sets $\subseteq \mathbb{N}$ is said to be* **hereditary under inclusion** *if every infinite subset of a set satisfying $\mathcal{P}$ also satisfies $\mathcal{P}$.*

It is easy to see that immunity is hereditary under inclusion.

How can we make it even harder for the members of a set to avoid our computable net, and so get stronger immunity properties? Well, the set A might be able to avoid every element of an infinite W_i being in A — but what if we replace each element of W_i by a finite set and challenge A to avoid at least one of those finite sets capturing an $x \in A$?

DEFINITION 12.2.9 *Let $\{D_i\}_{i \geq 0}$ be a computable list of all finite sets, and let f be computable.*

(1) We call a c.e. set $\{D_{f(i)}\}_{i \geq 0}$ of mutually disjoint finite sets $D_{f(i)}$ a **c.e. array**.

A c.e. set $\{W_{f(i)}\}_{i \geq 0}$ of mutually disjoint finite c.e. sets $W_{f(i)}$ is called a **weak c.e. array**.

(2) We say $A \subseteq \mathbb{N}$ **avoids** *$\{X_i\}_{i \geq 0}$ if for some i we have $X_i \cap A = \emptyset$.*

(3) If A avoids every c.e. array, A is **hyperimmune**.
And if A avoids every weak c.e. array, A is **hyperhyperimmune**.
If we remove the finiteness requirement on the sets $W_{f(i)}$ in the definition of hyperhyperimmune, we get A is **strongly hyperhyperimmune**.

EXERCISE 12.2.10 *Show that A hyperimmune implies A immune, A hyperhyperimmune implies A hyperimmune, and A strongly hyperhyperimmune implies A hyperhyperimmune.*

Another route to stronger immunity properties is to make A sparse enough for it to be impossible for us to computably cut it up into a finite number of infinite pieces:

DEFINITION 12.2.11   *(1) We say X **splits** A if both $A \cap X$ and $\overline{A} \cap X$ are infinite.*

*(2) If A is infinite and not split by any c.e. set, A is **cohesive**.*

*And if A is infinite and not split by any computable set, A is **r-cohesive** — where "r" is for "recursive" in the old Kleene terminology.*

EXERCISE 12.2.12   *Show that A cohesive implies A r-cohesive, and that A r-cohesive implies A strongly hyperhyperimmune.*

It turns out that some of these new definitions connect up with a familiar way of *pushing up* computability. Remember how the Ackermann function *dominated* all primitive recursive functions, and the Busy Beaver Function dominated all computable functions. We look at domination properties of functions Turing equivalent to A.

DEFINITION 12.2.13   *Let $a_0 < a_1 < a_2 < \ldots$ be a list of $A \subseteq \mathbb{N}$ in ascending order. Then the function p_A given by $p_A(x) = a_x$ is called the **principal function** for A.*

THEOREM 12.2.14
Let $A \subseteq \mathbb{N}$ be infinite. Then A is hyperimmune if and only if no computable function dominates p_A.

PROOF $(\Rightarrow)$ Assume some computable f dominates p_A. Say $f(n) \geq_A (n)$ for all $n \geq N$.

Then for each $n \geq N$ the finite interval $[n, f(n)] \subset \mathbb{N}$ contains some $x \in A$. It is easy to choose a c.e. array from amongst these intervals — say $[N, f(N)]$, $[f(N) + 1, f(f(N) + 1)], \ldots$, etc. Since A does not avoid such an array, we get A not hyperimmune.

$(\Leftarrow)$ On the other hand, say A fails to avoid the c.e. array $\{D_{g(i)}\}_{i \geq 0}$. Defining

$$f(x) = \max\{y \mid y \in D_{g(z)}, \text{ some } z \leq x\},$$

we get a computable f which dominates p_A.   ⬜

You can now use this to show that the nondeficiency set $\overline{D}$ in Example 12.2.6 is actually hyperimmune. But the domination properties of another function closely related to $A \in \Delta_2$ gives an even better result.

DEFINITION 12.2.15   *Let* $\{A^s\}_{s \geq 0}$ *be a* Δ_2-*approximating sequence for* $A \in \Delta_2$. *Then the* **computation function** C_A *for* $\{A^s\}_{s \geq 0}$ *is defined by*

$$C_A(x) = \mu s \geq x \, [\, A^s \! \restriction \! x = A \! \restriction \! x \,].$$

EXERCISE 12.2.16   *Given* Δ_2-*approximating sequence* $\{A^s\}_{s \geq 0}$ *for* A, *show that* $C_A \equiv_T A$.

The next lemma turns up everywhere:

THE DOMINATION LEMMA

   Let $A \in \Delta_2$ *have computation function* C_A. *Then* A *is computable if and only if there is some computable* f *dominates* C_A.

PROOF   ($\Rightarrow$) If A is computable, we can just take $A^s = A \! \restriction \! s$ for each s. Then C_A is just the identity function.

($\Leftarrow$) Say C_A is dominated by f computable, say $f(x) > C_A(x)$ for $x \geq N$. Define:

   $M(x) = $ the greatest $y \leq x$ such that $(\forall t \in [x, f(x)]) \, [\, A^t \! \restriction \! y = A^{f(x)} \! \restriction \! y \,]$.

Then notice:

   (1) Since $A \in \Delta_2$, we have M total and so computable,

   (2) If $x \geq N$ and $y < M(x)$, we have $A(y) = A^{f(x)}(y)$, and

   (3) For every y there is an $x \geq N$ such that $y < M(x)$, since for all x we have $C_A(x) \geq x$.

   So to compute $A(y)$ we computably find an $x \geq N$ such that $y < M(x)$, and compute $A(y) = A^{f(x)}(y)$.   ⬜

Notice, of course, that the Domination Lemma works perfectly well with a c.e. approximating sequence $\{A^s\}_{s \geq 0}$ for $A \in \Sigma_1$ — and rather more simply, with just "μs" in place of "$\mu s \geq x$" in the definition of C_A. As does the next little result, whose full significance will only be seen in the next section.

EXERCISE 12.2.17 (The Permitting Lemma) *Let $A, B \in \Delta_2$ have computation functions C_A, C_B. Show that if C_A dominates C_B, then $B \leq_T A$.*

If we define ψ *dominates* θ for partial functions ψ, θ to mean

$$(\exists n)(\forall x \geq n)\, [\, \theta(x) \downarrow \Rightarrow \psi(x) \downarrow \geq \theta(x)\,],$$

then we get a useful practical version of the Domination Lemma:

EXERCISE 12.2.18 (The Strong Domination Lemma) *Let $A \in \Delta_2$, and let W be c.e. Show that A is computable if and only if there is a p.c. f which dominates $C_A \restriction W$.*

We can use the Domination Lemma to prove:

THEOREM 12.2.19
Every nonzero $\mathbf{a} \leq \mathbf{0}'$ is hyperimmune.

PROOF   We build on the solution to Example 12.2.6.

Assume we have a Δ_2-approximating sequence $\{A^s\}_{s \geq 0}$ for the incomputable $A \in \mathbf{a} \leq \mathbf{0}'$. We show that $\mathrm{Graph}(C_A)$ is a hyperimmune set of degree $\mathbf{a}$.

Since C_A is total, $\mathrm{Graph}(C_A)$ is infinite.

By Exercise 12.2.16 we also have $C_A \equiv_T A$, giving $\mathrm{Graph}(C_A) \equiv_T A$.

Now to get a contradiction to $\mathrm{Graph}(C_A)$ being hyperimmune, let f computable dominate the principal function of $\mathrm{Graph}(C_A)$, for arguments $> N$, say.

But if $\langle x, s \rangle, \langle y, t \rangle \in \mathrm{Graph}(C_A)$, and $x < y$, then we must have $s \leq t$. So given $\langle x, s \rangle \in \mathrm{Graph}(C_A)$, we have $s \leq \langle x, s \rangle < f(x)$ for all $x > N$. That is, $C_A(x) \leq f(x)$ for all $x > N$.

But this means f dominates C_A, which, by the Domination Lemma, contradicts the incomputability of A. ⬚

EXERCISE 12.2.20 *Show that if $\mathbf{b} \geq \mathbf{a}$ hyperimmune $\mathbf{a}$ then $\mathbf{b}$ is also hyperimmune.*

[**Hint:** Code a hyperimmune $A \in \mathbf{a}$ into a suitably sparse set $\in \mathbf{b}$.]

It was Stanley Tennenbaum who suggested in 1961 that stronger and stronger immunity properties were actually taking everyone in the opposite direction to Post's problem. Here is the confirmation that strong domination properties certainly do that.

EXERCISE 12.2.21   *Show that if f dominates every p.c. function, then $\emptyset' \leq_T f$.*

[**Hint:** Consider the p.c. function $\psi(x) = \mu s \, (x \in K^s)$.]

Such functions do exist in $\emptyset'$:

EXERCISE 12.2.22   *Show that there does exist a function $f \in \emptyset'$ which dominates every p.c. function.*

[**Hint:** Consider $f(x) = \max \{ \varphi_i(x) \mid i \leq x \; \& \; \varphi_i(x) \downarrow \}$.]

Does it change the situation if you give sets satisfying immunity properties c.e. complements? Not much, we will discover.

DEFINITION 12.2.23   *A c.e. set A is **hypersimple, hyperhyper-simple, strongly hyperhypersimple** or **maximal** if its complement is hyperimmune, hyperhyperimmune, strongly hyperhyperimmune or cohesive, respectively.*

I should mention that you often see "hyper" or "hyperhyper" abbreviated to "h-" or "hh-" — for example hh-simple just means hyperhypersimple.

EXERCISE 12.2.24   *Show that a coinfinite c.e. set is hypersimple if and only if no computable f dominates $p_{\bar{A}}$.*

EXERCISE 12.2.25   *Show that every nonzero c.e. degree contains a hyper-simple set.*

Let us look at maximality, involving our strongest immunity property. For the m-degrees everything works nicely as usual.

EXAMPLE 12.2.26   *Show that if A is maximal, then the m-degree $\mathbf{a}_m$ of A is minimal in the structure $\mathcal{E}_m$ of the c.e. m-degrees, in that there is no m-degree $< \mathbf{a}_m$ other than $\mathbf{0}_m$.*

SOLUTION   Assume A is maximal, and $B \leq_m A$ via f.

Since $f(\mathbb{N})$ is c.e., either (a) $\bar{A} - f(\mathbb{N}) = \bar{A} - f(\bar{B})$ is finite, or (b) $\bar{A} \cap f(\mathbb{N}) = f(\bar{B})$ is finite.

If case (a) holds, $A \leq_m B$ via g, where g is computed as follows:

(1) Compute $g(x) = p_{\bar{B}}(0)$ for $x \in \bar{A} - f(\bar{B})$.

(2) Otherwise, start alternately enumerating members of A and $f(\mathbb{N})$.

(3) If x is enumerated into A first, compute $g(x) = p_B(0)$.

(4) But if we get $f(y) = x$ first, compute $g(x) = y$.

If case (b) holds, we get B computable — since if $f(\overline{B})$ is finite, $\overline{B} = f^{-1}(f(\overline{B}))$ is c.e. The result follows by the Complementation Theorem. $\Box$

But for the Turing case, there is again an unpromising link with domination properties:

THEOREM 12.2.27
If M is maximal, then $p_{\overline{M}}$ dominates every computable function.

PROOF Say $p_{\overline{M}}$ fails to dominate f computable.

Since $\overline{M}$ is hyperimmune, Theorem 12.2.14 tells us that f fails to dominate $p_{\overline{M}}$.

So we can assume there are infinitely many numbers x at which we get $f(x) < p_{\overline{M}}(x)$ but $p_{\overline{M}}(x+1) \leq f(x+1)$, in which case we get the interval $(f(x), f(x+1)]$ containing ≥ 2 members of $\overline{M}$.

Here is a crafty way of enumerating a set A which peels off just one of these members from each such interval, and so ends up splitting $\overline{M}$:

(1) Start enumerating M.

(2) Enumerate x into A if x appears in M.

(3) For each interval $(f(x), f(x+1)]$, at each stage $s > x$, enumerate into A the least member of $(f(x), f(x+1)] - M^s$, if it exists. $\Box$

It was Tony Martin — whom we will mention again in relation to Borel determinacy — who used domination to connect up lattice theoretic relations with another important strand of information content — the high/low hierarchy. The result? Some really bad news about immunity properties and Post's problem. But a massive step towards showing the high c.e. degrees are lattice theoretic.

THEOREM 12.2.28 (Martin, 1966)
There is an A-computable function that dominates every computable function if and only if $A' \geq_T \emptyset''$.

Hence, every maximal set is high.

PROOF I will use the fact — Exercise 10.5.18 — that Tot is Π_2-complete, and so $\equiv_T \emptyset''$.

($\Rightarrow$)   We assume Φ^A dominates every computable f and build a Δ_2^A-approximating sequence $\{\mathrm{Tot}^s\}_{s \geq 0}$ for Tot.

Define computably in A:

$$i \in \mathrm{Tot}^s \iff (\forall x \leq s)\,(x \in W_i\,[\Phi^A(s)]\,).$$

Then notice that if $i \in \mathrm{Tot}$, then $W_i = \mathbb{N}$, so that the computation function of W_i is computable, and so dominated by Φ^A. Hence $\mathrm{Lim}_s\,\mathrm{Tot}^s(i)$ exists — and is equal to 1 if and only if $i \in \mathrm{Tot}$.

($\Leftarrow$)   Conversely, assume we are given Δ_2^A-approximating sequence $\{\mathrm{Tot}^s\}_{s \geq 0}$ for Tot. We *want* to define

$$\Phi^A(x) = \max\,\{\varphi_i(x) \mid i \leq x \;\&\; i \in \mathrm{Tot}\,\},$$

but we need to do this computably in A. What we do is notice that for each i and x we can find computably in A a stage $s > x$ such that $\varphi_{i,s}(x) \downarrow \vee\, i \notin \mathrm{Tot}^s$. Call this stage $s(i,x)$. Now define computably in A:

$$\Phi^A(x) = \max\,\{\,\varphi_{i,s(i,x)}(x) \mid i \leq x \;\&\; \varphi_{i,s(i,x)}(x) \downarrow\,\}.$$

Then if φ_i is computable, we have for all large enough x that $i \leq x$ and $(\forall s \geq x)\,(i \in \mathrm{Tot}^s)$ — and so $\varphi_{i,s(i,x)}(x) \downarrow\, < \Phi^A(x)$.

So Φ^A dominates all computable functions.   ☐

There do turn out to be high degrees other than $\mathbf{0}'$, both c.e. and not c.e. However, if one wants to say more about *existence* of sets with these stronger immunity properties, it makes it a lot harder to do it with c.e. complements. For the moment we will be content with the following:

THEOREM 12.2.29
There exist $2^{\aleph_0}$ cohesive sets.

PROOF   Since cohesiveness is hereditary under inclusion, we only need find *one* cohesive set A — then there will be continuously many infinite subsets of A which are also cohesive.

We define a nest $A_0 \supseteq A_1 \supseteq \ldots$ of sets A_i, each A_{i+1} not split by W_i. We choose a distinct a_i from each A_i, and define $A = \{\,a_0, a_1, a_2, \ldots\,\}$.

The construction of A_i and a_i

Stage 0

Define $A_0 = \mathbb{N}$ and $a_0 = 0$.

Assume an infinite A_i already defined, not split by any W_j, $j < i$. And assume we have chosen an $a_j \in$ each A_j, $j \leq i$, all distinct.

Stage $i + 1$

Define

$$A_{i+1} = \begin{cases} A_i \cap W_i & \text{if } A_i \cap W_i \text{ infinite,} \\ A_i \cap \overline{W}_i & \text{otherwise,} \end{cases}$$

$$a_{i+1} = \text{ some } a \in A_{i+1} \text{ with } a \neq \text{ any } a_j, \, j \leq i.$$

Clearly this definition means the inductive assumptions are carried forward to the next stage.

Take $A = \{\, a_0, a_1, a_2, \ldots \,\}$. Then given W_i, we have A_{i+1} not split by W_i. And since $A - \{\, a_0, \ldots, a_i \,\} \subseteq A_{i+1}$, nor is A split by W_i. $\quad\square$

This proof is not very effective! We will do a lot better when we come to construct a maximal set.

EXERCISE 12.2.30 *Show that there is a cohesive set $\leq_T \emptyset'''$.*

EXERCISE 12.2.31 *Show that every infinite set has a cohesive subset. Deduce that there is a degree of a cohesive set above any given degree.*
[**Hint:** Use Exercise 11.3.26.]

EXERCISE 12.2.32 *Show that there is no cohesive set with a cohesive complement.*

There is a very interesting consequence of Theorem 12.2.29 for *automorphisms* of $\mathcal{E}$. We have a natural definition:

DEFINITION 12.2.33 *(1) An **automorphism** of $\mathcal{E}$ is a bijective function $\psi : \mathcal{E} \to \mathcal{E}$ such that for all c.e. sets A, B we have*

$$A \subseteq B \iff \psi(A) \subseteq \psi(B).$$

*(2) A set $\mathcal{A}$ of c.e. sets is **lattice invariant** if $\mathcal{A} = \psi(\mathcal{A})$ for every automorphism of $\mathcal{E}$.*

*(3) A set $\mathbf{A}$ of degrees is **lattice invariant** if $\mathbf{A}$ is the set of degrees of some lattice invariant set of c.e. sets.*

EXERCISE 12.2.34 *Show that if $\mathcal{A}$ is lattice theoretic, then $\mathcal{A}$ is lattice invariant.*

COROLLARY 12.2.35
There exist $2^{\aleph_0}$ automorphisms of $\mathcal{E}$.

PROOF The point is you can bijectively map a cohesive set A onto itself without having much effect on $\mathcal{E}$ — essentially, $\mathcal{E}$ lives inside $\overline{A}$.

If ψ permutes A but is the identity on $\overline{A}$, then $\psi(W_i)$, $\psi^{-1}(W_i)$ are each $= W_i$, modulo a finite set of elements, so are c.e. And

$$W_i \subseteq W_j \iff \psi(W_i) \subseteq \psi(W_j),$$

since ψ is a permutation of $\mathbb{N}$.

The result follows from the fact that there are continuously many permutations of A, which I will leave you to check.   ⏹

What we have started in this section is a very big topic. There is Post's problem for the various strong reducibilities, and the role of immunity properties there. And there is the relationship between Turing structure and lattice invariance. There is the whole question of exactly which immunity properties are lattice theoretic. Soare's book is the best source for further reading.

I will end the section by looking at the question of whether hyperhypersimplicity is lattice theoretic — it certainly does not appear to be.

DEFINITION 12.2.36 *If $S \subseteq \mathbb{N}$, write $W \restriction S = W \cap S$, and $\mathcal{E} \restriction S$ for $\mathcal{E}$ restricted to S, the set of all $W \restriction S$ with W c.e.*

EXERCISE 12.2.37 *Show that $\mathcal{E} \restriction S$ is a distributive lattice for all S.*

Hopefully you will remember the Definition 7.2.12 of a Boolean algebra, and the fact (Exercise 7.2.13) that $\mathcal{E}$ is not complemented, and so not a Boolean algebra.

PROPOSITION 12.2.38 (Lachlan, 1968)
Let $S \subseteq \mathbb{N}$ be infinite. If $\mathcal{E} \restriction S$ is a Boolean algebra then S is hyperhyperimmune.

PROOF Assume S is not hyperhyperimmune, failing to avoid the array $\{W_{f(i)}\}_{i \geq 0}$. We want to enumerate A so that $\overline{A} \restriction S \neq W_i$, every i, so $\mathcal{E} \restriction S$ is not complemented.

Idea of the proof: We want to choose some x for W_i, with the aim of enumerating x into A if x appears in $W_i \upharpoonright S$ — otherwise keeping x in $\overline{A} \upharpoonright S$ with hopefully $x \in \overline{W}_i$. To make sure the numbers x for different numbers i do not clash, we only enumerate $x \in W_i \cap W_{f(i)}$ into A for W_i. And to be sure such an x is $\in S$, we can make $x \in W_{f(i)} \cap S$!

So, we define $A = \cup\{W_i \cap W_{f(i)} \mid i \geq 0\}$. Then to see $W_i \upharpoonright S \neq \overline{A}$, just choose $x \in W_{f(i)} \cap S$.

So if $x \in \overline{A} \upharpoonright S$, we have $x \in S - W_i$ — and if $x \in S - \overline{A}$, so $x \in A \upharpoonright S$, we have $x \in W_i \upharpoonright S$. Hence, $W_i \upharpoonright S \neq \overline{A} \upharpoonright S$. $\qquad\Box$

The remarkable thing is that the converse holds. So for c.e. sets we get Lachlan's surprising result:

THEOREM 12.2.39

If A is a coinfinite c.e. set, A is hyperhypersimple if and only if $\mathcal{E} \upharpoonright \overline{A}$ is a Boolean algebra.

Hence, the property of being hyperhyperimmune is lattice theoretic.

You will need to know something about priority arguments and splitting to complete the proof. This will be the business of the rest of this chapter. In the next section I will tell you a little about how Post's problem was eventually solved, and the new techniques that these solutions depended on.

12.3 Approximation and Priority

Post's problem was eventually solved using a completely different approach. Computable approximations are nothing new to us — it is some time since we constructed a simple set in proving Theorem 6.2.3 — but we will now add the radical new ingredient of *injury*, with a correspondingly new and important role for *priority*. What we call *priority arguments* have become the cornerstone of much subsequent research, including the eventual discovery of a solution to Post's problem in the spirit in which Post asked it.

Probably the best way to learn about priority arguments is to just do one. Once again we will return to Theorem 10.6.3 of Kleene and Post, and see what we need to do to make the incomparable Turing degrees not just $\leq_T \mathbf{0}'$ — but c.e. This is going to be difficult because we will not be able to use any oracles.

> **THEOREM 12.3.1 (The Friedberg–Muchnik Theorem)**
> $\mathcal{E}$ is not linearly ordered — that is, there exist c.e. Turing degrees $\mathbf{a}$ and $\mathbf{b}$ such that $\mathbf{a} \not\leq \mathbf{b}$ and $\mathbf{b} \not\leq \mathbf{a}$.

PROOF   We start off as before, needing to construct sets A and B such that $A \not\leq_T B$ and $B \not\leq_T A$.

Again we break these conditions up into an infinite list of requirements:

$$\mathcal{R}_{2i} : \quad A \neq \Phi_i^B$$
$$\mathcal{R}_{2i+1} : \quad B \neq \Phi_i^A.$$

The way to get A and B c.e. though is to actually make the construction produce c.e. approximating sequences $\{A^s\}_{s \geq 0}$, $\{B^s\}_{s \geq 0}$ to A and B — and this is where we have to discard any oracles that might spoil the computability of these enumerations.

The construction will take place at stages $0, 1, \ldots, s+1, \ldots$ as before. At stage $s+1$ we will computably construct $A^{s+1}, B^{s+1} \supseteq A^s, B^s$, so as to help satisfy just one requirement, although this time we will not be able to decide ahead of time exactly *which* requirement. Without oracles we will blunder around making mistakes which later have to be put right.

The strategy for satisfying $\mathcal{R}_{2i}$

In isolation, $\mathcal{R}_{2i}$ is easy to satisfy, even without oracles. At a general stage $s+1$ we focus, in turn, on just one phase of the following:

(1) Choose a potential *witness* x to $A \neq \Phi_i^B$, where x is not yet in A — we aim to make $A(x) \neq \Phi_i^B(x)$.

(2) Do nothing more unless we get a stage $s+1$ at which $A^s(x) = 0 = \Phi_i^B(x)[s]$.

(3) In which case enumerate x into A^{s+1}.   And if $\Phi_i^B(x)[s]$ has use $z = \varphi_i^B(x)[s]$, preserve $B^s \restriction z = B \restriction z$ for evermore. We call this z a *B-restraint*.

The analysis of outcomes for the strategy

The only outcomes are:

$\boxed{w}$ The strategy waits forever at (2) for $A^s(x) = 0 = \Phi_i^B(x)[s]$ — in which case either $\Phi_i^B(x) \uparrow$ or $\Phi_i^B(x) = 1 \neq A(x)$. So $\mathcal{R}_{2i}$ is satisfied.

$\boxed{s}$ The strategy halts at (3) with $A(x) = A^{s+1}(x) = 1 \neq 0 = \Phi_i^B(x)[s]$. Since we preserve $B^s \restriction z = B \restriction z$, we have $\Phi_i^B(x) \neq A(x)$, so again $\mathcal{R}_{2i}$ is satisfied.

We can set this strategy out on a flow diagram:

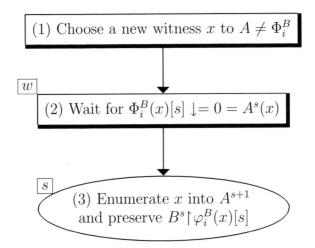

(1) Choose a new witness x to $A \neq \Phi_i^B$

w

(2) Wait for $\Phi_i^B(x)[s] \downarrow = 0 = A^s(x)$

s

(3) Enumerate x into A^{s+1}
and preserve $B^s \upharpoonright \varphi_i^B(x)[s]$

The strategy for $\mathcal{R}_{2i+1}$ below that for $\mathcal{R}_{2i}$

Discussion: How will $\mathcal{R}_{2i+1}$ spoil the strategy $\mathcal{R}_{2i}$?

Obviously, $\mathcal{R}_{2i+1}$ will want to pursue a similar strategy to that for $\mathcal{R}_{2i}$ — with A and B interchanged. $\mathcal{R}_{2i+1}$ may want to enumerate a number y into B and preserve a beginning of some A^t, while $\mathcal{R}_{2i}$ may want to enumerate x into A and preserve B^s. Conflicting aims!

We resolve these conflicts by giving $\mathcal{R}_{2i}$ and $\mathcal{R}_{2i+1}$ *priority* $2i$ and $2i+1$ — we say $\mathcal{R}_{2i}$ has *higher priority* than $\mathcal{R}_{2i+1}$. That is, we make $\mathcal{R}_{2i}$ and $\mathcal{R}_{2i+1}$ line up, like customers at a cafeteria counter.

Imagine you are customer $\mathcal{R}_{2i+1}$. $\mathcal{R}_{2i}$ always gets to choose first. You can choose too, so long as you do not obstruct $\mathcal{R}_{2i}$. You can choose whatever $\mathcal{R}_{2i}$ has previously passed over — can choose a witness $y \neq x$, or choose $y \in B$, even if $B^s \upharpoonright z = B \upharpoonright z$ has been preserved already by $\mathcal{R}_{2i}$, so long as $y > z$. And you can overtake when $\mathcal{R}_{2i}$ is waiting for some special item to be prepared — if $\mathcal{R}_{2i}$ is stuck at step (2) of her strategy, you can carry out (1) or (3) of your own version of $\mathcal{R}_{2i}$'s strategy.

Actually it's a bit more complicated than that. Choosing the witness $y \neq x$ is really just pre-selection — you can hungrily look further along the counter and decide on y — but if $\mathcal{R}_{2i}$ suddenly decides to pay, snatching a $z > y$ as she goes, you have to change your choice to a witness $y' > z$.

But things can get worse. You have overtaken, paid and left the counter — your version of outcome s — when up comes customer $\mathcal{R}_{2i}$ blazing mad. She accuses you of taking an item off the counter she had already chosen for herself. Your meal — your outcome s complete with preserved beginning of A^s — involves her witness x which she has decided is part of *her* outcome

s , and must be enumerated into A. It is not just your pride that is injured. Your meal is now ruined, and you return to the counter to line up again.

You are relieved to see that customer $\mathcal{R}_{2i}$ is no longer there — for the moment she appears to be sitting at her table looking very satisfied. But if she ever returns, she will be just ahead of you again!

The strategy:

(1) Choose a witness y to $B \neq \Phi_i^A$, where y has not yet appeared in the strategy — either as a witness or below any restraint previously set up. We say y is *fresh*.

(2) At each later stage $s+1$ at which $\mathcal{R}_{2i}$ is halted at (1) or (2) of its strategy, first check if (a) $y <$ a B restraint set up by $\mathcal{R}_{2i}$ — in which case we throw y away and go back to (1) to get a fresh witness. We say $\mathcal{R}_{2i+1}$ has been *injured*.

Otherwise, ask: (b) Is $B^s(y) = 0 = \Phi_i^A(y)[s]$?

If "no" go back to 2(a) at the next stage, if "yes" go straight to (3).

(3) Enumerate y into B^{s+1}. And set up an A restraint $w = \varphi_i^A(y)[s]$.

If $\mathcal{R}_{2i}$ *injures* $\mathcal{R}_{2i+1}$ at a later stage — that is, enumerates x into A with $x < w$ — throw y and w away, and return to (1) to start all over again.

The analysis of outcomes for the injurable strategy

For $\mathcal{R}_{2i}$ the outcomes are exactly as before. For $\mathcal{R}_{2i+1}$ they are very similar:

w The strategy waits forever at (2) for $B^s(y) = 0 = \Phi_i^A(y)[s]$, and $\mathcal{R}_{2i+1}$ is satisfied.

s The strategy halts at (3) with $B(x) = B^{s+1}(y) = 1 \neq 0 = \Phi_i^A(y)[s]$. We set up the A restraint w, and $\mathcal{R}_{2i+1}$ is satisfied.

But what about the injuries — surely they introduce another outcome in which $\mathcal{R}_{2i}$ keeps on injuring $\mathcal{R}_{2i+1}$?

Not at all!

$\mathcal{R}_{2i}$, of course, is *never* injured. It only has outcomes w and s, each of which mean $\mathcal{R}_{2i}$ *never again* chooses a fresh witness or sets up a new restraint.

So there is a stage after which $\mathcal{R}_{2i+1}$ too is *never* injured. This means that for $\mathcal{R}_{2i+1}$ also there are only two outcomes w and s.

The strategy is *finite injury*, and the use of priority has enabled us to satisfy both requirements.

The strategy for $\mathcal{R}_{2i+1}$ with all the other requirements

It is easy to see now that all the requirements can successfully pursue their own copies of the strategy.

$\mathcal{R}_0$ is never injured.

So after some stage s_0 $\mathcal{R}_1$ is never injured, and so never injures after some stage $s_1 \geq s_0$.

Inductively we get $\mathcal{R}_{2i+1}$ is not injured after some stage, and so gets satisfied via $\boxed{w}$ or $\boxed{s}$.

We have proved the theorem. □

REMARK 12.3.2 This proof is a prototype. It is full of intuitive explanation we can do without next time. There are presentational conventions we can use to cut down the waffle. I will introduce the modern method of laying out the outcomes to the strategies on a *tree*, which will clarify how strategies combine. If I were proving a new theorem in a research paper, I might want to include a more systematic verification that the strategies work.

But I will want to keep to the approach via analyses of strategies — or *basic modules* — for groups of requirements. There are more concise, more formal ways of doing the easier priority arguments: The Friedberg–Muchnik Theorem in a few lines, maybe. But this is not calculus. If you get used to that way of doing it, you will find it hard to come to grips with the harder proofs later on. □

EXERCISE 12.3.3 *Show that there exists an infinite independent set of c.e. degrees.*

[**Hint:** Make the sequence in Exercise 10.6.7 c.e.]

If you remember Remark 10.6.8, you will see how to use this last result to show the existential theory of $\mathcal{E}$ is decidable.

Permitting Techniques

The Permitting Lemma 12.2.17 gives us a way of constructing a Δ_2 degree **b** below another $\mathbf{a} \leq \mathbf{0}'$. If you want $B \in \mathbf{b}$ to satisfy other requirements, you usually have to use the Domination Lemma on the $A \in \mathbf{a}$ to show it does *permit* the strategies for these requirements. Permitting takes a specially important and simple form relative to c.e. sets. Let us show the Friedberg–Muchnik strategy can be permitted below a nonzero c.e. **c** in place of $\mathbf{0}'$.

COROLLARY 12.3.4 (of the proof of Theorem 12.3.1)
If $\mathbf{c} > \mathbf{0}$ *is c.e. then* $\mathcal{E}(\leq \mathbf{c})$ *is not linearly ordered — that is, there exist c.e. Turing degrees* $\mathbf{a}, \mathbf{b} < \mathbf{c}$ *such that* $\mathbf{a} \not\leq \mathbf{b}$ *and* $\mathbf{b} \not\leq \mathbf{a}$.

PROOF We need to build c.e. approximating sequences to sets $A, B \leq_T C \in \mathbf{c}$ such that $A \not\leq_T B$ and $B \not\leq_T A$. Assume we are given c.e. approximating sequence $\{C^s\}_{s \geq 0}$ for C.

As well as the requirements:

$$\mathcal{R}_{2i} : \quad A \neq \Phi_i^B$$
$$\mathcal{R}_{2i+1} : \quad B \neq \Phi_i^A,$$

we also have *overall requirements*:

$$\mathcal{S}^A : \quad (\forall x \geq 0)\,[\,C_A(x) \leq C_C(x)\,],$$
$$\mathcal{S}^B : \quad (\forall x \geq 0)\,[\,C_B(x) \leq C_C(x)\,].$$

By the Permitting Lemma $\mathcal{S}^A$ and $\mathcal{S}^B$ will give $A, B \leq_T C$.

The strategy for satisfying $\mathcal{R}_{2i}$

We just need to add the permitting. This will entail $\mathcal{R}_{2i}$ choosing multiple witnesses $x_1, x_2, \ldots$, each of which simultaneously pursues the strategy.

By convention $s + 1$ is always the current stage:

(1) Choose a fresh witness x_k to $A \neq \Phi_i^B$.

(2) Wait for a stage $s + 1$ at which $A^s(x_k) = 0 = \Phi_i^B(x_k)[s]$.

(3) In which case set up a B restraint $z = \varphi_{i,s}^B(x_k)$ — and wait for a stage at which $C^{s+1}\!\restriction x_k \neq C^s\!\restriction x_k$, when we can go to (4).

Simultaneously we return to (1) to start a copy of the strategy with a fresh witness x_{k+1}.

(4) Enumerate x_k into A^{s+1}. Halt the $\mathcal{R}_{2i}$ strategy on all witnesses.

The analysis of outcomes

This time we get:

$\boxed{w}$ The strategy waits forever at (2) — giving requirement $\mathcal{R}_{2i}$ satisfied via either $\Phi_i^B(x) \uparrow$ or $\Phi_i^B(x) = 1 \neq A(x)$.

$\boxed{s}$ The strategy halts at (4) with $A(x) = A^{s+1}(x) = 1 \neq 0 = \Phi_i^B(x)[s]$. The B restraint z in (3) ensures $\mathcal{R}_{2i}$ satisfied via $\Phi_i^B(x) \downarrow \neq A(x)$.

But there is also a *pseudo-outcome* — that is, an apparent outcome which turns out to be impossible because it is blocked by our special C.

$\boxed{i}$ The strategy gets to (3) with infinitely many witnesses $x_1, x_2, \ldots$, but none of the x_k's ever gets to (4).

This is how the kindly C permits an x_k to go to (4), so avoiding $\boxed{i}$:

To get a contradiction, assume each x_k, $k \geq 1$, reaches (3) at stage $\hat{s}(x_k)$. Since x_k never reaches (4)

$$C^{\hat{s}(x_k)}\!\restriction x_k = C\!\restriction x_k.$$

But then $\hat{s}$ dominates $C_C \upharpoonright \{x_1, x_2, \dots\}$. But by the Domination Lemma — in the strong form in Exercise 12.2.18 — this would give C computable, contrary to what we assumed.

Now say each requirement pursues its copy of the strategy for $\mathcal{R}_{2i}$. The strategy still has only finitary outcomes $\boxed{w}$ and $\boxed{s}$, so is finite injury again. So as before, every requirement eventually becomes satisfied. $\Box$

EXERCISE 12.3.5 *Let $\mathbf{a}$ be c.e. and $> \mathbf{0}$. Show that there exists an infinite independent set $\{\mathbf{a}_1, \mathbf{a}_2, \dots\}$ of c.e. degrees below $\mathbf{a}$.*

Deduce that the existential theories of $\mathcal{E}(\leq \mathbf{a})$ and $\mathcal{D}(\leq \mathbf{a})$ are decidable.

We can now use our new skills to show that the hierarchy $\{\mathbf{D}_n\}_{n>0}$ of n-c.e. degrees of Definition 12.1.2 does not collapse.

THEOREM 12.3.6

    *There exists a **properly d.c.e. degree**   — that is a degree $\mathbf{d} \in \mathbf{D}_2 - \mathcal{E}$.*

PROOF   We need to build a d.c.e. approximating sequence $\{D^s\}_{s \geq 0}$ to a set D such that $D \not\equiv_T W$ for any c.e. set W.

Presentational convention: It often helps to present the requirements without fussy indices and explicit priority ordering. In this case it is routine to get a standard listing of all triples of the form (W, Φ, Θ) with W c.e., and with Φ, Θ p.c. functionals to computably list all the requirements we need. This time a typical requirement on the list is:

$$\mathcal{R}_{W, \Phi, \Theta}: \quad D \neq \Phi^W \vee W \neq \Theta^D.$$

Discussion: Our first idea for satisfying $\mathcal{R}_{W, \Phi, \Theta}$ is to use our control over D in the equation $W = \Theta^D$ to preserve enough of W to keep some $\Phi^W(x)$ fixed — and then change $D(x)$ to make $D \neq \Phi^W$.

Of course this will not work — the $D(x)$ change is likely to clash with the D restraint.

The second idea is to do it anyway — do the restraint *except for* the putting of x in D. You never know, you might be lucky and W might not change to make $D = \Phi^W$ again. And if W does change — ha! ha! — the c.e. W has been tricked! We can *extract* x from D. And that seemingly useless restraint on D is now complete again, giving $\Theta^D =$ the old W — but $\neq$ to the new W. Our d.c.e. D can change back. The c.e. W cannot. $\mathcal{R}_{W, \Phi, \Theta}$ is satisfied.

We have been a bit vague about what beginnings of D and W are involved here. The detailed strategy will make that clear.

The strategy for satisfying $\mathcal{R}_{W,\Phi,\Theta}$

Presentational conventions: We assume all expressions mentioned in the strategy are approximations at the current stage $s + 1$, so often leave out mention of s, etc., when not really necessary. Where it is clear what set argument a use function φ^W or θ^D has, we omit it.

(1) Choose a fresh witness x to $D \neq \Phi^W \vee W \neq \Theta^D$.

(2) Wait for a stage at which $D(x) = 0 = \Phi^W(x)$ and $W \upharpoonright \varphi(x) = \Theta^D \upharpoonright \varphi(x)$.

(3) In which case enumerate x into D^{s+1} and restrain $D^{s+1} \upharpoonright \theta(\varphi(x))[s]$ — and wait for a stage at which $D(x) = \Phi^W(x)$ again, when we go to (4).

(4) Extract x from D^{s+1}.

Here is our flow diagram with some outcomes attached:

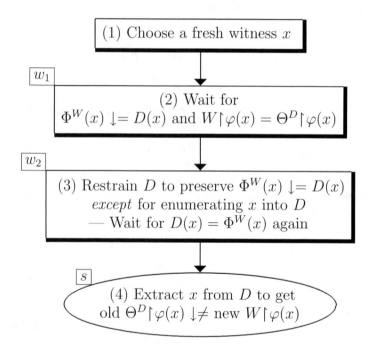

The analysis of outcomes

This time we get:

$\boxed{w_1}$ The strategy waits forever at (2) — giving the satisfaction of $\mathcal{R}_{W,\Phi,\Theta}$ via one of $\Phi^W(x)$, $W \upharpoonright \varphi(x)$ or $\Theta^D \upharpoonright \varphi(x)$ not being defined, or via $\Phi(x) \downarrow \neq 0 = D(x)$, or $W \upharpoonright \varphi(x) \downarrow \neq \Theta^D \upharpoonright \varphi(x) \downarrow$.

$\boxed{w_2}$ The strategy halts at (3) with equation $D(x) = \Phi^W(x)$ never holding again, giving $\mathcal{R}_{W,\Phi,\Theta}$ satisfied.

$\boxed{s}$ The strategy reaches (4) and terminates.

Say the strategy previously visited (3) at a stage $t + 1 <$ the current stage $s + 1$. The restraint imposed then means we have

$$D^{s+1} \upharpoonright \theta(\varphi(x))[t] = D^t \upharpoonright \theta(\varphi(x))[t],$$

so that

$$\Theta^D \upharpoonright \varphi(x)[t] = \Theta^D[s+1] \upharpoonright \varphi(x)[t].$$

But $W^s \upharpoonright \varphi(x)[t] \neq W \upharpoonright \varphi(x)[t]$, since $\Phi^W(x)[s] = 1 \neq D^t(x)$. So

$$W^s \upharpoonright \varphi(x)[t] \neq \Theta^D[s+1] \upharpoonright \varphi(x)[t].$$

Hence $W \neq \Theta^D$, and $\mathcal{R}_{W,\Phi,\Theta}$ is satisfied.

We can now put the strategies together to prove the theorem much as before.

Say each requirement pursues its copy of the strategy for $\mathcal{R}_{W,\Phi,\Theta}$. Each such strategy has only finitary outcomes $\boxed{w_1}$, $\boxed{w_1}$ and $\boxed{s}$. So as before, every requirement has a stage after which its strategy is never injured, and passes on to lower priority requirements a stage after which it never injures.

Since every strategy succeeds, every $\mathcal{R}_{W,\Phi,\Theta}$ becomes satisfied. ◻

To be sure that the n-c.e. hierarchy does not collapse at levels $n > 2$, you need to prove:

EXERCISE 12.3.7 (The n-C.E. Hierarchy Theorem) *Show that for each $n \geq 2$ there exists a* **properly n-c.e. degree** $\mathbf{a} \in \mathbf{D}_n - \mathbf{D}_{n-1}$.

[**Hint:** Construct an n-c.e. $A \not\equiv_T$ any (n−1)-c.e. U. Model your strategy on that for Theorem 12.3.6. By changing $A(x)$ n times — instead of $D(x)$ twice in Theorem 12.3.6 — make U unable to defeat the strategy.]

You can also combine the strategy for the n-c.e. Hierarchy Theorem with permitting.

EXERCISE 12.3.8 *Show that there is a properly d.c.e. degree* $\mathbf{d} <$ *any given c.e. degree* $\mathbf{a} > \mathbf{0}$.

[**Hint:** You need to ask for separate A permissions for *both* $x \in D^{s+1} - D^s$ in (3) *and* $x \in D^s - D^{s+1}$ in (4) of the strategy for Theorem 12.3.6. As before fresh witnesses need choosing while all existing witnesses are waiting for permissions.]

You can make the degree properly d.c.e. in this exercise properly n-c.e. by building a representative according to the strategy for Exercise 12.3.7, suitably permitted. You can then prove:

EXERCISE 12.3.9 *Show that if* **a** *is properly n-c.e., then there exists a properly n-c.e. degree* **b** < **a**.

[**Hint:** Use Exercise 12.1.6.]

The proof of the Friedberg–Muchnik Theorem and the invention of the priority method took the subject in a new direction. It led to powerful techniques for bringing out the complexity of the Turing universe — but without Post's emphasis on natural information content. There is plenty of information content in the sets constructed, of course. The problem is to extract it in some generally meaningful form. It would be many years before these different approaches would be brought together again.

12.4   Sacks' Splitting Theorem and Cone Avoidance

Now for another classic theorem and some new techniques. The next example prepares the ground.

EXAMPLE 12.4.1 *Show that if* $A = A_0 \sqcup A_1$ *is a splitting of* A *into disjoint c.e. sets, then (a)* $A \leq_T A_0 \oplus A_1$, *and (b)* $A_0, A_1 \leq_T A$.

SOLUTION   (a) is immediate.

For (b) the following algorithm computes A_0 from A. To decide if $x \in A_0$:

(1) Ask whether $x \in A$. If not, $x \notin A_0$.

(2) If $x \in A$, enumerate A_0 and A_1. If x appears first in A_0, $x \in A_0$. Otherwise x appears in A_1 first telling us $x \notin A_0$.   ⬚

Notice that in Friedberg's Splitting Theorem — Exercise 6.2.10 — the splitting of A is into incomputable c.e. sets which *are computable from* A. What we can do now is make the splittings *Turing incomparable*. The theorem is also about *cone avoidance*.

DEFINITION 12.4.2 *(1) We say that two Turing degrees* **a** *and* **b** *are* **incomparable** *if* **a** $\not\leq$ **b** *and* **b** $\not\leq$ **a**, *and write* **a**|**b**.

(2) If $\mathcal{C} \subseteq \mathcal{D}$, *we call* $\mathcal{C}(\geq \mathbf{a})$ *the* **cone above a in** $\mathcal{C}$ — *and constructing an* $\mathbf{a} \in \mathcal{C} - \mathcal{C}(\geq \mathbf{a})$ *we call* **upper cone avoidance**.

We similarly define the **cone below a in** $\mathcal{C}$, *and* **lower cone avoidance**.

Notice that **a**|**b** if and only if **b** avoids the cones above and below **a**. With degrees of splittings, upper cone avoidance is the problem.

THEOREM 12.4.3 (The Sacks Splitting Theorem)
Let A *be an incomputable c.e. set, and let* $\emptyset <_T D \leq_T \emptyset'$.

Then there exists a splitting $A = B \sqcup C$ *of* A *into disjoint low c.e. sets* $B, C \leq_T A$ *such that* $D \not\leq_T B$ *or* C.

PROOF This is what we need to produce:

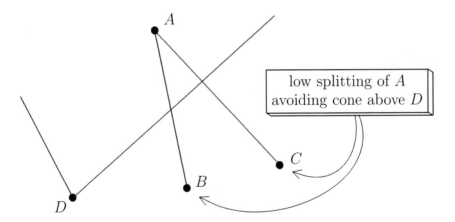

Assume we are given c.e. $\{A^s\}_{s \geq 0}$ for A, and Δ_2-approximating sequence $\{D^s\}_{s \geq 0}$ for D. We need to build c.e. $\{B^s\}_{s \geq 0}$, $\{C^s\}_{s \geq 0}$ for B and C to first of all satisfy the overall requirement:

$$\mathcal{S}: \quad A = B \sqcup C.$$

There is really only one strategy for $\mathcal{S}$. If x is enumerated into A, we must immediately enumerate x into just one of B or C.

We also need all requirements of the form:

$$\mathcal{N}_x^B : \quad \text{Lim}_s \, B'(x)[s] \text{ exists,}$$

$$\mathcal{N}_x^C : \quad \text{Lim}_s \, C'(x)[s] \text{ exists,}$$

to make $B', C' \in \Delta_2$, and so $\leq_T \emptyset'$.

And for upper cone avoidance we need for each p.c. Φ:

$$\mathcal{R}_\Phi^B : \quad D \neq \Phi^B,$$

$$\mathcal{R}_\Phi^C : \quad D \neq \Phi^C,$$

to make $D \not\leq_T B$ or C.

Presentational conventions: (1) Write $j_s^B(x)$ for the B use of $B'(x)$ at stage $s+1$. Notice — since $x \in B' \iff \Phi_x^B(x) \downarrow$, we have $j_s^B(x) = \varphi_x^B(x)[s]$.

(2) A useful bit of notation is

$$\ell_s(D, \Phi^B) = \mu x \, (D^s \restriction x \neq \Phi^B \restriction x[s])$$

for the *length of agreement* of D, Φ^B at stage $s+1$. We often just write ℓ_s.

Obviously we will only have to bother about $\mathcal{R}_\Phi^B$ if ℓ_s becomes unbounded while constructing B, so we will want to watch for $\mathcal{R}_\Phi^B$ *expansionary* stages at which ℓ_s becomes $> \ell_t$ for any $t < s$.

We will write

$$\text{Use}_s(D = \Phi^B) = \mu z \, (\Phi^{B \restriction z} \restriction \ell \downarrow [s])$$

for the B *use* of $\ell(D, \Phi^B)$ at stage $s+1$.

Discussion: What a lot of requirements! And none of them quite like what we have already dealt with. Of course, we only need to describe strategies for one $\mathcal{N}$ requirement and one $\mathcal{R}$ requirement.

And the strategy for $\mathcal{N}_x^B$, say, is really just the restraint half of the Friedberg–Muchnik strategy:

Say $B'(x)[s] \downarrow$. Restrain $B \restriction j_s^B(x)$ at all later stages, giving $\text{Lim}_s \, B'(x)$ exists $= B'(x)[s]$.

The strategy for $\mathcal{R}_\Phi^B$ is the novel one, the method of *Sacks restraints*. What is different about Sacks restraints? You use normal restraints to preserve a disagreement you want, as in the Friedberg–Muchnik Theorem. Sacks restraints you use to preserve an *agreement* you *do not* want!

What we do is monitor the length of agreement of D, Φ^B. At every $\mathcal{R}_\Phi^B$ expansionary stage we decide to permanently restrain the B use of $\ell(D, \Phi^B)$. The effect of infinitely many such restraints is to make $D = \Phi^B$ with Φ^B computable — which an incomputable D turns into a pseudo-outcome. Since the strategy makes this the only route to $D = \Phi^B$, it satisfies $\mathcal{R}_\Phi^B$.

But how do we protect these strategies, based purely on restraints, from that for $\mathcal{S}$ — which forces us to injure restraints by enumerating members x of A into B or C?

First notice that if we can protect the restraints for the $\mathcal{R}$ requirements, we can certainly do it for the $\mathcal{N}$ requirements. And the strategy for just $\mathcal{R}_\Phi^B$ with $\mathcal{S}$ is easy — enumerate every $x \in A$ into C.

The crucial situation is $\mathcal{R}_\Phi^C$ below $\mathcal{R}_\Phi^B$ with $\mathcal{S}$. No problem! Pursue the strategy for $\mathcal{R}_\Phi^B$ with $\mathcal{S}$ with no regard for $\mathcal{R}_\Phi^C$, until we run out of $\mathcal{R}_\Phi^B$ expansionary stages — the pseudo-outcome defeated. Eventually all new members x of A are greater than the greatest B restraint, and can be safely enumerated into B. There are only finitely many injuries to $\mathcal{R}_\Phi^C$, which takes its turn to be satisfied.

This completes the proof of the theorem. Not convinced? Oh, alright, here are some details.

Since you make me do something I don't think is necessary, I will take it as an opportunity to bring in an important new idea — that of a *tree of outcomes*.

The strategy for satisfying $\mathcal{R}_\Phi^B$ with $\mathcal{S}$

Background activity: If $x \in A^{s+1} - A^s$ — that is, $x \in A^{\text{at } s+1}$ — enumerate x into C^{s+1}.

(1) Wait for an $\mathcal{R}_\Phi^B$ expansionary stage $s + 1$.

(2) Impose a restraint on $B \restriction \text{Use}_s(D = \Phi^B)$.

Return to (1).

Analysis of outcomes

$\boxed{w}$ At all stages $>$ than some s^* we wait at (1). $\mathcal{R}_\Phi^B$ is satisfied.

$\boxed{i}$ $\mathcal{R}_\Phi^B$ visits (2) infinitely often.

Given x, we compute $D(x) = \Phi^B(x)$ as follows. First look for a stage $s + 1$ at which $\mathcal{R}_\Phi^B$ visits (2) with $\ell_s(D, \Phi^B) > x$.

Then the restraint on $B \restriction \text{Use}_s(D = \Phi^B)$ is never injured and so ensures $\Phi^B(x) = \Phi^B(x)[s]$. We must then have $\Phi^B(x)[s] = \text{Lim}_s D^s(x) = D(x)$, since otherwise there are only finitely many $\mathcal{R}_\Phi^B$ expansionary stages, contradicting our being in outcome $\boxed{i}$.

But D is *not* computable. So outcome $\boxed{i}$ is impossible.

Presentational conventions: We can lay out all the outcomes for a list of requirements on a *tree of outcomes*.

Listing our $\mathcal{R}$ requirements $\mathcal{R}_0, \mathcal{R}_1, \ldots$ and ignoring the $\mathcal{N}$ requirements, our tree in this case might look like:

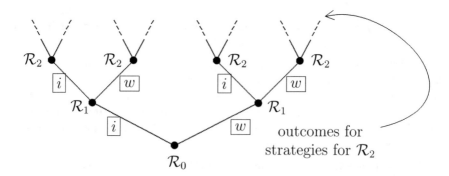

outcomes for
strategies for $\mathcal{R}_2$

As you can see, the requirements and their strategies are located at nodes, and the branches represent outcomes of those strategies. In more complicated proofs — but not here — the strategies might depend on the position on the tree. Which is why we also refer to the *tree of strategies*.

By the way, my trees grow upwards. But if you prefer higher priority requirements to be higher on the page, downward growing trees are quite acceptable!

The strategy for satisfying $\mathcal{R}_\Phi^C$ below $\mathcal{R}_\Phi^B$ having outcome $\boxed{w}$

Background activity: Say $x \in A^{\mathrm{at}\, s+1}$. If $x <$ a B restraint, enumerate x into C^{s+1} — otherwise enumerate x into B^{s+1}.

(1) Wait for a $\mathcal{R}_\Phi^C$ expansionary stage $s + 1$.

(2) Impose a restraint on $C{\upharpoonright}\operatorname{Use}_s(D = \Phi^C)$.

Return to (1).

Analysis of outcomes

Only the background activity has changed. This makes no difference to outcome $\boxed{w}$. We just have to check that $\boxed{i}$ remains a pseudo-outcome.

Choose a stage s^* such that no $s + 1 > s^*$ is $\mathcal{R}_\Phi^B$ expansionary, and no $x \in A^{\mathrm{at}\, s+1}$ is $<$ a restraint defined by $\mathcal{R}_\Phi^B$.

Then the previous analysis of outcome $\boxed{i}$ for $\mathcal{R}_\Phi^B$ applies here, except that when we compute $D(x) = \Phi^C(x)$ by taking a stage $s + 1$ at which $\mathcal{R}_\Phi^C$ visits (2) with $\ell_s > x$, we must take $s + 1 > s^*$.

The strategy for all the requirements is now clear. All we do is adjust the background activity to deal with more requirements. Say $x \in A^{\mathrm{at}\, s+1}$. Let z be the restraint of highest priority with $x < z$. We must include the restraints set up by the $\mathcal{N}$ requirements now. If z is a B restraint enumerate x into C, and otherwise enumerate x into B. Then inductively each requirement ceases to be injured, and so ceases to injure, and all the requirements are satisfied via successful strategies. ∎

COROLLARY 12.4.4

Let $\mathbf{a} > \mathbf{0}$ *be c.e., and* $\mathbf{0} < \mathbf{d} \leq \mathbf{0}'$.

Then there is a splitting $\mathbf{a} = \mathbf{b} \cup \mathbf{c}$ *of* $\mathbf{a}$ *into incomparable low c.e. degrees* $\mathbf{b}$ *and* $\mathbf{c}$ *avoiding the cone above* $\mathbf{d}$.

PROOF Assume $A \in \mathbf{a}$. If $\mathbf{d} \leq \mathbf{a}$, choose $D \in \mathbf{d}$, and if $\mathbf{d} \nleq \mathbf{a}$, choose $D \in \mathbf{a}$. Take $\mathbf{b} = \deg(B)$, $\mathbf{c} = \deg(C)$, where B, C are as in the Sacks Splitting Theorem.

Then $A = B \oplus C$ by Example 12.4.1, so $\mathbf{a} = \mathbf{b} \cup \mathbf{c}$. And since $D \nleq_T B$ or C, $\mathbf{d} \nleq \mathbf{b}$ or $\mathbf{c}$, and $\mathbf{a} \nleq \mathbf{b}$ or $\mathbf{c}$ so $\mathbf{b} | \mathbf{c}$. And the lowness of $\mathbf{b}, \mathbf{c}$ follows immediately from that for B and C. $\qquad\Box$

The Sacks Splitting Theorem has many applications. There are a number of simple consequences worth mentioning. For instance, notice that the low degrees are not closed under the join operation. In fact all the nonzero c.e. degrees are generated by the c.e. low degrees with join — even $\mathbf{0}'$ is the join of two low c.e. degrees.

Another corollary is that there is a c.e. degree incomparable with every $\mathbf{d}$, $\mathbf{0} < \mathbf{d} < \mathbf{0}'$ — just take a c.e. splitting $\mathbf{b}, \mathbf{c}$ of $\mathbf{0}'$ avoiding the cone above $\mathbf{d}$. At most one of $\mathbf{b}, \mathbf{c}$ is $\leq \mathbf{d}$, since otherwise $\mathbf{d} \geq \mathbf{b} \cup \mathbf{c} = \mathbf{0}'$.

Also notice that we can relativise the theorem to any given oracle, taking care to code the oracle into the splitting.

EXERCISE 12.4.5 *Let* $\mathbf{a} > \mathbf{e}$ *be c.e. in* $\mathbf{e}$, *and* $\mathbf{e} < \mathbf{d} \leq \mathbf{e}'$.

Show that there is a splitting $\mathbf{a} = \mathbf{b} \cup \mathbf{c}$ *of* $\mathbf{a}$ *into incomparable degrees* $\mathbf{b}, \mathbf{c} \geq \mathbf{e}$ *avoiding the cone above* $\mathbf{d}$, *with* $\mathbf{b}, \mathbf{c}$ *c.e. in* $\mathbf{e}$, *and low over* $\mathbf{e}$ *(that is,* $\mathbf{b}' = \mathbf{c}' = \mathbf{e}'$*)*.

EXERCISE 12.4.6 *Show that every* $\mathbf{a} \geq \mathbf{0}'$ *is the join of two degrees* $\mathbf{a}_1, \mathbf{a}_2 < \mathbf{a}$ *with* $\mathbf{a}_1' = \mathbf{a}_2' = \mathbf{a}$.

EXERCISE 12.4.7 *Show that every n-c.e. degree* $\mathbf{a} > \mathbf{0}$ *is the join of two degrees* $\mathbf{a}_1, \mathbf{a}_2 < \mathbf{a}$.

In fact, one can split any n-c.e. $\mathbf{a} > \mathbf{0}$ into two incomparable n-c.e. degrees — but not with upper cone avoidance, it turns out.

EXERCISE 12.4.8 *Given* $\mathbf{d} \geq \mathbf{a}$ *and* $\mathbf{c}$ *which is CEA* $\mathbf{a}$, *show that* $\mathbf{d} \cup \mathbf{c}$ *is CEA* $\mathbf{d}$.

Deduce that if $\mathbf{d} < \mathbf{0}''$ *then* $\mathbf{0}'' = \mathbf{b}_1 \cup \mathbf{b}_2$ *for incomparable degrees* $\mathbf{b}_1, \mathbf{b}_2 > \mathbf{d}$ — *we say* $\mathbf{0}''$ *is* **splittable over d**.

[**Hint:** Consider $\mathbf{d} \cup \mathbf{0}'$.]

By the way, you can split every c.e. $\mathbf{a}$ into a pair of low c.e. degrees by just splitting $\mathbf{0}'$:

EXERCISE 12.4.9 *Show that there exist low c.e. degrees* $\mathbf{b}, \mathbf{c}$ *such that every c.e.* $\mathbf{a} = \mathbf{b}_0 \cup \mathbf{c}_0$ *with* $\mathbf{b}_0, \mathbf{c}_0$ *low c.e. degrees* $\leq \mathbf{b}, \mathbf{c}$ *respectively.*

[**Hint:** Get a low splitting B, C of K_0. Use it to find a low splitting $B_0, C_0 \leq_m B, C$ for each $W_i = \{x \mid \langle x, i \rangle \in K_0\}$.]

12.5   Minimal Pairs and Extensions of Embeddings

I said earlier that results about the structure of $\mathcal{D}$ were essentially *about* Turing definability. And that is true if you are interested in the wider consequences of computability for the foundations of science. But degree structure can be about other general issues. For those interested in finding out how decidable the elementary theories of $\mathcal{D}$ or its substructures are, degree structure is about *extensions of embeddings*.

In Remark 10.6.8 I described how being able to embed any finite partial ordering in $\mathcal{D}$ gave us the decidability of the Σ_1 theory of $\mathcal{D}$. And we just saw — Exercise 12.3.5 — how we can do the same for the Σ_1 theory of $\mathcal{E}$.

Embeddings alone are no good for deciding 2-quantifiers sentences. Allowing $\mathbf{0}$ in our language, a φ of the form:

$$(\forall \mathbf{x}_0)(\exists \mathbf{x}_1)\,[\,\mathbf{0} < \mathbf{x}_0 \Rightarrow \mathbf{0} < \mathbf{x}_1 < \mathbf{x}_0\,], \tag{12.1}$$

says about $\mathcal{D}$: "Given the p.o. $\mathbf{0} < \mathbf{x}_0$, you can *extend* it to $\mathbf{0} < \mathbf{x}_1 < \mathbf{x}_0$." In general, an $\forall\exists$ — or Π_2 — sentence about a degree structure needs you to decide whether some given finite p.o. P can be *extended* in the structure to $Q_i \supseteq P$ for one of a list $Q_0, \ldots, Q_k$ of finite partial orderings. The Sacks Splitting Theorem is more than strong enough to decide the φ of (12.1) in $\mathcal{E}$ — it is true. φ is false in $\mathcal{D}$ — see Section 13.4.

So from this point of view, structural results are *about* extensions of embeddings. They can tell us how we can extend orderings — and identify *obstacles* to extensions. An obvious obstacle to extending embeddings is that given $\mathbf{a} > \mathbf{b}, \mathbf{c}$ we may not be able to extend to $\mathbf{a} > \mathbf{x} > \mathbf{b}, \mathbf{c}$ — since we may have $\mathbf{a} = \mathbf{b} \cup \mathbf{c}$. Also we cannot insert degrees $< \mathbf{0}$ or — in $\mathcal{E}$ — $> \mathbf{0}'$.

Shoenfield's Conjecture from 1965 famously asserted that in $\mathcal{E}$ these were the *only* obstacles to extensions of a P involving just one $Q \supseteq P$. That is,

$\mathcal{E}$ is an upper semi-lattice densely ordered much like the rationals between 0 and 1 form a densely ordered linear ordering. The conjecture was more important for its intention than its correctness! There were two particular sub-conjectures — there are no c.e. *minimal pairs*, and you can always *cup* $\mathbf{b} > \mathbf{0}$ to a $\mathbf{a} > \mathbf{b}$ in $\mathcal{E}$ — both of which turned out to be completely wrong.

The next definition is stated for $\mathcal{E}$, but can be used with minor modifications in other local degree structures, such as $\mathcal{D}_e(\leq \mathbf{0}'_e)$ or $\mathbf{D}_n$.

DEFINITION 12.5.1 *(1) Degrees* $\mathbf{a}, \mathbf{b} > \mathbf{0}$ *in* $\mathcal{E}$ *form a* **minimal pair** *if there is no* $\mathbf{x} \in \mathcal{E}$ *with* $\mathbf{a}, \mathbf{b} \geq \mathbf{x} > \mathbf{0}$.

(2) $\mathbf{a} > \mathbf{0}$ *is* **cappable** *if there is a* $b > \mathbf{0}$ *with* $\mathbf{0} = \mathbf{a} \cap \mathbf{b}$ *in* $\mathcal{E}$. *Otherwise* $\mathbf{a}$ *is* **noncappable**.

(3) $\mathbf{b}$ *is* **cuppable** *to* $\mathbf{a} > \mathbf{b}$ *if there is a* $\mathbf{c} < \mathbf{a}$ *with* $\mathbf{a} = \mathbf{b} \cup \mathbf{c}$ *in* $\mathcal{E}$. $\mathbf{a} < \mathbf{0}'$ *is* **cuppable** *if it is cuppable to* $\mathbf{0}'$, *and otherwise is* **noncuppable**.

We will frame the proof of the next theorem as a *tree construction*, actually using the tree of outcomes to help us frame the strategies.

THEOREM 12.5.2 (Lachlan, 1966; Yates, 1966)
There exists a minimal pair $\mathbf{a}, \mathbf{b}$ *of c.e. degrees.*

PROOF We need to build $A \in \mathbf{a}$ and $B \in \mathbf{b}$ via c.e. approximations $\{A^s\}_{s \geq 0}$, $\{B^s\}_{s \geq 0}$. To make A and B not computable we satisfy:

$$\mathcal{P}_W^A : \quad A \neq \overline{W},$$
$$\mathcal{P}_W^B : \quad B \neq \overline{W},$$

for each c.e. W. And to make $\mathbf{a}, \mathbf{b}$ a minimal pair we satisfy:

$$\mathcal{R}_{\Theta, \Psi} : \quad (\Theta^A \text{ total} = \Psi^B) \implies \Theta^A \text{ computable},$$

for each pair Θ, Ψ of p.c. functionals.

Similarly to before, define the length of agreement as

$$\ell_s = \ell_s(\Theta^A, \Psi^B) = \mu x \, (\Theta^A \restriction x \, [s] \neq \Psi^B \restriction x \, [s]),$$

along with the $\mathcal{R}_{\Theta, \Psi}$ expansionary stages $s + 1$ at which ℓ_s reaches a new maximum value. We write $m_s = \max \{\ell_t \, | \, t \leq s\}$. And, similarly to before, let

$$\text{Use}_s = \text{Use}_s(\Theta^A = \Psi^B) = \mu z \, [\Theta^{A \restriction z} \restriction \ell[s] = \Psi^{B \restriction z} \restriction \ell[s]].$$

Discussion: The strategy for the $\mathcal{P}$ requirements — say $\mathcal{P}^A$ — is to choose a witness $x \notin A$ and to enumerate it into A^{s+1} if $x \in W^{\text{at } s+1}$.

For $\mathcal{R}_{\Theta,\Psi}$ the idea is to ensure that if ℓ_s is unbounded then we always protect at least one side of $\Theta^A = \Psi^B$ for arguments $x < m_s$. This means if $\Theta^A \restriction m$ changes, then $\Psi^B \restriction m$ is preserved *until* the next $\mathcal{R}_{\Theta,\Psi}$ expansionary stage. This will satisfy $\mathcal{R}_{\Theta,\Psi}$ while ensuring that although $\mathcal{R}_{\Theta,\Psi}$ may use infinitely many restraints, all the restraints are temporary.

The strategy for $\mathcal{R}_{\Theta,\Psi}$ with the $\mathcal{P}$ requirements

Background activity: Assume that only one of the $\mathcal{P}$ requirements enumerates a witness into A or B at stage $s + 1$.

(1) Restrain $A \restriction \text{Use}_s$ and $B \restriction \text{Use}_s$.
Wait for a new expansionary stage, and then go to (2).

(2) Throw away all restraints for $\mathcal{R}_{\Theta,\Psi}$.
Wait for a stage $t + 1$ at which $\ell_t < \ell_s$, in which case return to (1).

Analysis of outcomes

$\boxed{w}$ There are only finitely many expansionary stages.
Then $\mathcal{R}_{\Theta,\Psi}$ is trivially satisfied since $\Theta^A \neq \Psi^B$. The strategy only acts finitely often, and if it halts at (1) there may be a permanent restraint.

$\boxed{i}$ There are infinitely many expansionary stages.
Assume that the *injury set* I of witnesses enumerated into A or B by $\mathcal{P}$ requirements above $\mathcal{R}_{\Theta,\Psi}$ is finite — as it will turn out. Let s^* be such that $A^{s^*}(z) = A(z)$ and $B^{s^*}(z) = B(z)$ for every $z \in I$.

Say Θ^A is total. Here is an algorithm showing Θ^A is computable:

To compute $\Theta^A(x)$, look for the least stage $s+1 > s^*$ at which $\ell_s > x$. Then $\Theta^A(x) = \Theta^A(x)[s]$. This will follow from the fact that at every expansionary stage $t + 1 \geq s + 1$ we have

$$\Theta^A[s] \restriction \ell_s = \Theta^A[t] \restriction \ell_s = \Psi^B[t] \restriction \ell_s. \tag{12.2}$$

This is because if $t + 1$ is expansionary, and $\ell_{t'} < \ell_t$ at some least stage $t' + 1 > t + 1$, only one of $\Theta^A[t'] \restriction \ell_s$, $\Psi^B[t'] \restriction \ell_s$ can have changed from the value in Equation (12.2). And the other will be restrained via (1) until the next expansionary stage $> t' + 1$ when *both* values are restored.

Presentational conventions: (1) Remember — we will situate strategies for the nth requirement at nodes on level n of the tree of outcomes/strategies, and use branches to represent the outcomes.

It is a matter of taste *how much information* you put on your tree in the way of outcomes. You can put all sorts of detail about restraints and status of witnesses, breaking up outcomes into sub-outcomes. I will just put what has a significant impact on the strategies lower down the tree.

(2) *How do we use the tree?*

At the end of the construction only one outcome at each node will be real — a *true* outcome. The true outcomes will define a *true path* — a tongue-in-cheek bit of terminology due to Leo Harrington who gave the tree method its present-day form.

We make the tree useful *during the construction* by approximating the true path via a true path at each stage $s + 1$. To do this we inductively work our way down the approximated true path, applying a suitable strategy at each node, and deciding on a corresponding true outcome at that stage to extend the true path.

(3) Notice — just as we *vertically* prioritise requirements, we must define a *horizontal* order $<_L$ ("to the left of") on the set of outcomes at each node. The ordering $<_L$ on the outcomes induces a lexicographical ordering on the nodes of the tree. The true path is usually the path of "leftmost" nodes visited infinitely often during the building activity. The $<_L$ ordering is not arbitrary like the priority ordering. We carefully situate less fragile outcomes — say ones which are only true if eventually always on the current true path — to the right. We usually throw away all restraints and witnesses to the right of the current true path.

The tree of strategies

The $\mathcal{P}$ nodes will have just outcome $\boxed{s}$ and not branch — we do not need to tell the $\mathcal{R}$ strategy about the injury set I, it will discover enough itself. The $\mathcal{R}$ nodes will branch left along outcome $\boxed{i}$ and right along outcome $\boxed{w}$.

In building A and B the outcome *true at stage $s + 1$* will be $\boxed{i}$ if the $\mathcal{R}$ strategy is at (2) and $\boxed{w}$ if (1) applies. So the tree looks very much like the diagram we saw earlier, although you must imagine the non-branching $\mathcal{P}$ nodes inserted.

The strategy for satisfying $\mathcal{P}_W^A$ below $\mathcal{R}_{\Theta,\Psi}$

Background activity: Assume the $\mathcal{P}^B$ requirements below $\mathcal{P}_W^A$ are busy when $\mathcal{P}_W^A$ is not, enumerating numbers into B.

(1) Choose a fresh witness x.

(2) Wait for $x \in W$.

(3) Enumerate x into A.

Discussion: First, just a comment on the power of the tree presentation. Notice how it *looks* like the same $\mathcal{P}$ strategy below $\boxed{w}$ or $\boxed{i}$. But not so. The $\mathcal{P}$ strategy below $\boxed{i}$ only applies when $\mathcal{R}_{\Theta,\Psi}$ implements (2) — so the strategy is to wait for expansionary stages. While the $\mathcal{P}$ strategy below $\boxed{w}$ is to choose a fresh $x >$ any restraint whenever the current true path moves right to $\boxed{w}$.

Now to serious matters. The tree will help us adopt appropriate strategies, *so long as* the true path at stage $s + 1$ is close to the eventual true path. The problem is not identifying true outcomes as leftmost ones, it is *orchestrating* those outcomes into a leftmost path.

To be specific, say we have $\mathcal{R}_{\Theta, \Psi}$ below $\mathcal{R}_{\Theta_0, \Psi_0}$, both with outcome $\boxed{i}$ on the eventual true path. Obviously each requirement will visit $\boxed{i}$ on the current true path infinitely often. But will they ever do it simultaneously? If not, the correct strategy for a $\mathcal{P}_W^A$ below $\mathcal{R}_{\Theta, \Psi}$ and $\mathcal{R}_{\Theta_0, \Psi_0}$ may never be implemented. If communication of information is the strength of the tree framework, flexibility of timing is not. The method is truly orchestral. Everyone has to be in the room at the same time. And they had better watch the conductor!

The solution which works here is a sort of *relativity* which clarifies as powerfully in its small way as does its material world namesake. Not only does an $\mathcal{R}$ strategy only *act* when it is on the current true path, it only *observes* its expansionary stages relative to those stages!

Presentational conventions: Given a node on the tree T of strategies, we can notate it using the string α of outcomes on the path above it, listed in order of priority. We can then index with α all expressions evaluated at node α — for example, $\mathcal{P}_W^A$ below just $\boxed{i}$ for $\mathcal{R}_{\Theta, \Psi}$ and $\boxed{w}$ for $\mathcal{R}_{\Theta_0, \Psi_0}$ would be at node $\alpha = \boxed{w}\,\boxed{i}$, and we write $T(\alpha) = \mathcal{P}_\alpha$ for $\mathcal{P}_W^A$ at that node.

If α is on the true path at stage $s + 1$, we call $s + 1$ an α-*stage*. Say $T(\alpha) = \mathcal{R}_{\Theta, \Psi}$. We can now give the real definition of the length function:

$$\ell_{\alpha, s+1} = \begin{cases} \mu x \left(\Theta^A \!\restriction\! x[s] \neq \Psi^B \!\restriction\! x[s] \right) & \text{if } s + 1 \text{ is an } \alpha \text{ stage,} \\ \ell_{\alpha, s} & \text{otherwise.} \end{cases}$$

As previously, the α-*expansionary* stages $s + 1$ are those at which $\ell_{\alpha, s}$ reaches a new maximum.

Having put so much effort into our tree framework, here is the payoff.

The strategy for satisfying $\mathcal{P}_W^A$ below $\mathcal{R}_{\Theta, \Psi}$ below $\mathcal{R}_{\Theta_0, \Psi_0}$

Assume $\mathcal{R}_{\Theta, \Psi}$, $\mathcal{R}_{\Theta_0, \Psi_0}$ and $\mathcal{P}_W^A$ located at nodes α, α_0 and β. The α and α_0 strategies are just the $\mathcal{R}$ strategy — with obvious notational adjustments — as already described. Since the $\mathcal{P}$ nodes do not branch, and the $\mathcal{R}$ strategies do not interact, the α and α_0 strategies do not depend on the tree.

The strategy for $\mathcal{P}_\beta$ is

(1) Choose a fresh witness x.

(2) Wait for $x \in W$.

(3) Enumerate x into A.

But that is just the same too, surely? Again — not so. It only acts at β-stages. We need to look at β more closely to see what the outcomes really are. If one or more of $\mathcal{R}_\alpha, \mathcal{R}_{\alpha_0}$ have a $\boxed{w}$ outcome above β, we just put together our previous analyses of $\boxed{w}$, $\boxed{i}$ in the obvious way.

If both $\mathcal{R}_\alpha$ and $\mathcal{R}_{\alpha_0}$ have outcomes $\boxed{i}$, the β strategy succeeds because the only β-stages are also α and α_0 stages, which are α- and α_0-expansionary, so that restraints due to α, α_0 are thrown away.

Now we have the full picture. You should now try and verify to your own satisfaction:

EXERCISE 12.5.3 *Let the* **leftmost path** *be the set of α for which there are infinitely many α-stages, and only finitely many α'-stages for $\alpha' <_L \alpha$.*

Show inductively that for each α on the leftmost path (a) there is a string $\alpha^\frown \boxed{x}$ on the leftmost path, with $\boxed{x} = \boxed{w}$, $\boxed{i}$ or $\boxed{s}$, and (b) the α strategy satisfies $T(\alpha)$ with outcome $\boxed{x}$.

In other words, you should try putting together what we have done in proving that the strategies do satisfy the requirements, located as I have described on the tree. As a result you will have verified that the leftmost path is indeed a true path. $\qquad\qquad\qquad\qquad\qquad\qquad\qquad\qquad\qquad\Box$

REMARK 12.5.4 The original proof did not use a tree. It solved the problem of non-synchronous expansionary stages by letting a witness progress past higher priority $\mathcal{R}$ requirements *sequentially*. An $\mathcal{R}$ requirement with true outcome $\boxed{w}$ might permanently obstruct a finite number of witnesses which arrived asking for the $\mathcal{R}$ restraints to be thrown away. Or one with true outcome $\boxed{i}$ would allow all to pass on to a higher priority $\mathcal{R}$ — or into A or B if there is no such $\mathcal{R}$ — at each of its expansionary stages. An $\mathcal{R}$ would have to have knowledge of the finite number of witnesses permanently restrained above it, as well as the finite injury set, to be sure of being satisfied by outcome $\boxed{i}$.

This very different way of doing things — good on flexible timing, weak on communicating information between strategies — gives what is called the *pinball model* of priority. You have to be old enough, or have a good sense of cultural history, to benefit from the analogy! The method is almost as neglected now as the museum piece from which it gets its name. If you want to disprove Shoenfield's Conjecture that you can always cup, you really should use a pinball construction. I wish I had space to show you the proof.

Although the tree framework for priority goes back to Friedberg's use of e-states in the fifties, it really only surfaced in something like its present form in some very hard proofs of Lachlan in the early seventies. It has now become essential knowledge for anyone wanting to follow latest developments in computability theory. That is why it was well worth weighting down the last two proofs with a tree framework they did not really need. $\qquad\Box$

The minimal pair construction is a special case of a *lattice embedding*.

DEFINITION 12.5.5   *A lattice $\langle L, \leq, \vee, \wedge \rangle$ is **embeddable** in $\mathcal{E}$ if there is a bijection $\psi : L \to \mathcal{L} \subset \mathcal{E}$ such that for all $a, b \in L$*

$$a \leq b \iff \psi(a) \leq \psi(b),$$
$$\psi(a \vee b) = \psi(a) \cup \psi(b) \text{ and}$$
$$\psi(a \wedge b) = \psi(a) \cap \psi(b).$$

*We say that the embedding **preserves least element** if L has a least element 0 and $\psi(0) = \mathbf{0}$. Similarly one can define ψ **preserves greatest element** if 1 exists in L and $\psi(1) = \mathbf{0'}$.*

So a lattice embedding is an embedding which not only preserves the ordering $\leq$, but also the join and meet operations $\vee$ and $\wedge$.

This is the *diamond lattice*:

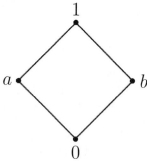

EXAMPLE 12.5.6   *Show that the diamond lattice is embeddable in $\mathcal{E}$ preserving least element.*

SOLUTION   Assume the diamond lattice notated as in the diagram. Take the minimal pair $\mathbf{a}, \mathbf{b}$ from Theorem 12.5.2, and define ψ by: $\psi(0) = \mathbf{0}$, $\psi(1) = \mathbf{a} \cup \mathbf{b}$, $\psi(a) = \mathbf{a}$ and $\psi(b) = \mathbf{b}$.   ⛝

EXERCISE 12.5.7   *Show that the diamond lattice is embeddable into the low c.e. degrees preserving least element.*

[**Hint:** Show that one can carry out the proof of Theorem 12.5.2 with added requirements to make $A \oplus B$ low.]

I could have stated the definition of lattice embedding for other degree structures. But lattice embeddings into $\mathcal{E}$ are specially interesting since they are a basic part of the Σ_2 theory of $\mathcal{E}$.

EXERCISE 12.5.8 *Show that the statement of Theorem 12.5.2 is an $\exists\forall$ sentence of the elementary theory of $\mathcal{E}$.*

The problem of exactly which lattices are embeddable in $\mathcal{E}$ — the *lattice embedding problem* for $\mathcal{E}$ — is a particularly tough one. Until it is better understood there is not much hope of knowing whether the Π_2 theory of $\mathcal{E}$ is decidable or not.

Here is a nice connection with the enumeration degrees below $\mathbf{0}'_e$.

EXERCISE 12.5.9 *Show that if A, B are low Π_1 sets, and $G \in \Sigma_2$ with $G \leq_e A$ and B, then there is a $C \in \Pi_1$ such that $G \leq_e C \leq_e A$ and B.*

Deduce that the diamond is lattice embeddable into the Π_1 enumeration degrees preserving least element.

[**Hint:** Let $G = \Psi_i^A = \Psi_j^B$ and define

$$\langle x, t \rangle \in C \iff (\forall s \geq t)\,[\,x \in \Psi_i^A[s] \vee x \in \Psi_j^B[s]\,],$$

and use the fact that any set c.e. in A or B must be Δ_2. Then consider the natural embedding of the low diamond from Exercise 12.5.7.]

You can see in fact that any lattice embedding into the low c.e. degrees is a lattice embedding in the Π_1 enumeration degrees, under the natural embedding of $\mathcal{D}$ into $\mathcal{D}_e$.

You can use the same sorts of methods we used in embedding the diamond to embed *any* countable distributive lattice with least element 0 into the low c.e. degrees preserving 0. Here are two basic nondistributive lattices which are embeddable preserving least element.

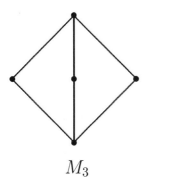

M_3

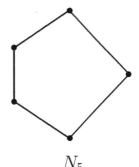

N_5

And here are two non-embeddable lattices.

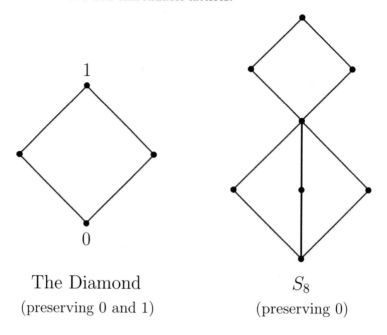

The Diamond

(preserving 0 and 1)

S_8

(preserving 0)

EXERCISE 12.5.10 *Show that there is a minimal pair of properly n-c.e. degrees for each $n \geq 2$.*

So the research programme Shoenfield set in motion still has a long way to go. At least one loose end was tied up by Slaman and Soare in 2000, when they proved a suitably amended version of Shoenfield's faulty conjecture.

12.6   The Π_3 Theory — Information Content Regained

Since the Π_2 theory of $\mathcal{E}$ turns out to be so complicated, you might reasonably expect Σ_3 statements about the c.e. degrees to be *very* hard to decide. Also, you might be thinking that we have come a long way from Post's great aim of using information content to reveal Turing definable relations.

The following statement is actually quite easy to prove. And if you look more closely at the proof, you can extract some general information content which dramatically brings the different approaches to Post's problem back together.

> **THEOREM 12.6.1 (Yates, 1966)**
> *There exists a noncappable c.e. degree* $\mathbf{a} < \mathbf{0}'$.

PROOF Actually we will prove a Π_4 statement. Assume given $C \in \Delta_2$ with $\emptyset <_T C$. We satisfy all requirements of the form:

$$\mathcal{N}_\Phi : \quad C = \Phi^A \Rightarrow C \text{ computable}$$

$$\mathcal{R}_{W,U} : \quad W \text{ computable } \vee (\exists D = \Gamma^A = \Lambda^W)\, [\overline{D} \neq U\,],$$

with D, W, U c.e. sets, Φ, Γ, Λ p.c. functionals, and D, Γ, Λ built by us.

Notice that if all the $\mathcal{N}$ requirements are satisfied, $C \not\leq_T A$, so $\emptyset' \not\leq_T A$. And if $\mathcal{R}_{W,U}$ is satisfied for all U, either W is computable, or $\emptyset <_T D \leq_T A, W$, so the degree of W cannot cap the degree of A.

Presentational convention: Following Lachlan, we can think of functionals such as Γ (or Λ) as c.e. sets of *axioms* of the form "$\Gamma^\sigma(z) = \delta$" with $\delta \leq 1$. So we build Γ by defining such axioms for it at stages $s + 1 > 0$.

In this proof you will notice that the use functions γ, λ of Γ and Λ are very simple, so Γ and Λ are being used to present familiar permitting arguments.

The strategy for $\mathcal{N}_\Phi$

We use Sacks restraints again.

(1) Wait for a $\mathcal{N}_\Phi$ expansionary stage $s + 1$.

(2) Impose a restraint on $A {\restriction} \mathrm{Use}_s(C = \Phi^A)$.

Return to (1).

Analysis of outcomes

$\boxed{w}$ At all stages $>$ than some s^* we wait at (1). $\mathcal{N}_\Phi$ is satisfied.

$\boxed{i}$ $\mathcal{N}_\Phi$ visits (2) infinitely often.

As in the Sacks Splitting Theorem, we argue that the restraints enable us to compute increasingly long beginnings of C, and since C is not computable, this outcome never happens.

The strategy for $\mathcal{R}_{W,U}$

Background activity: Maintain the equation $D(z) = \Gamma^A(z) = \Lambda^W(z)$ for all $z \leq s$ at stage $s + 1$ by defining $\Gamma^{A {\restriction} z+1}(z)[s] = \Lambda^{W {\restriction} z}(z)[s] = D^s(z)$ whenever needed.

(1) Choose a fresh witness x for $\overline{D} \neq U$.

(2) Wait for $x \in U$.

(3) Then wait for $W^{s+1} \restriction x \neq W^s \restriction x$, while concurrently returning to (1) to start with a fresh witness.

(4) When $W^{s+1} \restriction x \neq W^s \restriction x$, so $\Lambda^W(x)[s+1] \uparrow$, *immediately* enumerate x into A^{s+1}, so $\Gamma^A(x)[s+1] \uparrow$.

And define $\Gamma^{A \restriction x+1}(x)[s+1] = \Lambda^{W \restriction x}(x)[s+1] = D^s(x) = 1$.

Throw away all the other witnesses.

Analysis of outcomes

$\boxed{w}$ Some witness x waits forever at (2).

Then $x \in \overline{D} - U$, so $R_{W,U}$ is satisfied.

No more witnesses arrive at (3), so no more witnesses are chosen.

$\boxed{s}$ Some witness x arrives at (4).

Then $D(x) = \Gamma^A(x) = \Lambda^W(x)$ is established at $s+1$ and maintained at all later stages. So $x \in D \cap U$ giving $R_{W,U}$ is satisfied again.

Once again, no more witnesses are chosen for $R_{W,U}$.

$\boxed{i}$ Every witness arrives at (3) but fails to reach (4).

Let $x_0, x_1, \ldots$ be a list of all witnesses, and let $s(x) = $ the least stage $>$ that at which the least $x_i > x$ arrives at (3). Then s dominates the computation function C_W of W. So by the Domination Lemma, W is computable.

Now notice — All outcomes involve at most finitary injury or obstruction to other requirements. So we can inductively argue as usual that each strategy is finitely injured, and so finitely injures. $\Box$

EXERCISE 12.6.2   *Show that every noncappable c.e. degree is noncappable in the n-c.e. degrees* $\mathbf{D}_n$.

REMARK 12.6.3 (Prompt simplicity and noncappable degrees)   We asked remarkably little of A to make its degree noncappable. In the proof of Theorem 12.6.1 we asked that in (4) of the strategy for noncapping, if we got infinitely many witnesses x permitted by W at some stage $s+1$, then at least one such x was *immediately* permitted by A at stage $s+1$.

It turns out that with a little refinement, this is *exactly* the information content of a noncappable degree. For historical reasons, the term used to describe an appropriate A is *promptly simple* — but a more user-friendly version close to what I described is *promptly permitting*. Anyway, the promptly simple degrees **PS** are precisely the noncappable degrees **NC**.

You can use this new type of information content — not quite the sort Post had in mind, but a big step towards it — to show some very nice *alge-*

braic properties of the classes $\mathbf{NC} = \mathbf{PS}$ and $\mathbf{M} =$ the cappable degrees. In particular $\mathbf{NC}$ is a *filter*:

(a) $\mathbf{NC}$ is *closed upwards* — that is, $\mathbf{a} \in \mathbf{NC}$ and $\mathbf{a} \leq \mathbf{b} \Rightarrow \mathbf{b} \in \mathbf{NC}$ (this is immediate, of course), and

(b) Given $\mathbf{a}, \mathbf{b} \in \mathbf{NC}$, we also have $\mathbf{a} \cap \mathbf{b} \in \mathbf{NC}$ if it exists.

And $\mathbf{M}$ is an *ideal* in $\boldsymbol{\mathcal{E}}$:

(a) $\mathbf{M}$ is *closed downwards* — that is, $\mathbf{a} \in \mathbf{M}$ and $\mathbf{a} \geq \mathbf{b} \Rightarrow \mathbf{b} \in \mathbf{M}$ (immediate again), and

(b) Given $\mathbf{a}, \mathbf{b} \in \mathbf{M}$, we also have $\mathbf{a} \cup \mathbf{b} \in \mathbf{M}$.

You should see Soare's book for more details. ⬚

Where do the cuppable degrees fit into this picture? This Σ_3 statement, disproving another ingredient of Shoenfield's Extension of Embeddings Conjecture, is harder to prove:

THEOREM 12.6.4

There exists a noncuppable c.e. degree $\mathbf{a} > \mathbf{0}$.

PROOF We want A c.e. with $A >_T \emptyset$ and $K \not\leq_T A \oplus U$ each c.e. $U <_T K$. So we satisfy all requirements of the form

$$\mathcal{P}_W : \quad \overline{A} \neq W$$
$$\mathcal{R}_{\Theta,U} : \quad D = \Theta^{A,U} \implies K = \Gamma^U$$

where we build the c.e. sets A, D, and the p.c. functional Γ.

Strategy for $\mathcal{P}_W$

What could be simpler than the simple set strategy again?

(1) Choose a fresh witness x.

(2) Wait for $x \in W$.

(3) Enumerate x into A.

Analysis of outcomes

$\boxed{w}$ Wait forever at (2). $x \in \overline{W} - A$, $\mathcal{P}_W$ satisfied.

$\boxed{s}$ Reach (3). $x \in A \cap W$, $\mathcal{P}_W$ satisfied.

Strategy for $\mathcal{R}_{\Theta,U}$

The idea is to maintain $K(y) = \Gamma^U(y)$ by keeping some $\theta(z) < \gamma(y)$. If $K(y)$ changes, we force a $U {\restriction} \gamma(y)$ change by enumerating z into D while at the same time preserving $A \upharpoonright \theta(z)$.

(1) Given y, choose a fresh *agitator* z (apt description) for y, if no such z already exists.

Wait for $\ell_s = \ell_s(D = \Theta^{A,U}) > z$.

Define $\Gamma^U(y) = K(y)$ with $\gamma(y) > \theta(z)$. Restrain $A \upharpoonright \theta(z)$.

(2) If $U \upharpoonright \theta(z)$ changes, throw away the A restraint and return to (1).

(3) If $y \in K^{\text{at } s+1}$, define $D(z) \neq \Theta^{A,U}$.

Wait for a $U \upharpoonright \theta(z)$ change.

(4) Rectify $K(y) = \Gamma^U(y)$, and maintain this at later stages.

Analysis of outcomes

$\boxed{w_y}$ We wait forever at (1) for y. $D \neq \Theta^{A,U}$, so $\mathcal{R}_{\Theta,U}$ is satisfied.

$\boxed{i_y}$ The strategy loops infinitely often through (2) and back to (1) on behalf of the agitator z for y. This means that $\Theta^{A,U}(y) \uparrow$, and $\mathcal{R}_{\Theta,U}$ is satisfied again.

$\boxed{w_2}$ We wait forever at (3). Then $D \neq \Theta^{A,U}$ again, giving $\mathcal{R}_{\Theta,U}$ satisfied.

$\boxed{s_1}$ The infinitary outcome $K = \Gamma^U$ is successfully progressed at y.

The tree of outcomes

As for the Minimal Pair Theorem 12.5.2 the $\mathcal{P}$ nodes will just have a combined outcome $\boxed{s}$ and will not branch.

This allows us to combine $\boxed{w_1}$, $\boxed{w_2}$ into $\boxed{w}$, and then order the outcomes for $\mathcal{R}_{\Theta,U}$ very sensibly:

$$\boxed{i} <_L \boxed{s_1} <_L \boxed{w}.$$

Notice — These are only the outcomes that z "sees" (more relativity) — to make this clear we could call them $\boxed{i}(z)$, $\boxed{s_1}(z)$ and $\boxed{w}(z)$. The full tree would include outcomes

$$\boxed{i}(0) <_L \boxed{i}(1) <_L \cdots <_L \boxed{s_1}(0) <_L \ldots \boxed{s} <_L \boxed{w}(0) <_L \ldots.$$

The eventual true path will extend some $\boxed{i}(z)$, $\boxed{w}(z)$, or the outcome $\boxed{s_1}$ as the limit of outcomes $\boxed{s_1}(0)$, $\boxed{s_1}(1)$, $\ldots$.

Strategy for $\mathcal{P}_W$ below $\mathcal{R}_{\Theta,U}$

In order for $\mathcal{P}_W$ to carry out its strategy, it must also "take charge" of the higher priority $\mathcal{R}_{\Theta,U}$ strategy. $\mathcal{P}_W$ sees the outcomes the agitator z of its current witness x sees. $\boxed{i}$ is *true* at $s+1$ if the $\mathcal{P}_W$ strategy loops through (2) of the $\mathcal{R}_{\Theta,U}$ strategy, $\boxed{s_1}$ at $\mathcal{R}_{\Theta,U}$ expansionary stages, and $\boxed{w}$ if $\mathcal{P}_W$ is stuck at (1) or (3).

The $\mathcal{P}_W$ strategy is just as before below outcome $\boxed{i}$ or $\boxed{w}$. In the former case it acts when $\mathcal{R}_{\Theta,U}$ discards its A restraint, and in the latter it acts above the finitary permanent A restraint.

The only problem $\mathcal{R}_{\Theta,U}$ with true outcome $\boxed{s}$ presents for $\mathcal{P}_W$ is that $\boxed{s_1}(0)$, $\boxed{s_1}(1)$, ... may entail infinitely many A restraints. So we can only ask $\mathcal{P}_W$ to protect finitely many such outcomes.

This means all we need do is share out the A restraints amongst the $\mathcal{P}$ requirements, instead of locating them all at $\mathcal{R}_{\Theta,U}$.

Specifically, choosing x fresh ensures $x > \mathrm{Use}_s(D = \Theta^{A,U})$. So $\mathcal{P}$ acting at expansionary stages — and imposing A restraints to protect all *existing* agitators when x is chosen from lower priority requirements — will have done its bit to progress outcome $\boxed{s}$. That is, if z is the greatest agitator such that x is chosen $> \theta(z)$, $\mathcal{P}_W$ protects outcomes $\boxed{s_1}(0), \dots \boxed{s_1}(z)$. Of course, x will be thrown away if a U change later brings x into conflict with $\boxed{s_1}(0), \dots \boxed{s_1}(z)$. But if infinitely many such x get thrown away, then $\boxed{i}(z')$, some $z' \le z$, is the true outcome.

Strategy for $\mathcal{P}_W$ below $\mathcal{R}_{\Theta,U}$ below $\mathcal{R}_{\Theta',U'}$

To put all the strategies together, there are timing problems to deal with as in building a minimal pair. With more than one $\mathcal{R}$ requirement above $\mathcal{P}_W$, $\mathcal{P}_W$ must observe expansionary stages relatively again, so that $\mathcal{P}_\alpha$ on the true path will get infinitely many α-true stages again. Also — given $\boxed{i}(z)$ for $\mathcal{R}_{\Theta,U}$ below $\boxed{i}(z')$ for $\mathcal{R}_{\Theta',U'}$, both on the true path, $U \upharpoonright \theta(z)$ changes must be observed by $\mathcal{P}_\alpha$ relative to stages at which A restraints for $\mathcal{R}_{\Theta',U'}$ are thrown away.

You should now have enough to complete the proof. As before you should check:

EXERCISE 12.6.5 *There does exist an infinite left-most path which is travelled without bound during the building of A. And the strategies along this true path do satisfy the requirements.*

If you want to do this formally, you should use an induction along the true path. You should really attempt to do that, as you will learn a lot from it — certainly more than from such details passively received. $\quad\square$

REMARK 12.6.6 It is, of course, too much to expect that **PS** should coincide with the cuppable degrees. You only have to observe how the promptness of the A changes in the above proof is inhibited by the temporary restraints, used to try and make the agitators stay honest, to give up on that hope.

But here is a thought. Say we get rid of the infinitary outcome $\boxed{i}$ by only bothering with U for which $\mathrm{Lim}_s\,\Theta^{A,U}(x)[s]$ always exists. We can do this by only considering U low. Then A being promptly simple *is* enough.

And this is where the picture becomes even more pleasing. Define **LC** = the set of c.e. degrees $\mathbf{a}$ cuppable by a low c.e. $\mathbf{b}$ — $\mathbf{a} \cup \mathbf{b} = \mathbf{0}'$ and $\mathbf{b}' = \mathbf{0}'$. Then

$$\mathbf{LC} = \mathbf{PS} = \mathbf{NC}.$$

What an extraordinary result. It is not surprising that prompt simplicity appears to be a crucial notion for the Π_2 theory of $\mathcal{E}$. $\qquad\Box$

EXERCISE 12.6.7 (Arslanov)   *Let $\mathbf{0} < \mathbf{a}$ c.e. Show that there is a d.c.e. degree $\mathbf{d}$ such that $\mathbf{a} \cup \mathbf{d} = \mathbf{0}'$.*

Deduce that every d.c.e. $\mathbf{a} > \mathbf{0}$ is cuppable in $\mathbf{D}_2$.

[**Hint:** Build E c.e., $D \in \mathbf{D}_2$ and Γ p.c. satisfying

$$\mathcal{S}: \quad K = \Gamma^{A,D}$$
$$\mathcal{N}_\Phi: \quad E \neq \Phi^D$$

for all p.c. Φ, and where $\Gamma^{A,D}$ uses the two oracles A and D to make $K \leq_T A \oplus D$. For $\mathcal{S}$, if $y \in K^{\mathrm{at}\,s+1}$, rectify $K(y) = \Gamma^{A,D}(y)$ by changing $A{\upharpoonright}\gamma(y)$ or $D{\upharpoonright}\gamma(y)$.

Use a *moving marker* strategy for $\mathcal{N}_\Phi$. Choose a *threshold* z and a witness x for $\mathcal{N}_\Phi$. Move all $\gamma(y)$, $y \geq z$, above $\varphi(x)$ whenever $\Phi^D(x) {\downarrow}^{\mathrm{at}\,s+1}$, so we can rectify $K(y) = \Gamma^{A,D}(y)$ for $y \geq z$ without injuring the computation of $\Phi^D(x)$. Do this first using an enumeration into D when $\Phi^D(x) {\downarrow}^{\mathrm{at}\,s+1}$. Use an eventual $A{\upharpoonright}\gamma(z)$ change — guaranteed by $A >_T \emptyset$ — to rectify $K(y) = \Gamma^{A,D}(y)$, $y \geq z$, while restoring $D{\upharpoonright}\varphi(x)$, and enumerate x into E.]

EXERCISE 12.6.8   *Show every nonzero $\mathbf{a} \in \mathbf{D}_2$ is low cuppable in $\mathbf{D}_2$.*

It is now time to look at infinite injury. If you successfully negotiated Theorem 12.6.4, including Exercise 12.6.5, you have nothing to fear.

12.7   Higher Priority and Maximal Sets

Our last section in this chapter, a chapter which is already the longest in the book, does a number of good things. Most importantly, it proves a link between degree class and information content in the spirit of Post — we discover the lattice invariance of the high c.e. degrees. Also very useful, it introduces you to infinite injury priority arguments. And we learn some more about the lattice $\mathcal{E}$ of c.e. sets, in particular that it has *maximal* elements, so is not densely ordered. We will start with the promised construction of a maximal set (Definition 12.2.23), which prepares the ground for Theorem 12.7.6 which follows it.

THEOREM 12.7.1

There exists a maximal set A.

PROOF   We need to find a constructive version of Theorem 12.2.29.

The 1958 proof of Richard Friedberg used an early form of tree argument, called an *e-state construction*. I will present it in the tree framework we have been using. The requirements to satisfy are of the form

$$S: \quad \overline{A} \text{ is infinite,}$$

$$\mathcal{R}_W: \quad \overline{A} - W \text{ is finite } \vee \; \overline{A} \cap W \text{ is finite,}$$

with W c.e.

The strategy for S

At stage $s+1$ every level $n+1$ of the tree will have a *resident* number a_n^s. To start with, level $n+1$ will have resident $a_n^0 = n$. During the construction we may throw some segment of residents $[a_n^s, a_{n+k-1}^s]$ into A, defining $a_i^{s+1} = a_{i+k}^s$ for $i \geq n$ — we say a_{n+k}^s *evicts* a_n^s.

We satisfy S by making $\text{Lim}_s\, a_n^s$ exist for all n.

The strategy for $\mathcal{R}_W$

Assume the priority listing puts $\mathcal{R}_W$ at level $n+1$ of the tree. $\mathcal{R}_W$ will try to ensure all residents of levels $> n$ are in W. Let

$$\ell_s = \ell_s(\overline{A} - W \text{ finite}) = \mu i \geq n\, [a_i^s \notin W].$$

(1) Wait for an $a_i^s \in W$ with $i > \ell_s$.

(2) Throw the interval $[a_{\ell_s}^s, a_{i-1}^s]$ of residents into A^{s+1}. a_i^s evicts a_{ℓ_s}.

Analysis of outcomes

$\boxed{0}$ The strategy visits (2) infinitely often. ℓ_s is unbounded with s, and no member of the interval $[a_n^s, a_{\ell_s}^s]$ ever gets evicted by the strategy. Since $[a_n^s, a_{\ell_s}^s] = [a_n, a_{\ell_s}] \subset \overline{A} \cap W$ for all s, we get both S and $\mathcal{R}_W$ satisfied via $\overline{A} - W$ finite.

$\boxed{1}$ The strategy gets stuck at (1). Then all evictions cease after some stage s^*, and all limits a_i exist again, satisfying S. And every $a_i \notin W$ for $i \geq \ell_{s^*}$, so $\mathcal{R}_W$ is satisfied via $\overline{A} \cap W$ finite.

The tree of outcomes

It is now clear the tree is binary branching with outcomes $\boxed{0} <_L \boxed{1}$ at each node.

Outcome $\boxed{0}$ for $\mathcal{R}_\alpha = \mathcal{R}_W$ is *true at stage $s+1$* if $s+1$ is *expansionary* — that is, ℓ_s increases at stage $s+1$. Otherwise $\boxed{1}$ is true at stage $s+1$.

We say α is *true for a_i^s at stage $s+1$* if for each $\beta \subset \alpha$, with $\mathcal{R}_\beta = \mathcal{R}_{W'}$ say, we have $a_i^s \in W_s' \iff \beta^\frown\boxed{0} \subseteq \alpha$.

All the strategies together

When we include the strategies at every level, the procedure is to give the leftmost possible path through the tree the opportunity to evict numbers. So we have the possibility of a_n^s being evicted through a number of different $\mathcal{R}$ requirements. The important thing is that a_n^s can only be evicted by requirements at levels above $n+1$, and for each of those, each time they evict they go through an expansionary stage — so for each such requirement, ℓ_s is eventually $> n$, or there are no more evictions. So $\mathcal{S}$ is still satisfied, and then the satisfaction of each $\mathcal{R}_W$ follows as before, with true outcomes those along the leftmost path, where each α on the true path will be true for every large enough a_i. ◻

Actually, you will notice that although the $\mathcal{R}$ requirements are active enough for us to call this an infinite injury argument, there are no restraints for an $\mathcal{R}_W$ to injure. If you feel cheated, try the following.

EXERCISE 12.7.2 (Sacks, 1964)   *Show that there exists a maximal set of degree $< 0'$.*

[**Hint:** Add the requirements

$$\mathcal{N}_\Phi : \quad E \neq \Phi^A,$$

where you build E. Below outcome $\boxed{0}$ for $\mathcal{R}_W$ you need to use A restraints $< \ell_s$. Below outcome $\boxed{1}$ you use A restraints after all evictions have finished.]

While you are at it, you might like to prove the sister result of this. It is the one which really meant an end to Post's approach to incompleteness in the c.e. degrees via immunity properties.

EXERCISE 12.7.3 (Yates, 1965)   *Show that there exists a maximal set $A \in 0'$.*

[**Hint:** Use Exercise 12.2.21, and in the proof of Theorem 12.7.1, make $a_n > \varphi_i(n)$ whenever $\varphi(n)\downarrow$ and $n > i$.]

We are almost ready for our main result. But we also need a constructive version of the earlier Theorem 12.2.28 of Martin. It was R.W. Robinson who first had the bright idea of making computation functions do the dominating, so providing the basis for the *high permitting* method.

DEFINITION 12.7.4 *We say a c.e. approximating sequence $\{A^s\}_{s \geq 0}$ for A is a Σ_1 high approximating sequence if the corresponding computation function C_A dominates all computable functions.*

LEMMA 12.7.5
Given any high c.e. $\mathbf{a}$ there is a $B \in \mathbf{a}$ with Σ_1 high approximating sequence.

PROOF   Using Theorem 12.2.28, let $A \in \mathbf{a}$ be c.e. with Θ^A dominating all computable functions.

We build D c.e. satisfying the requirements

$$\mathcal{S}: \quad D \leq_T A$$

$$\mathcal{R}_i: \quad \varphi_i \text{ computable} \implies C_{D \oplus A} \text{ dominates } \varphi_i.$$

We can then take $B = D \oplus A \geq_T A$ with all the right properties.

The strategy for $\mathcal{R}_i$

We use $\{\langle i, x \rangle \mid x \geq 0\}$ to satisfy $\mathcal{R}_i$. The idea is to enumerate $\langle i, x \rangle$ into D late enough to ensure C_D — or rather $C_{D \oplus A}$ — dominates φ_i on a whole segment of arguments.

(1) For each k, wait for $s + 1$ such that $\varphi_i(y) \downarrow [s] < s$ for all y with $2\langle i, k \rangle \leq y \leq 2\langle i, k+1 \rangle + 1$.

(2) Enumerate $\langle i, k \rangle$ into D^{s+1}. Write $s_i(k) = $ this $s + 1$.

Analysis of outcomes

$\boxed{w}$ We get stuck at (1) for some k. Then φ_i is not total, so $\mathcal{R}_i$ is satisfied.

$\boxed{s}$ We reach (2) for every k. So φ_i is total and $C_{D \oplus A}(y) > \varphi_i(y)$ for every $y > \langle i, 0 \rangle$.

The strategy for $\mathcal{S}$

In (2), we only allow $\langle i, k \rangle$ to be enumerated into D^{s+1} if

$$\Theta^A(k)[s] \downarrow \ \& \ (\forall t < s)\,[\, A{\upharpoonright}\theta(k)[s] \neq A{\upharpoonright}\theta(k)[t]\,].$$

Outcome

If φ_i is computable, then so is s_i. Let

$$\widehat{C}_A(k) = \mu s\,[\,\Theta^A(k)[s] \downarrow \,\&\, A{\upharpoonright}\theta_s(k) = A{\upharpoonright}\theta(k)[s]\,].$$

Then $\widehat{C}_A$ dominates s_i — since otherwise we could compute $\Theta^A(k)$ for infinitely many k using $s_i(k)$, contradicting Θ^A dominating every computable function.

This means the $\mathcal{S}$ strategy only obstructs the $\mathcal{R}_i$ strategy from enumerating into D finitely many members of $\{\langle i, x\rangle \,|\, x \geq 0\}$.

The lemma follows.   ⬚

We really are ready now to prove:

THEOREM 12.7.6 (Martin, 1966)
Every high c.e. degree contains a maximal set. Hence the set of high c.e. degrees is lattice invariant.

PROOF   Let **h** be high, with H c.e. $\in$ **h**. By Lemma 12.7.5 we can assume a Σ_1 high approximating sequence $\{H^s\}_{s \geq 0}$ for H.

All we have to do is add high permitting below H to the construction of the maximal A in Theorem 12.7.1, along with some extra "reverse permitting" to make $H \leq_T A$.

The $\mathcal{R}_W$ strategy with high permitting

If we only allow an eviction of an a_i^s at stage $s+1$ if $H^{s+1}{\upharpoonright}i \neq H^s{\upharpoonright}i$, then $A \leq_T H$ — since we get $a_i^s \in \overline{A} \Leftrightarrow H^{s+1}{\upharpoonright}i = H{\upharpoonright}i$. We now let

$$\ell_s = \mu i \geq n\,[a_i^s \notin W \,\&\, a_i^s \text{ is not due for eviction}],$$

where (see below) a_i^s becomes due for eviction before being permitted by H to actually be evicted.

(1) Wait for an $a_i^s \in W$ with $i > \ell_s$.

(2) All residents in the interval $[a_{\ell_s}^s, a_{i-1}^s]$ become *due for eviction*, and a_i^s starts *waiting to evict* them.

Write $\ell_s = k$.

Wait for a later stage $s+1$ at which $H^{s+1}{\upharpoonright}k \neq H^s{\upharpoonright}k$.

(3) Throw the interval $[a_k^s, a_{i-1}^s]$ of residents into A^{s+1}. a_i^s evicts a_k.

The outcomes

$\boxed{0}$ As previously, the strategy visits (2) infinitely often. ℓ_s is unbounded with s, and no member of the interval $[a_n^s, a_{\ell_s}^s]$ ever gets evicted by the strategy. Since $[a_n^s, a_{\ell_s}^s] = [a_n, a_{\ell_s}]$ for all s, we get $\mathcal{S}$ satisfied as before.

For $\mathcal{R}_W$, we must examine the high permitting. For each a_i^s, let $s^*(i) =$ the stage at which a_i^s becomes due for eviction — so $s^*(i) \downarrow$ for all but a finite number of $i \geq 0$. So C_H dominates s^*, giving $H^{s+1} \restriction i \neq H^s \restriction i$ for some $s > s^*$ for all but finitely many numbers i. The outcome is now much as before.

We have $[a_n^s, a_{\ell_s}^s] = [a_n, a_{\ell_s}] \subset \overline{A} \cap W$ for all $s >$ some t, and get $\mathcal{R}_W$ satisfied via $\overline{A} - W$ finite again.

$\boxed{1}$ The strategy gets stuck at (1). The analysis is just as before, with $\mathcal{S}$ satisfied and $\mathcal{R}_W$ satisfied via $\overline{A} \cap W$ finite.

The reverse permitting

As well as the evictions, we will want to throw extra residents a_i^s into A to get $H \leq_T A$. We just have to be careful not to sabotage the $\mathcal{R}$ strategies by throwing in all the residents waiting to evict before they get H permission to do the evicting. We add on at stage $s + 1$:

Let α_i^s be the longest string of outcomes true for a_i^s at stage $s + 1$. Assume $H^{s+1} \restriction z \neq H^s \restriction z$ with z least. Throw a_z^s into A^{s+1} — unless α_z^s is waiting to evict some a_i^s and $\alpha_z^s \leq_L \alpha_{z+1}^s$, in which case throw a_{z+1}^s into A.

The outcomes

Since we have

$$H^{s+1} \restriction z = H \restriction z \iff (a_z^{s+1}, a_{z+1}^{s+1}) = (a_z, a_{z+1}),$$

we get $H \leq_T A$.

And since we choose to throw a_{z+1}^s into A if α_z^s is waiting to evict some a_i^s and $\alpha_z^s \leq_L \alpha_{z+1}^s$, we manage to protect the $\mathcal{R}$ strategies along the eventual true path. $\qquad\Box$

EXERCISE 12.7.7 *Given a Δ_2-approximating sequence $\{A^s\}_{s \geq 0}$ for a set $A \leq_T \emptyset'$ for which C_A dominates all computable functions, we say that $\{A^s\}_{s \geq 0}$ is a Δ_2 **high approximating sequence**.*

Show that $\mathbf{a} \leq \mathbf{0}'$ is high if and only if there is an $A \in \mathbf{a}$ with Δ_2 high approximating sequence for A.

EXERCISE 12.7.8 *Given a Σ_2-approximating sequence $\{A^s\}_{s \geq 0}$ for the set $A \leq_e \overline{K}$, define the corresponding **computation function** by*

$$C_A(x) = \mu s \geq x \, [\, A^s \restriction x \subseteq A \,].$$

(a) Show that if C_A is dominated by some computable f, then A is c.e.

(b) If we also have a Σ_2-approximating sequence $\{B^s\}_{s \geq 0}$ for B, show that if C_A dominates C_B, then $B \leq_e A$.

*(c) If C_A dominates all computable functions, we say $\{A^s\}_{s \geq 0}$ is a Σ_2 **high approximating sequence** for A.*

Show that an enumeration degree $\mathbf{a} \leq \mathbf{0}'_e$ is high if and only if there is an $A \in \mathbf{a}$ with Σ_2 high approximating sequence.

REMARK 12.7.9 (Classifying priority arguments) There is a much more informative classification of priority arguments than the old division into finite and infinite injury. If you only think in terms of injuries, where does that put the minimal pair or noncuppable degree constructions? The construction of a minimal pair, or a noncuppable degree, is not finite injury in the sense that some requirements injure restraints infinitely often. But it cannot be framed as a finite injury argument either, since some requirements generate infinitely many restraints. The discovery in the 1970s of priority constructions of a new scale of difficulty demanded a new way of measuring complexity of constructions. And that came out of Harrington's work clarifying the new tree constructions.

If you put a finite injury argument on a tree, you can identify the true path using a $\emptyset'$ oracle — so call the construction a $\emptyset'$ *priority argument*. Similarly, an infinite injury construction — or the minimal pair or noncupping constructions — turns out to be $\emptyset''$ *priority arguments*. You need $\emptyset''$ to compute their true paths. And — the origin of all this work on trees — Lachlan's "monster" priority construction (I will leave you to find out what that is!) involves finite injury along a $\emptyset''$ true path, so becomes a $\emptyset'''$ *priority argument* in the new framework.

What is $\emptyset^{(4)}$ priority needed for? The jury is still out on that one. An interesting footnote is that Lempp and Lerman have found a way of analysing $\emptyset^{(n)}$ priority arguments in terms of lower priority. ⌷

Although I have spent a lot of time on priority arguments and their applications, I have not managed to more than scratch the surface of what is probably the deepest and most technically complex area of the whole subject. Rather than try to be comprehensive — a ridiculous aim! — I have tried to introduce you to an *approach* which will mentally prepare you for further reading or research. If you want to find out more, you should go to the specialist books of Soare or Odifreddi. But you should be warned — some of the most important and interesting results are still buried in barely readable research papers or, worse, still in privately circulated rough notes.

Chapter 13

Forcing and Category

Forcing was invented by Paul Cohen and first used in 1963 to show that *Cantor's Continuum Hypothesis* (CH) cannot be decided from what we know to be true about sets. Hopefully you will remember CH from our earlier Remarks 3.1.16 and 8.2.9. More precisely, what Cohen did was devise a way of building a model of the usual axioms of set theory — that is, of the system ZFC of Zermelo–Fraenkel set theory with the Axiom of Choice — and *forcing* it to be a model in which CH was not true. Since Gödel had already built a model of ZFC — specifically the universe of sets *constructible* from the usual set theoretic operations — in which CH *is* true, neither CH nor its negation could possibly be provable within ZFC.

Forcing can be applied in more homely contexts. Solomon Feferman noticed as early as 1965 that you could force relative to the first order language of arithmetic. And it soon became obvious that there were nice connections between forcing and other elegant frameworks — such as those using measure and category to put Kleene and Post's discoveries in context.

The importance of forcing and other such ways of looking at things is partly *presentational*. Computability theory is plagued with technical messiness and ad hoc methods. These frameworks can unify whole areas of theory, and focus our attention on the essential similarities between what at first seem diverse ways of proving different results.

13.1 Forcing in Computability Theory

Let us first look at Feferman's take on forcing in arithmetic. In Section 13.3 we will *really* bring things down to earth — in fact, you could even jump to that section if you wanted to without too many problems.

What Feferman did was use the language of arithmetic, with some added set constant symbols, to describe properties of sets of numbers. Interpreted relative to true arithmetic, we can get various possible models for the set constants. The ones we will focus on will be those built *generically* in a very basic way using finitary *forcing* to decide what is true about them in the language.

So the basic language we will use will be the usual first order language $\mathcal{L}_{\mathrm{PA}}$ for $\mathfrak{N} = \langle \mathbb{N}, ', +, \times, = \rangle$ described in Section 3.1. We will also allow set constants $S_0, S_1, \ldots$, and atomic wfs of the form $t \in S_i$, with t a term of $\mathcal{L}_{\mathrm{PA}}$ — call this extended language $\mathcal{L}_{\mathrm{PA}}^{(s)} = \mathcal{L}_{\mathrm{PA}}(S_0, S_1, \ldots)$. "$s$" is for "sets" of course. For convenience, I will not distinguish unnecessarily between the language and its interpretation. For instance, I will allow a statement of the form $k \in S_i$ both as a wf of $\mathcal{L}_{\mathrm{PA}}^{(s)}$ and as a statement about the interpretations of k and S_i.

A *forcing condition* Q is a finite set of membership statements of the form $k \in S_i$ or $k \notin S_i$.

If a sequence $Q_0 \subset Q_1 \subset \ldots$ of conditions is *complete* — that is, if for each k, i either $(k \in S_i) \in \cup_{j \geq 0} Q_j$ or $(k \notin S_i) \in \cup_{j \geq 0} Q_j$ — then we can extract from it a fixed sequence of sets $S_0, S_1, \ldots$. This is how we get a generic $Q_0 \subset Q_1 \subset \ldots$, which will be complete with corresponding generic $S_0, S_1, \ldots$.

DEFINITION 13.1.1 (The forcing relation) *Let Q be a forcing condition, and ψ a wf of $\mathcal{L}_{\mathrm{PA}}^{(s)}$. We inductively define $Q \Vdash \psi$ — that is, Q* **forces** *ψ — by*

   (1) $Q \Vdash$ each member of Q,

   (2) $Q \Vdash$ each true atomic wf of PA,

   (3) $Q \Vdash \varphi \vee \psi \iff Q \Vdash \varphi$ or $Q \Vdash \psi$,

   (4) $Q \Vdash \neg\varphi \iff$ there is no $Q' \supseteq Q$ for which $Q' \Vdash \varphi$,

   (5) $Q \Vdash (\exists x)\varphi(x) \iff Q \Vdash \varphi(k)$ for some k.

   We say $Q_0, Q_1, \ldots$ is **generic** *if $\cup_{i \geq 0} Q_i \Vdash \varphi$ or $\Vdash \neg\varphi$, each wf φ — in which case we also call $\cup_{i \geq 0} Q_i$* **generic**.

   And $S_0, S_1, \ldots$ is **generic** *if $D = \{k \in S_i \,|\, k \in S_i\} \cup \{k \notin S_i \,|\, k \notin S_i\}$ is generic. In particular, we can get the notion of S being* **generic** *by limiting ourselves to just one set constant S.*

Notice — I have used here the well-known fact that any wf is logically equivalent to one using only connectives $\neg$ and $\vee$ and quantifier $\exists$. D is for the *diagram* of $S_0, S_1, \ldots$, by the way. We usually identify a given $S_0, S_1, \ldots$ with its diagram for the purposes of forcing. So we write $S_0, S_1, \cdots \Vdash \psi$ for $D \Vdash \psi$.

This is the basic Existence Theorem for generic sets:

THEOREM 13.1.2 (Feferman, 1965)
There exists a generic $S_0, S_1, \dots$.

PROOF Inductively construct a generic $Q_0, Q_1, \dots$.

Assume a listing $\{\psi_i\}_{i \geq 0}$ of all wfs of $\mathcal{L}_{\mathrm{PA}}^{(s)}$ — easily got from a suitable Gödel numbering, for instance. Let $Q_0 = \emptyset$. Assume $Q_0, \dots, Q_m$ already defined with $Q_{i+1} \Vdash \psi_i$ or $\Vdash \neg \psi_i$ each $i < m$.

If $Q_m \Vdash \neg \psi_m$, just take $Q_{m+1} = Q_m$. Otherwise, by part (4) of Definition 13.1.1 we can take some $Q_{m+1} \supseteq Q_m$ such that $Q_{m+1} \Vdash \psi_m$. $\quad\square$

Here is another basic property of the forcing relation: in the limit, forced statements are true.

Let $\mathfrak{M} = \mathfrak{M}(\mathrm{arith}, S_0, S_1, \dots) = \langle \mathfrak{N}, S_0, S_1, \dots \rangle$ be the standard model $\mathfrak{N}$ of PA with the sets $S_0, S_1, \dots$. If ψ of $\mathcal{L}_{\mathrm{PA}}^{(s)}$, we write $S_0, S_1, \cdots \models \psi$ for $\mathfrak{M}(\mathrm{arith}, S_0, S_1, \dots) \models \psi$. Then:

THEOREM 13.1.3
Let $S_0, S_1, \dots$ be generic. Then for each wf ψ of $\mathcal{L}_{\mathrm{PA}}^{(s)}$

$$S_0, S_1, \cdots \models \psi \iff S_0, S_1, \cdots \Vdash \psi.$$

PROOF We show by induction on the complexity of ψ that $S_0, S_1, \cdots \models \psi \iff$ some finite subset Q of D forces ψ.

The base of the induction is immediate from Definition 13.1.1.

We have $D \Vdash \psi$ atomic $\iff$ ψ is a true atomic statement about $\mathfrak{N}$, or ψ is a true statement of the form $k \in S_i$ — in either case this being if and only if $\mathfrak{M} \models \psi$.

For the inductive step I will just do $\neg \psi$, and leave you to fill in the details for $\varphi \vee \psi$ and $(\exists x) \varphi(x)$.

($\Leftarrow$) Say $Q \Vdash \neg \psi$, for some finite $Q \subset D$.

I will show that no $Q' \subset D$ forces ψ — from which it will follow by the inductive hypothesis that $S_0, S_1, \cdots \not\models \psi$, so that $S_0, S_1, \cdots \models \neg \psi$, which is what we want.

Say $Q' \subset D$. Then $Q' \cup Q \supseteq Q$, giving $Q' \cup Q \Vdash \neg \psi$.
But we also have $Q' \cup Q \supseteq Q'$ — so $Q' \not\Vdash \psi$.

($\Rightarrow$) Say $S_0, S_1, \cdots \models \neg \psi$.
Since D is generic, $D \Vdash \neg \psi$ or $D \Vdash \psi$.

Now say $Q \Vdash \psi$, some finite $Q \subset D$ — then by the inductive hypothesis, $S_0, S_1, \cdots \models \psi$, a contradiction.

Now it is your turn.

EXERCISE 13.1.4 *Assume* $S_0, S_1, \cdots \models \varphi \iff D \Vdash \varphi$ *and* $S_0, S_1, \cdots \models \psi \iff D \Vdash \psi$. *Show that:*

(a) $S_0, S_1, \cdots \models \varphi \vee \psi \iff D \Vdash \varphi \vee \psi$, *and*

(b) $S_0, S_1, \cdots \models (\exists x)\varphi(x) \iff D \Vdash (\exists x)\varphi(x)$.

You have now completed the induction. ▯

It now follows that whatever properties generic sets may have, being describable in elementary arithmetic is not one of them.

COROLLARY 13.1.5

If S is generic then S is not arithmetical.

PROOF Let S be generic, and let $\psi(x)$ be a wf of PA such that

$$S \models [\psi(x) \Leftrightarrow x \in S].$$

So by Theorem 13.1.3 $S \Vdash [\psi(x) \Leftrightarrow x \in S]$, and hence for some string $\sigma \subset S$ we have

$$\sigma \Vdash [\psi(x) \Leftrightarrow x \in S].$$

Choose a generic $\widehat{S} \supset \sigma$ such that $\widehat{S} \neq S$. Then $\widehat{S} \models \neg[\psi(x) \Leftrightarrow x \in S]$ — and by Theorem 13.1.3 again

$$\tau \Vdash \neg[\psi(x) \Leftrightarrow x \in S],$$

for some $\tau \subset \widehat{S}$,

But $\tau \supset \sigma$, contradicting the definition of $\Vdash$. ▯

EXERCISE 13.1.6 *Show that no generic set can be cohesive.*

The next example gives an easy application of the corollary and of Theorem 13.1.3.

> **DEFINITION 13.1.7** *Let $R(S)$ be a property of sets $S \subseteq \mathbb{N}$.*
> *We say R is **arithmetical** if it can be expressed in the language $\mathcal{L}_{\mathrm{PA}}(S)$.*
> *If R is arithmetical, we say that the collection $\mathcal{C}$ of sets S satisfying R is* **arithmetical**.

Intuitively — just as a set is arithmetical if it can be described in first order arithmetic, just the same applies to a relation on sets. Notice that you can also relativise this definition to a given set A. So a collection $\mathcal{C}$ of sets is *arithmetical in A* if it can be described in $\mathcal{L}_{\mathrm{PA}}(S, A)$.

EXERCISE 13.1.8 *Show that the set of all cohesive sets is arithmetical.*

> **EXAMPLE 13.1.9** *Show that the relation*
>
> $$\mathcal{A}(S) \iff_{\mathrm{defn}} S \text{ is arithmetical}$$
>
> *is not arithmetical.*

SOLUTION Assume otherwise, and take S generic.

By Corollary 13.1.5 S is not arithmetical. Then $S \models \neg\mathcal{A}(S_0)$, giving $S \Vdash \neg\mathcal{A}(S_0)$. Hence, $\sigma \Vdash \neg\mathcal{A}(S_0)$ for some $\sigma \subset S$.

But there does exist an arithmetical set $\widehat{S} \supset \sigma$ — which means that $\widehat{S} \models \mathcal{A}(S_0)$, giving $\tau \Vdash \mathcal{A}(S_0)$ for some $\tau \subset \widehat{S}$.

But we can choose a $\pi \supset \sigma$ and $\supset \tau$, giving $\pi \Vdash \mathcal{A}(S_0)$ *and* $\pi \Vdash \neg\mathcal{A}(S_0)$, a contradiction. $\qquad\Box$

REMARK 13.1.10 Just as we developed the arithmetical hierarchy for the arithmetical sets of numbers, we can do the same for the set of all arithmetical sets of sets. And just as before, the bottom of this hierarchy can be characterised as its computable members — where we can define a collection $\mathcal{C}$ of sets to be *computable* if there is an algorithm which enables us to decide whether $A \in \mathcal{C}$ via a computation using a finite number of oracle questions to A. We can make this more precise by saying $\mathcal{C}$ is computable if there is some p.c. Φ such that $\lambda X\, \Phi^X(0)$ is total, and $X \in \mathcal{C} \Leftrightarrow \Phi^X(0) = 1$. Then as before we inductively get the notion of a set $\mathcal{C} \subseteq 2^\omega$ being Σ_n^0 or Π_n^0.

You can also define the arithmetical sets of *functions* $\mathbb{N} \to \mathbb{N}$ similarly to how we did for sets, and describe the corresponding hierarchy of arithmetical sets of functions.

We will look more closely at the level corresponding to one universal quantifier when we look at Π_1^0 classes of functions and sets in Chapter 15.   ⬚

EXERCISE 13.1.11 *Let $\mathcal{C} \subseteq 2^\omega$ be a collection of sets, such that $S \in \mathcal{C} \Leftrightarrow (\exists x \in \mathbb{N})\, R(S, x)$, where the relation R of S, x can be written in $\mathcal{L}_{\mathrm{PA}}(S)$ without quantifiers.*

Show that $I_{\mathcal{C}} = \{i \,|\, W_i \in \mathcal{C}\}$ is a Σ_2 set.

What do you get if you *fix* some of the sets $S_0, S_1, \ldots$ in the definition of genericity and ask the remaining sets to do the forcing?

Let us consider forcing conditions Q which now consist of membership statements about only S_0 — that is, ones of the form $k \in S_0$ or $k \notin S_0$. Say we then choose fixed sets $S_1, S_2, \ldots$, and define the forcing relation $Q \Vdash \psi \in \mathcal{L}_{\mathrm{PA}}^{(s)}$ as in Definition 13.1.1 — except that we modify (2) to

(2)′ $Q \Vdash$ each true atomic wf of $\mathcal{L}_{\mathrm{PA}}(S_1, S_2, \ldots)$.

Then we get the definition of an $S = S_0$ which is *generic relative to* $S_1, S_2, \ldots$ — that is, is $(\oplus_{i>0} S_i)$-*generic*.

EXERCISE 13.1.12   *Given $A_0, A_1, \ldots$, show that there exists an $(\oplus_{i \geq 0} A_i)$-generic set S.*

EXERCISE 13.1.13   *Let S be $(\oplus_{i \geq 0} A_i)$-generic. Show that for each wf ψ of $\mathcal{L}_{\mathrm{PA}}(S, A_0, A_1, \ldots)$, we have*

$$S, A_0, A_1, \cdots \models \psi \iff S \Vdash \psi.$$

13.2   Baire Space, Category and Measure

Different sorts of mathematicians often look at the same structure in very different ways. An algebraist looking at the real numbers $\mathbb{R}$ sees a *field*, whereas an analyst sees a *metric space*. A computability theorist mainly sees *information content*. New insights sometimes come from adopting another's viewpoint. In this section we will look at computability through a topologist's eyes.

You may have been lucky enough to see some topology already. If not, here is a quick introduction. You can skip it if you are already an expert!

Basic topological notions

Topology extracts from the real numbers the "closeness" relation between reals. It does this using a *metric* or "distance" function based on the distance $|x - y|$ between two reals — or by describing the *open subsets*, these being an abstraction of the open intervals $(a, b) \subset \mathbb{R}$. The resulting abstract notions of *metric space* or *topology* have many applications.

Here is an abstract definition based on describing the open sets:

DEFINITION 13.2.1   *Let M be a set, and $\mathcal{T}$ a collection of subsets of M. We say $\mathcal{T}$ is a* topology *for M if*

(1) $\emptyset, M \in \mathcal{T}$,

(2) $X, Y \in \mathcal{T} \implies X \cap Y \in \mathcal{T}$, and

(3) For each $\mathcal{C} \subseteq \mathcal{T}$, we have $\cup_{X \in \mathcal{C}} X \in \mathcal{T}$.

X is **open** *in $\mathcal{T}$ if $X \in \mathcal{T}$ and* **closed** *if $\overline{X} \in \mathcal{T}$.*

EXERCISE 13.2.2   *Show that the set 2^X of all subsets of a set X is a topology — called the* **discrete topology** *for X.*

So how exactly does $\mathbb{R}$ fit into this definition?

We say a subset of $\mathbb{R}$ is *open* if it is the union (possibly infinite) of open intervals $(a, b) \subseteq \mathbb{R}$. Then the open sets of $\mathbb{R}$ form a topology, called the *usual topology* for $\mathbb{R}$. This way of getting a topology can be generalised:

DEFINITION 13.2.3   *If every $X \in$ a topology $\mathcal{T}$ is a union $X = \cup_{Y \in \mathcal{C}} Y$ of members of $\mathcal{B} \subseteq \mathcal{T}$, then we say $\mathcal{B}$ is a* **base** *for the topology $\mathcal{T}$.*

So the set of open intervals of reals is a base for the usual topology for $\mathbb{R}$.

In fact, you can get a base for the usual topology for $\mathbb{R}$ using the metric $d(x, y) = |x - y|$ — you just need to take all *neighbourhoods* of the form $N_\epsilon(a) = \{x \mid d(a, x) < \epsilon\}$, with $a \in \mathbb{R}$ and $\epsilon > 0$.

This too generalises:

DEFINITION 13.2.4   *A* **metric** *for a set M is a function $d : M^2 \to \mathbb{R}^+$ satisfying:*

    (a) $d(x, y) = 0 \Leftrightarrow x = y$,

    (b) $d(x, y) = d(y, x)$, and

    (c) (the triangle inequality) $d(x, z) \leq d(x, y) + d(y, z)$.

    We call $\langle M, d \rangle$ a **metric space**.

The metric d for the metric space M defines a topology on M just as we got the usual topology for $\mathbb{R}$ — the topology is the one with base all neighbourhoods $N_\epsilon(a)$ with $a \in M$.

EXERCISE 13.2.5   *Show that $d(z_1, z_2) = |z_1 - z_2|$ defines a metric on $\mathbb{C}$ (= the set of all complex numbers).*

EXERCISE 13.2.6   *Show that a subset U of a metric space M is open if and only if for every $a \in U$ there exists a neighbourhood $N_\epsilon(a) \subset U$.*

Now for something a little more subtle. We often need a metric space M with no "holes" in it. We need a way of saying that if you get closer and closer to some location, that location does exist in the space M.

DEFINITION 13.2.7   *(1) A sequence $a_0, a_1, \cdots \in M$ is a* **Cauchy sequence** *if for every $\epsilon > 0$, there is an $N \in \mathbb{N}$ such that $d(a_i, a_j) < \epsilon$ for every $i, j > N$.*

    (2) We say $a_0, a_1, \cdots \in M$ **converges to** *$a \in M$ if for every $\epsilon > 0$, there is an $N \in \mathbb{N}$ such that $d(a, a_i) < \epsilon$ for every $i > N$.*

    (3) M is **complete** *if every Cauchy sequence converges to a member of M.*

Clearly $\mathbb{R}$ with the usual metric is complete. As is $\mathbb{C}$ with the metric given in Exercise 13.2.5. But:

EXERCISE 13.2.8   *Show that the set $\mathbb{Q}$ of all rational numbers with the usual metric is a metric space, but is not complete.*

[**Hint:** Consider the sequence of rationals given by $a_1 = 1$, $a_{i+1} = \dfrac{a_i}{2} + \dfrac{1}{a_i}$.]

The last basic notion we want before returning to computability is that of a *continuous* function f between metric spaces. Intuitively, such a function maps points x, y which are "close to each other" to points $f(x), f(y)$ which are also "close to each other". A neat way of expressing this topologically is:

DEFINITION 13.2.9 *Let M_1, M_2 be metric spaces. Then $f : M_1 \to M_2$ is* **continuous** *if for each set U open in M_2, $f^{-1}(U)$ is open in M_1.*

Since the above definition does not use the metric, it gives a notion of continuity between *any* sets having topologies.

EXERCISE 13.2.10 *Given a set X we define the* indiscrete topology *on X to be $\mathcal{T} = \{\emptyset, X\}$.*

Verify that $\mathcal{T}$ is a topology for X, and characterise the continuous functions $X \to X$ for this topology.

Baire space

Our main concern is the computability of functions and sets over the natural numbers.

We can get a base for a topology for ω^ω by taking all neighbourhoods of the form $N_\sigma = \{ f \in \omega^\omega \mid \sigma \subset f \}$, with $\sigma \in \omega^{<\omega}$. Clearly this base is closed under finite intersections, and the open sets are all unions of such neighbourhoods. This topology can also be got from a metric on ω^ω — for example, you can take

$$
d(f, g) = \begin{cases} 0 & \text{if } f = g, \\ \dfrac{1}{\mu x\,[\,f(x) \neq g(x)\,] + 1} & \text{otherwise.} \end{cases}
$$

We will call the metric space we get *Baire space* although it is really just a special case of a more general class of topologies.

You can, of course, topologise 2^ω (= the set of all characteristic functions) in just the same way. This special case goes back to Cantor and is called *Cantor space*. Anyway, the above metric turns both ω^ω and 2^ω into complete metric spaces.

EXERCISE 13.2.11 *Verify that the Baire space ω^ω is complete.*

EXERCISE 13.2.12 *Define $N_{D,E} = \{ A \subseteq \mathbb{N} \mid A \subseteq D \,\&\, A \cap E = \emptyset \}$, for each pair D, E of finite sets of numbers.*

*Show that the set of all such neighbourhoods is a base for a topology for 2^ω.
Show that this topology is the same as the Baire topology for 2^ω.*

Our applications will depend on a theorem called the *Baire Category Theorem*. To state this we need a couple more topological notions. We will state them for the Baire version of 2^ω, although they apply to ω^ω or any other metric space.

DEFINITION 13.2.13 *(1) $U \subset 2^\omega$ is* **nowhere dense** *if given any $\sigma \in 2^{<\omega}$, there is a $\tau \supseteq \sigma$ such that $N_\tau \subseteq 2^\omega - U$.*

(2) $U \subseteq 2^\omega$ is **meager** *if U is the union of a countable set of nowhere dense sets. And, of course, U is* **comeager** *if it is the complement of a meager set.*

Traditionally, meager sets are called *category one*, and comeager sets *category two*.

There are various equivalent ways of describing nowhere dense sets.

DEFINITION 13.2.14 *If $S \subseteq$ a metric space M, the* **interior** *$Int(S)$ of S is $\cup\{U \subseteq S \mid U$ is open $\}$. And the* **closure** *$Cl(S)$ of S is $M - Int(\overline{S})$.*

Then:

LEMMA 13.2.15

Let $S \subseteq$ a metric space M. Then S is nowhere dense if and only if the closure of S has an empty interior.

PROOF ($\Rightarrow$) Say S is nowhere dense.

Then given N_σ there is an $N_\tau \subseteq N_\sigma \cap \overline{S}$. So $N_\tau \subseteq Int(\overline{S})$, giving $N_\tau \cap Cl(S) = \emptyset$.

Hence $N_\sigma \not\subseteq Cl(S)$ — which means $Cl(S)$ has empty interior.

($\Leftarrow$) Conversely, say $Cl(S)$ has empty interior.

That means given an N_σ, there is an $X \in N_\sigma - Cl(S)$. So $X \in N_\sigma \cap Int(\overline{S})$, which means there must be some $N_\tau \subseteq N_\sigma \cap \overline{S}$.

So S is nowhere dense. ▯

EXERCISE 13.2.16 *Show that any countable $U \subset 2^\omega$ is meager.*

EXERCISE 13.2.17 *Show that if U is a countable union of meager sets, then U is meager.*

EXERCISE 13.2.18 *Show that if U is the intersection of a countable number of comeager sets, then U is comeager.*

Intuitively, there are *more* sets in a comeager set than there are in a meager set. A concrete expression of this is that a meager set may be empty, whereas:

THEOREM 13.2.19 (The Baire Category Theorem)
If $U \subseteq 2^\omega$ is comeager, then U is non-empty.

PROOF Assume $U \subseteq 2^\omega$ is comeager, where $2^\omega - U = \cup_{i \geq 0} X_i$, each X_i nowhere dense. Then $U = \cap_{i \geq 0}(2^\omega - X_i)$.

We build a sequence $\sigma_0 \subset \sigma_1 \subset \ldots$ as follows:

Let $\sigma_0 = \emptyset$. Assume we have already chosen $N_{\sigma_i} \subset \cap_{j<i}(2^\omega - X_j)$.

Choose $\tau \supset \sigma_i$ with $N_\tau \subset (2^\omega - X_i)$, and so $\subset \cap_{j<i+1}(2^\omega - X_j)$ — and define $\sigma_{i+1} = \tau$. Since 2^ω is complete, $\sigma_0 \subset \sigma_1 \subset \ldots \subset$ some $f \in U$. ☐

Notice that the Baire Category Theorem applies here because of the completeness of the topology for 2^ω. Baire spaces in general are those topologies for which the Baire Category Theorem holds.

You can now extend the notions of meager and comeager to sets of degrees, in various degree structures.

DEFINITION 13.2.20 *Let $\mathbf{D}$ be a degree structure over 2^ω. We say a set $S \subseteq \mathbf{D}$ is **meager** if $\{X \mid X \in$ some $\mathbf{a} \in \mathbf{D}\}$ is meager.*
*And $S \subseteq \mathbf{D}$ is **comeager** if $\{X \mid X \in$ some $\mathbf{a} \in \mathbf{D}\}$ is comeager.*

Here is a first application:

EXAMPLE 13.2.21 *Let $\mathbf{a}$ be a non-zero Turing degree.*
Show that $\mathcal{D}(\geq \mathbf{a})$ is meager. Deduce that there exists a degree $\mathbf{b} | \mathbf{a}$.

SOLUTION   Let $A \in \mathbf{a}$. Define for each $i \geq 0$

$$S_i = \{\, X \in 2^\omega \mid A = \Phi_i^X \,\}.$$

We show: | S_i is nowhere dense |

Given $\sigma \in 2^{<\omega}$, we look for a $\tau \supseteq \sigma$ and $x \geq 0$ such that $\Phi_i^\tau(x) \downarrow \neq A(x)$.

Case I. τ exists.
This means for every $\tau' \supseteq \tau$, we have $\Phi_i^{\tau'}(x) \downarrow \neq A(x)$ — giving $\overline{S}_i \supseteq N_\tau$.

Case II. Otherwise.
We show that for each $X \in N_\sigma$, we have Φ_i^X not total, giving $\overline{S}_i \supseteq N_\sigma$, in this case.
Assume otherwise.
Then for each $x \geq 0$ there is a $\tau \supseteq \sigma$ such that $\Phi_i^\tau(x) \downarrow$. And since Case I does not hold, $\Phi_i^\tau(x) = A(x)$.
But then it is easy to see that A must be computable, a contradiction — since in this case we can compute $A(x)$ by computably finding the least (s, τ) such that $\tau \supseteq \sigma$ and $\Phi_{i,s}^\tau(x) \downarrow$. And then $A(x) = \Phi_{i,s}^\tau(x)$.
So each S_i is nowhere dense, giving $\{\, X \in 2^\omega \mid A \leq_T X \,\} = \cup_{i \geq 0} S_i$ meager. Hence $\mathbf{\mathcal{D}}(\geq \mathbf{a})$ is meager.
But $\mathbf{\mathcal{D}}(\leq \mathbf{a})$ is also meager by Exercise 13.2.16. So the set of degrees comparable with $\mathbf{a}$ is meager by Exercise 13.2.17.
It follows that the set of degrees incomparable with $\mathbf{a}$ is comeager, and so, by the Baire Category Theorem, is non-empty.   ◻

We can now see that "most" Turing degrees are not complete.

COROLLARY 13.2.22
The set $\mathbf{\mathcal{D}}(\geq \mathbf{0}')$ of complete Turing degrees is meager.

PROOF   Just take $\mathbf{a} = \mathbf{0}'$ in Example 13.2.21.   ◻

EXERCISE 13.2.23   *Let $\mathbf{\mathcal{C}} \subseteq \mathbf{\mathcal{D}}$ be a countable collection of non-zero Turing degrees.*
Show that there is a $\mathbf{b} \in \mathbf{\mathcal{D}}$ which is incomparable with every $\mathbf{a} \in \mathbf{\mathcal{C}}$.

Similar results could have been stated for almost all of the degrees structures we have seen so far — they only depend on the underlying reducibility being captured by a countable set of operators which are *continuous*.

PROPOSITION 13.2.24
Every p.c. functional Φ is continuous over 2^ω.

PROOF We just need to show that for every open $U \subseteq 2^\omega$ we have that $\Phi^{-1}(U) = \{X \mid \Phi^X \in U\}$ is open. By Exercise 13.2.6, it suffices to show:

$$(\forall A \in \Phi^{-1}(U))\,(\exists \tau)\,[\,A \in N_\tau \subseteq \Phi^{-1}(U)\,]$$

So say $A \in \Phi^{-1}(U)$, so that $\Phi^A \in U$.

Then for some σ we have $\Phi^A \in N_\sigma \subseteq U$ — where $\sigma \subset \Phi^A$. This means there is some $\tau \subset A$ such that $\sigma \subset \Phi^X$ for every $X \in N_\tau$ — that is, $N_\sigma \subseteq \Phi(N_\tau$.

But that means $A \in N_\tau \subseteq \Phi^{-1}(N_\sigma) \subseteq \Phi^{-1}(U)$ — which is what we wanted to show. □

EXERCISE 13.2.25 *Show that every enumeration operator Ψ is continuous as a function over 2^ω.*

We said earlier that the forcing and category frameworks were related. Here is a useful connection between the two:

THEOREM 13.2.26
An arithmetical collection $\mathcal{C}$ of sets is comeager if and only if $\mathcal{C}$ contains every generic set.

PROOF Assume $\mathcal{C}$ is arithmetical, so $X \in \mathcal{C}$ can be expressed as a wf $\psi(X)$ of $\mathcal{L}_{\mathrm{PA}}^{(s)}$.

($\Rightarrow$) Let S be generic, with $S \notin \mathcal{C}$. So $S \models \neg\psi(S)$, giving $S \Vdash S\neg\psi(S)$.

This means there is some $\sigma \subset S$ for which $\sigma \Vdash S\neg\psi(S)$. But this gives a $N_\sigma \subset \overline{\mathcal{C}}$, so that $\mathcal{C}$ is not comeager.

($\Leftarrow$) Assume that every generic $S \in \mathcal{C}$.

By Lemma 13.2.15, if $\mathcal{C}$ is not comeager, then there is some $N_\sigma \subseteq \mathrm{Cl}(\overline{\mathcal{C}})$.

Take a generic $S \supset \sigma$, so $S \in N_\sigma$.

But $S \models \psi(S)$, giving $S \Vdash \psi(S)$ — and so $\tau \Vdash \psi(S)$ for some τ with $\sigma \subseteq \tau \subset S$.

But that means $N_\tau \subseteq N_\sigma \cap \mathcal{C}$, contradicting $N_\sigma \subseteq \mathrm{Cl}(\overline{\mathcal{C}})$. □

This gives another proof of Example 13.1.9.

REMARK 13.2.27 Notice that this result says that if S is generic, then it is to be found in every comeager arithmetical set $\mathcal{C}$. But such a $\mathcal{C}$ is *natural* — describable in arithmetic — and very *typical*, in that most sets are in $\mathcal{C}$.

So generic sets seek out normality, in a sense, and that is why they are called generic. $\quad\square$

EXERCISE 13.2.28 *Let $\mathcal{C}$ be a collection of sets arithmetical in A. Show that $\mathcal{C}$ is comeager if and only if $\mathcal{C}$ contains every A-generic set.*

EXERCISE 13.2.29 *We say $\mathbf{b}$ is $\mathbf{a}$-generic if it contains an A-generic set B with $A \in \mathbf{a}$.*

Show that if $\mathbf{a} > \mathbf{0}$ then every $\mathbf{a}$-generic $\mathbf{b}$ is incomparable with $\mathbf{a}$.

Lebesgue Measure

We can also look at Baire space for 2^ω *measure theoretically*. Intuitively, we measure a set A of binary reals by estimating how much of the interval $[0,1] = N_\emptyset$ it covers. We do this by covering A with sets we can easily measure — sets made up of countable unions of open intervals N_σ — and taking the infimum of the measure of such covers.

Here are the basics of the *Lebesgue measure μ* for Cantor space — also called the *product measure* on $\{0,1\}^\omega$.

DEFINITION 13.2.30 (Henri Lebesgue, 1901) (1) Let $A \subseteq 2^\omega$. We say $\{N_\sigma\}_{\sigma \in \Gamma}$ is a **covering** of A if $A \subseteq \cup_{\sigma \in \Gamma} N_\sigma$.

(2) The Lebesgue **outer measure** $\mu^* : 2^\omega \to \mathbb{R}^{\geq 0}$ is given by:

$$\mu^*(A) = \operatorname{Inf}\left\{ \sum_{\sigma \in \Gamma} 2^{-|\sigma|} \mid \{N_\sigma\}_{\sigma \in \Gamma} \text{ a covering of } A \right\}.$$

(3) A is Lebesgue **measurable** if for each $X \subseteq 2^\omega$ we have

$$\mu^*(X) = \mu^*(X \cap A) + \mu^*(X \cap \overline{A}).$$

If A is measurable, the Lebesgue **measure** of A is $\mu(A) = \mu^*(A)$.

Notice that the Borel subsets of 2^ω, which we get by closing up the set of open intervals N_τ under complements and countable unions, are all measurable. We say that the *Borel σ-algebra* turns 2^ω into a *measure space*.

There are various basic properties of the measure μ which we can verify, and which all fit in with what we intuitively expect a measure to deliver.

EXERCISE 13.2.31 *Show that the Lebesgue measure μ on measurable subsets of 2^ω satisfies:*

(1) $\mu(\emptyset) = 0$.

(2) $f \in 2^\omega \Rightarrow \mu(\{f\}) = 0$.

(3) μ is monotone — *that is,* $A \subseteq B \Rightarrow \mu(A) \leq \mu(B)$.

(4) $\mu(N_\sigma) = 2^{-|\sigma|}$.

(5) μ is countably additive — *that is, for any* $\{A_n\}_{n \geq 0}$ *we have*

$$\mu\left(\bigcup_{n \geq 0} A_n\right) \leq \sum_{n \geq 0} \mu(a_n).$$

And if the A_n's are mutually disjoint we have

$$\mu\left(\bigcup_{n \geq 0} A_n\right) = \sum_{n \geq 0} \mu(a_n).$$

(6) A *countable* $\Rightarrow \mu(A) = 0$.

EXERCISE 13.2.32 *Show that a set A is measurable if and only if its complement is measurable.*

The results of applying measure theory to sets of degrees gives roughly the same picture that we got from the category methods. The sets of degrees we decided were less common because they were meager — such as the complete degrees — also turn out to have measure zero. And similarly those like the set $\{\mathbf{x} \mid \mathbf{x} \text{ is incomparable with some } \mathbf{a} > \mathbf{0}\}$ which were found to be comeager also have measure one. But, in general, the notions of meager and having measure zero are not the same.

What certainly *is* different is that the corresponding measure theoretic results are much messier to prove. But you will come across measure again later when we look at the notion of a random real.

13.3 *n*-Genericity and Applications

Forcing is a powerful technique for building models satisfying special requirements. But to people like us interested in models whose constructive properties give them more everyday relevance, forcing seems a very blunt instrument indeed. Even Feferman's more finely honed version delivers generic sets which are not even arithmetical. So what chance of using forcing locally?

In 1969 Peter Hinman looked at fragments of full forcing in arithmetic and found some useful applications. And in 1977 Carl Jockusch gave us the user-friendly version that we now use all the time.

Jockusch's definition replaces arithmetical relations with sets of strings. I will use our usual convention of suppressing the coding in computing relative to strings. In this case it means we can use $\{W_i\}_{i \geq 0}$ as a standard listing of c.e. sets of strings, with all the usual properties.

DEFINITION 13.3.1   *(1) We say $A \subseteq \mathbb{N}$ is **1-generic** if for every c.e. set X of strings, either*

   (a) $(\exists \tau \subset A)[\tau \in X]$, or

   (b) $(\exists \tau \subset A)(\forall \sigma \supseteq \tau)[\sigma \notin X]$.

   *We say A — and any $\tau' \supseteq$ such a τ — **forces** X. We write $A \Vdash X$ — or $\tau' \Vdash X$ — as appropriate.*

   *(2) We say **a** is **1-generic** if **a** contains a 1-generic A.*

REMARK 13.3.2   In general we can define **n-generic** in the same way. We just make the set X in the above definition Σ_n instead of Σ_1.

Of course, the bigger the n in the definition, the more powerful the forcing notion — but the less constructive the n-generic sets. It turns out that 1-genericity is already powerful enough to capture all the Kleene–Post type constructions we introduced in Chapter 10. And it does come with some genuinely local generic sets.   ⬚

Although the definition above looks a little different from Definition 13.1.1, you will not be surprised by:

EXERCISE 13.3.3   *Show that every generic set is n-generic, each $n \geq 1$.*
[**Hint:** For $n = 1$ show that for each c.e. X the class of sets A satisfying (a) or (b) of Definition 13.3.1 is arithmetical and hence is forced by S — in the right way!]

If you find Definition 13.3.1 a little abstract, here is a picture to keep in mind.

I have displayed the strings in a natural way on a tree (all my trees grow upwards by the way). The boxed nodes correspond to strings which — in this example — are in the c.e. set X we are trying to force. I have tried to illustrate how a set A — corresponding to an infinite path through the tree —

can *avoid* being 1-generic. To do this it must avoid strings satisfying clause (a) of Definition 13.3.1, but without ever giving up (a) as an option, and ending in clause (b).

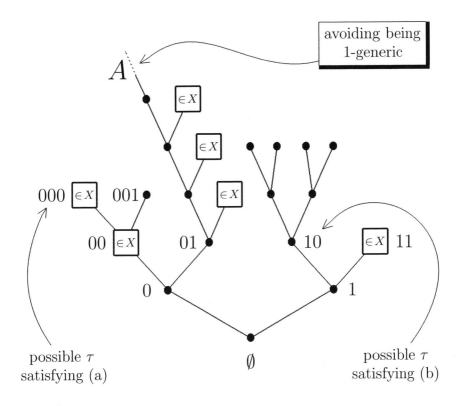

This next basic result corresponding to Feferman's Theorem 13.1.2 gives us a bound on the 1-generic set.

THEOREM 13.3.4 (The Existence Theorem for 1-Generic Sets)
There exists a 1-generic set $A \leq_T \emptyset'$.

PROOF We construct strings $\sigma_0 \subset \sigma_1 \subset \ldots \subset A = \bigcup_{i \geq 0} \sigma_i$ so that for each $i \geq 0$ we have $\sigma_{i+1} \Vdash W_i$.

The idea is that at stage $i + 1$ of the construction σ_{i+1} grabs a string in W_i if it possibly can — and if it cannot it blames σ_i for already having satisfied clause (b) of the definition of forcing W_i.

The construction of A

Stage 0.

Define $\sigma_0 = \emptyset$.

Stage $i + 1$.

Ask the crucial question: $\boxed{(\exists \tau \supset \sigma_i) \,[\tau \supseteq \text{ some } \tau' \in W_i]?}$

Case I: *Yes* — Choose such a τ, and define $\sigma_{i+1} = \tau$.

> **Outcome:** $\sigma_{i+1} \supset \sigma_i$ and $\sigma_{i+1} \Vdash W_i$ via (a) of Definition 13.3.1 — so $A \Vdash W_i$.

Case II: *No*— Define $\sigma_{i+1} = \sigma_i{}^\frown 0$.

> **Outcome:** $\sigma_{i+1} \supset \sigma_i$ and $\sigma_{i+1} \Vdash W_i$ via (b) of Definition 13.3.1 — so again we have $A \Vdash W_i$.

Checking the construction works:

(1) $A \Vdash W_i$ for each i, so A is 1-generic.

(2) $A \leq_T \emptyset'$ — to see this, notice the construction of A only needs an oracle for answering our crucial question above. But this question can be written

$$(\exists \tau \supset \sigma_i)(\exists s) \,[\tau \supseteq \text{ some } \tau' \in W_{i,s}]?$$

So it is uniformly answerable using an oracle for the Σ_1 set

$$\{\langle \sigma, i \rangle \mid (\exists \tau, \tau', s)[\sigma \subset \tau' \subseteq \tau \,\&\, \tau' \in W_{i,s}]\},$$

which is $\leq_T \emptyset'$.　　　　　　　　　　　　　　　　　　　　　□

EXERCISE 13.3.5 (The Existence Theorem for n-Generic Sets)
Show that there exists an n-generic set $A \leq_T \emptyset^{(n)}$.

Now that we know 1-generic sets exist locally, we can use them to satisfy all sorts of requirements. Or rather — we can show that in proving the above theorem we forced a whole lot of c.e. sets which turn out to express all sorts of interesting properties. At a practical level, proving just one existence theorem saves us repeating the proof in many different guises. We just say, "Ha, you want to build a set with *that* property — here is one I made earlier, it is called a 1-generic set." The only tricky part is to find the right c.e. set X to force to get the property wanted.

EXAMPLE 13.3.6 *Show that if A is 1-generic then A is not computable.*

SOLUTION Define $X = \{\sigma \mid \varphi_i$ and σ disagree on some argument $y\}$. Then

$$\sigma \in X \iff (\exists s, y)[\underbrace{y < |\sigma| \ \& \ \varphi_{i,s}(y) \downarrow \neq \sigma(y)}],$$

$$\text{computable relation}$$

so $X \in \Sigma_1$ and so is c.e.

This means that if A is 1-generic $A \Vdash X$. So by Definition 13.3.1 there is a $\tau \subset A$ such that $\tau \Vdash X$.

Case (a). $\tau \in X$ — then $\tau(y) \neq \varphi_i(y)$ some $y < |\tau|$. So $A \neq \varphi_i$ since $\tau \subset A$.

Case (b). For all $\sigma \supset \tau$ we have $\sigma \notin X$. But then we cannot have $\varphi(|\tau|) \downarrow$, since otherwise

$$\sigma = \tau^\frown (1 \dot- \varphi_i(|\tau|)) \in X.$$

So φ is not total and so cannot be the characteristic function of A. $\quad$ ▯

We can prove a lot more. Just to illustrate the power of the technique, let us reframe and extend the proof of Theorem 10.6.3 using the forcing method.

THEOREM 13.3.7

If $\mathbf{a}$ is 1-generic, then there exist incomparable degrees $\mathbf{a}_0, \mathbf{a}_1 < \mathbf{a}$ such that $\mathbf{a}_0 \cup \mathbf{a}_1 = \mathbf{a}$.

PROOF We first need to decide how we will use a single 1-generic set $A \in \mathbf{a}$ to give us *two* sets A_0 and A_1 such that $A_0 \not\leq_T A_1$ and $A_1 \not\leq_T A_0$.

What we do is take $A_0 = \{x \mid 2x \in A\}$, $A_1 = \{x \mid 2x + 1 \in A\}$ — that is, $A = A_0 \oplus A_1$.

Notation: Given $\sigma \in 2^{<\omega}$, we can write $\sigma = \sigma_0 \oplus \sigma_1$, where $\sigma_0(x) = \sigma(2x)$ for $2x < |\sigma|$, and $\sigma_1(x) = \sigma(2x + 1)$ for $2x + 1 < |\sigma|$.

A look at the proof of Theorem 10.6.3 should make some sense of the choice of

$$X = \{\sigma \mid (\exists x < |\sigma_0|)[\Phi_i^{\sigma_1}(x) \downarrow \neq \sigma_0(x)]\}$$

to force. Intuitively — if $\sigma \in X$ and $A_0, A_1 \supseteq \sigma_0, \sigma_1$, then $A_0 \neq \Phi_i^{A_1}$.

We first notice $\boxed{X \text{ is c.e.}}$ since

$$\sigma \in X \iff (\exists s)\,(\exists x < |\sigma_0|)\,\underbrace{[\,\Phi_{i,s}^{\sigma_1}(x) \downarrow \neq \sigma_0(x)\,]}_{\text{computable relation}},$$

giving $X \in \Sigma_1$.

So $\boxed{A \Vdash X}$ — so there is a $\sigma \subset A$ such that $\sigma \Vdash X$.

> **LEMMA 13.3.8**
> Either $\sigma \in X$ or $\Phi_i^{A_1}$ is not total.

PROOF Assume that $\Phi_i^{A_1}$ is total, and that $\Phi_i^{A_1}(|\sigma_0|) \downarrow = \delta$, say.

Then $\Phi_i^{\tau_1}(|\sigma_0|) = \delta$, some $\tau \supset \sigma$, with $|\tau_0| > |\sigma_0|$.

But since $\sigma_0(|\sigma_0|) \uparrow$, we can choose τ with $\tau_0(|\sigma_0|) \neq \delta$, so $\Phi_i^{\tau_1}(x) \downarrow \neq \tau_0(x)$ with $x = |\sigma_0|$ — which means $(\forall \tau \supset \sigma)[\tau \notin X]$ cannot hold. And this means that $\sigma \Vdash X$ through part (a) of Definition 13.3.1.

So $\sigma \in X$, as required.   ⬚

But $A_0 \neq \Phi_i^{A_1}$ whichever holds — either $\sigma \in X$ or $\Phi_i^{A_1}$ not total.

We handle $A_1 \neq \Phi_i^{A_0}$ in just the same way.

The theorem follows from Exercise 10.3.10, taking $\mathbf{a}_i = \deg(A_i)$, each $i \leq 1$.

⬚

REMARK 13.3.9 In fact we can show that $\mathbf{a}_0$ and $\mathbf{a}_1$ are *also* 1-generic, and the theorem can be iterated. So the structure of $\mathcal{D}(\leq \mathbf{a})$, $\mathbf{a}$ 1-generic, is very rich. And so is the structure of $\mathcal{D}(\leq \mathbf{0}')$ since we can choose our 1-generic $\mathbf{a} \leq \mathbf{0}'$.   ⬚

EXERCISE 13.3.10 *For any string σ let $\bar{\sigma}$ be given by*

$$\bar{\sigma} = \begin{cases} 0 & \text{if } \sigma(x) = 1, \\ 1 & \text{if } \sigma(x) = 0. \end{cases}$$

Show that if W_i is a c.e. set of strings, then so is $\{\bar{\sigma} \mid \sigma \in W_i\}$.

Hence, show that $A \subseteq \mathbb{N}$ is 1-generic if and only if $\overline{A}$ is 1-generic.

EXERCISE 13.3.11 *Remember that $A \subseteq \mathbb{N}$ is **immune** if and only if A is infinite and contains no infinite c.e. subsets.*

Show that for each i, $Y_i = \{\sigma \mid (\exists x)[\sigma(x) = 0 \,\&\, x \in W_i]\}$ is c.e.

Show that if A is 1-generic, then A forces each such Y_i, and hence that A is immune.

Deduce that no 1-generic set can be c.e.

EXERCISE 13.3.12   *We say that $A \subseteq \mathbb{N}$ is* **bi-immune** *if both A and $\overline{A}$ are immune.*

Show that there exists a bi-immune set $\leq_T \emptyset'$.

EXERCISE 13.3.13   *Show that every 1-generic set is hyperimmune.*

Deduce that the set of hyperimmune sets is comeager.

EXERCISE 13.3.14   *Show that if A is n-generic then A contains no infinite Σ_n subset. Deduce that there is no $(n+1)$-generic degree below $\mathbf{0}^{(n)}$.*

Of course, Exercise 13.3.11 still leaves open the possibility that the *degree* of the 1-generic A is c.e.

Can a 1-generic $\mathbf{a} = \mathbf{0}'$? Far from it!

DEFINITION 13.3.15   *We say A* **forces its jump** *if for all $m, n \in \mathbb{N}$ there is a $\sigma \subset A$ such that either*

(a) $\Phi_m^\sigma(n) \downarrow$, *or*

(b) $(\forall \tau \supseteq \sigma)\,[\Phi_m^\tau(n) \uparrow]$.

Of course, the similarity between (a) and (b) here and (a) and (b) of Definition 13.3.1 is no coincidence! This definition turns out to be just a particularly important instance of the forcing of c.e. sets of strings in general.

Before verifying this, let us see why sets which force their jump are useful.

LEMMA 13.3.16

Let $\mathbf{a} = \deg(A)$. Then if A forces its jump, then $\mathbf{a}' = \mathbf{a} \cup \mathbf{0}'$ — in which case we say $\mathbf{a}$ is **generalised low**.

Hence, if $\mathbf{a} \leq \mathbf{0}'$ then $\mathbf{a}$ is low.

PROOF   Let $\mathbf{a} = \deg(A)$ where A forces its jump.

(1) | Prove $\mathbf{a} \cup \mathbf{0}' \leq \mathbf{a}'$ |

As we observed earlier, the Jump Theorem 10.4.14 gives $\mathbf{a} \leq \mathbf{a}'$ and $\mathbf{0}' \leq \mathbf{a}'$.

(2) | Prove $\mathbf{a}' \leq \mathbf{a} \cup \mathbf{0}'$ |

I will give an algorithm to decide, using oracles for A and $\emptyset'$, whether $\langle x, y \rangle \in A'$ or not for all $x, y \in \mathbb{N}$.

1. Enumerate $\{\sigma \mid \sigma \subset A\}$, shorter strings first — using an oracle for A.

2. For each such σ, ask whether

   (a) $(\exists s)\, \Phi^{\sigma}_{x,s}(y) \downarrow$ — a Σ_1 question, and so answerable using the oracle for $\emptyset'$ — or

   (b) $(\forall \tau \supseteq \sigma, s \geq 0)\, \Phi^{\tau}_{x,s}(y) \uparrow$ — a Π_1 question, so $\leq_T \emptyset'$ again.

3. Using the fact that A forces its jump, take the first σ satisfying (a) or (b) in 2, and define

$$s_0 = \begin{cases} \text{the } s \text{ in (a)} & \text{if (a) holds for } \sigma \\ 0 & \text{if (b) holds.} \end{cases}$$

Then s_0 is computable from A and $\emptyset'$, and $\langle x, y \rangle \in A' \iff \Phi^{\sigma}_{x,s_0}(y) \downarrow$.

The lemma follows immediately. ▯

We are now ready to prove:

THEOREM 13.3.17
If A is 1-generic, then A forces its jump.

PROOF   Say A is 1-generic. Let

$$X_{\langle x,y \rangle} = \{\sigma \mid \Phi^{\sigma}_x(y) \downarrow\}$$
$$= \{\sigma \mid (\exists s)\Phi^{\sigma}_{x,s}(y) \downarrow\}.$$

Since $X_{\langle x,y \rangle}$ is Σ_1, and so c.e., we have $A \Vdash X_{\langle x,y \rangle}$.
So there exists a $\sigma \subset A$ such that *either*
(a) $\sigma \in X_{\langle x,y \rangle}$ — that is, $\Phi^{\sigma}_x(y) \downarrow$, *or*
(b) $(\forall \tau \subset \sigma)\,[\tau \notin X_{\langle x,y \rangle}]$ — that is, $(\forall \tau \subset \sigma)\,[\Phi^{\tau}_x(y) \uparrow]$.
So A does force its jump. ▯

Surprisingly, one can get a converse to this theorem — A is 1-generic *if and only if* A forces its jump.

We now discover how far $\mathbf{0}'$ is from being 1-generic:

COROLLARY 13.3.18

Every 1-generic $\mathbf{a}$ *is generalised low.*
In particular, all the 1-generic degrees $\leq \mathbf{0}'$ *are low.*

PROOF If $A \in \mathbf{a}$ is 1-generic, then Theorem 13.3.17 tells us A forces its jump — in which case Lemma 13.3.16 makes $\mathbf{a}$ generalised low.

Now say $\mathbf{a} \leq \mathbf{0}'$. Then $\mathbf{a}' = \mathbf{a} \cup \mathbf{0}' = \mathbf{0}'$, so $\mathbf{a}$ is low. $\qquad \square$

This corollary suggests that the structure of $\mathcal{D}(\leq \mathbf{0}')$ is very rich, even at the level of the low degrees.

EXERCISE 13.3.19 *Show that no degree* $\geq \mathbf{0}'$ *can be 1-generic.*

What can we do with 2-generic sets that we cannot do with 1-generic sets? Quite a lot it turns out. I will give just one example.

THEOREM 13.3.20

If $\mathbf{a}$ *is 2-generic, there is a minimal pair* $\mathbf{a}_0, \mathbf{a}_1 < \mathbf{a}$.

PROOF Let $A = A_0 \oplus A_1$ where $A \in \mathbf{a}$ is 2-generic.

Since A is also 1-generic, of course, we have by the proof of Theorem 13.3.7 that A_0, A_1 are Turing incomparable, so not computable. We just need to show A_0, A_1 satisfies the additional requirements:

$$\mathcal{R}_i : \quad \text{If } \Phi_i^{A_0} = \Theta_i^{A_1} \text{ then } \Phi_i^{A_0} \text{ is computable,}$$

where $\{\Phi_i, \Theta_i\}_{i \geq 0}$ is a standard list of all pairs of p.c. functionals.

Here is the cleverly designed set we need A to force:

$$X_i = \{\sigma \mid (\exists x)\,[\,\Phi_i^{\sigma_0}(x) \downarrow \neq \Theta_i^{\sigma_1}(x) \downarrow\,]\}$$
$$= \{\sigma \mid (\exists x)(\forall s)\,[\,\Phi_{i,s}^{\sigma_0}(x) \downarrow \neq \Theta_{i,s}^{\sigma_1}(x) \downarrow\,]\}.$$

Clearly $X_i \in \Sigma_2$, so $A \Vdash X_i$. So there is some least $\sigma \subset A$ with $\sigma \Vdash X_i$.

Case I. $\sigma \in X_i$.

Then for some x we have $\Phi_{i,s}^{\sigma_0}(x) \downarrow \neq \Theta_{i,s}^{\sigma_1}(x) \downarrow$, giving $\Phi_i^{A_0}(x), \Theta_i^{A_1}(x)$ defined and unequal. So $\mathcal{R}_i$ is satisfied.

Case II. For all $\tau \supseteq \sigma$, we have $\tau \notin X_i$.

This means that

$$(\forall x, s \geq 0, \tau \supseteq \sigma)\left[\Phi_{i,s}^{\tau_0}(x) \downarrow \,\& \, \Theta_{i,s}^{\tau_1}(x) \downarrow \Rightarrow \Phi_{i,s}^{\tau_0}(x) = \Theta_{i,s}^{\tau_1}(x)\right]$$

Assuming $\Phi_i^{A_0}, \Theta_i^{A_1}$ total, we show how to compute $\Phi_i^{A_0}(x)$ for each x. We just look for the least $\tau \supseteq \sigma$, $s \geq 0$ such that $\Phi_{i,s}^{\tau_0}(x) \downarrow = \Theta_{i,s}^{\tau_1}(x)$ — then

> Claim: $\Phi_{i,s}^{\tau_0}(x) = \Phi_i^{A_0}(x)$

Otherwise we can find a $\pi \subset A$, such that $\Phi_i^{\pi_0}(x) \downarrow = \Phi_i^{A_0}(x)$ — where we can also assume $\pi \supseteq \sigma$ and $\pi_1 \supseteq \tau_1$. But then $\Phi_i^{\pi_0}(x) \downarrow \neq \Theta_i^{\pi_1}(x)$, contradicting our being in Case II.

So in either case $\mathcal{R}_i$ is satisfied. $\qquad \Box$

The above proof certainly seems to need 2-genericity. But you will have to take my word for it that there is a 1-generic that bounds no minimal pair. Such results tend to be very hard to prove. One of the best and most difficult is Christine Haught's theorem that the 1-generic degrees below $\mathbf{0}'$ are *closed downwards* — meaning that below $\mathbf{0}'$ if $\mathbf{0} < \mathbf{b} \leq \mathbf{a}$ 1-generic, with $\mathbf{a} < \mathbf{0}'$, then $\mathbf{b}$ is 1-generic.

A good exercise on n-genericity is to extend the proof of Corollary 13.3.18.

EXERCISE 13.3.21 *Show that if $\mathbf{a}$ is n-generic, then $\mathbf{a}^{(n)} = \mathbf{a} \cup \mathbf{0}^{(n)}$ — that is, $\mathbf{a}$ is **generalised low**$_n$.*

Forcing in the partial degrees

Can we apply forcing techniques in the enumeration degrees? If you look back to the proof of Theorem 11.4.2, you will see something which looks very much like the usual Kleene–Post type forcing. In fact:

> **DEFINITION 13.3.22** *We say that an e-degree $\mathbf{a}_e$ is **set 1-generic** if it contains a 1-generic set.*

And:

EXERCISE 13.3.23 *Show that if* $\mathbf{a}_e \in \mathcal{D}_e$ *is set 1-generic, then* $\mathbf{a}_e$ *is quasi-minimal.*

But when we say that an e-degree is plain *1-generic*, we actually mean something different. There is a notion of forcing specifically geared to partial *functions*, first used by John Case. It is most conveniently defined first for the partial degrees. The first adjustment we need to make is to the kind of strings we force with, and the inclusion relation between them.

DEFINITION 13.3.24 *(1) A* **partial string** σ *is a finite sequence of symbols from* $\mathbb{N} \cup \{\uparrow\}$. *We write* $S^* = (\mathbb{N} \cup \{\uparrow\})^{<\omega}$ *for the set of all partial strings.*

(2) We say $\tau \in S^*$ **functionally extends** σ — *written* $\sigma \subseteq \tau$ — *if*

$$\forall x < |\sigma| \, [\tau(x) = \sigma(x) \lor \sigma(x) = \uparrow].$$

The intended interpretation of $\tau(x) = \uparrow$ is of course $\tau(x) \uparrow$. In particular, we identify a function $f \in \{1, \uparrow\}^\omega$ with the semicharacteristic function we get by making $f(x) \uparrow$ for each $x \in f^{-1}(\uparrow)$. We also notice that the definition of $\sigma \subseteq \tau$ is in keeping with the usual definition of $f \subseteq g$ between partial functions. For consistency, we define the *domain* $\mathrm{Dom}(\tau)$ of τ to be the set of $x < |\tau|$ for which $\tau(x) \neq \uparrow$.

Most of what follows is due to Kate Copestake from 1988.

DEFINITION 13.3.25 *A partial function* $\psi \in (\mathbb{N} \cup \{\uparrow\})^\omega$ *is* **1-generic** *if for every c.e. set* $X \subseteq S^*$, *either*

(a) $(\exists \sigma \subset \psi) \, [\sigma \in X]$, *or*

(b) $(\exists \sigma \subset \psi)(\forall \tau \in S^*) \, [\sigma \subseteq \tau \Rightarrow \tau \notin X]$.

A partial degree, or e-degree, $\mathbf{a}$ *is* **1-generic** *if there is a 1-generic* $\psi \in \mathbf{a}$.

Much of the basic development goes as expected.

EXERCISE 13.3.26 *Show that there exists a 1-generic* ψ *of e-degree* $\leq \mathbf{0}'_e$.

Such a 1-generic ψ is not total, of course.

EXERCISE 13.3.27 *Show that every 1-generic e-degree is quasi-minimal.*

[**Hint:** Ask the 1-generic ψ to force the c.e. sets of partial strings of the form

$$X_i = \{\sigma \in S^* \mid (\exists x, y_1 \neq y_2)\,[\,\langle x, y_1 \rangle, \langle x, y_2 \rangle \in \Psi_i^\sigma\,]\},$$

with the intention of forcing Ψ_i^ψ to not be the coded graph of a function (by making Ψ_i^ψ not single-valued).]

Here is something a little new.

EXERCISE 13.3.28   *Show that if the partial function ψ is 1-generic, then it has no p.c. extension.*

[**Hint:** Ask ψ to force each set $X_i \subseteq S^*$ given by

$$\tau \in X \iff (\exists x < |\tau|)\,[\,\varphi_i(x) \downarrow \neq \tau(x)\,].\,]$$

Again, we can get a notion of *n-generic* by just replacing the c.e. set X in Definition 13.3.25 by an $X \in \Sigma_n$. And we can prove similar properties:

EXERCISE 13.3.29   *Show that if ψ is n-generic then $Graph(\psi)$ does not contain an infinite Σ_n subset.*

Deduce that there is no $(n+1)$-generic e-degree below $\mathbf{0}_e^{(n)}$.

[**Hint:** Use Theorem 11.4.3.]

By now you are asking why we bother with two different notions of genericity for the e-degrees — although they look different, they surely lead to the same set of e-degrees? Well, that is the surprising thing — 1-generic e-degrees are *never* set 1-generic. This follows from the next three exercises. I have put them in ascending order of difficulty.

EXERCISE 13.3.30   *Show that if ψ is 1-generic then $Dom(\psi) <_e \psi$.*

[**Hint:** To show $\psi \neq \Psi_i^{Dom(\psi)}$ for any i, ask ψ to force each $X_i \subseteq S^*$ given by

$$\tau \in X_i \iff (\exists x < |\tau|, y)\,[\,\tau(x) \neq y \ \& \ \langle x, y \rangle \in \Psi_i^{Dom(\tau)}\,].\,]$$

EXERCISE 13.3.31   *Show that if A is a 1-generic set and $\psi <_e A$, then ψ has a p.c. extension and so is not 1-generic.*

[**Hint:** Ask A to force the set X_i of binary strings given by

$$\tau \in X_i \iff (\exists x, y, z)\,[\,\langle x, y \rangle, \langle x, z \rangle \in \Psi_i^{\tau^+} \ \& \ y \neq z\,].\,]$$

EXERCISE 13.3.32 *Show that A is an n-generic set $\iff A = \text{Dom}(\psi)$ for some n-generic function ψ.*

[**Hint:** ($\Leftarrow$) Assume A does not force some Σ_n set $X \in 2^\omega$, and show ψ does not force $\widehat{X} = \{\tau \in S^* \mid \text{Dom}(\tau) \in X\}$, where $\text{Dom}(\tau) = \chi_{\text{Dom}(\tau)} \restriction |\tau|$.

($\Rightarrow$) Given A n-generic, build ψ as the union of partial strings ψ_i forcing the ith Σ_n set of partial strings, with $A = \text{Dom}(\psi)$.]

The following corollary summarises what we have proved:

COROLLARY 13.3.33 (Copestake, 1988)
Every 1-generic e-degree $\mathbf{a}$ bounds a set 1-generic $\mathbf{b} < \mathbf{a}$, where no $\mathbf{c} \leq \mathbf{b}$ is 1-generic.

PROOF Say we have $\psi \in \mathbf{a}$ which is 1-generic.

By Exercise 13.3.30 we have a $\mathbf{b} = \deg_e(\text{Dom}(\psi)) < \mathbf{a}$ — and by Exercise 13.3.32 $\mathbf{b}$ is set 1-generic.

Now if $\mathbf{c} \leq \mathbf{b}$, $\mathbf{c}$ cannot be 1-generic by Exercise 13.3.31. $\qquad\square$

EXERCISE 13.3.34 *Show that the set of total e-degrees is meager.*

[**Hint:** Show that **TOT** is arithmetical and contains no e-degree of a generic set.]

13.4 Forcing with Trees and Minimal Degrees

Strings are just one type of constraint on a set. There are others we can force with. Here is a rather more complicated but very powerful forcing instrument, which has very many applications. It involves specialising our idea of what a *tree* should be. Where forcing with strings gave us a grip on the Σ_1 theory of $\mathcal{D}$, forcing with trees, it turns out, gives us a grip on the Σ_2 theory of $\mathcal{D}$.

DEFINITION 13.4.1 *(1) Let $\sigma, \tau, \pi \in S$.*
We say $\sigma, \tau \supset \pi$ split π if $\sigma \mid \tau$.

*(2) We say $T : S \to S$ is a **tree** if for all $\sigma \in S$, $i \leq 1$*

$$T(\sigma^\frown i) \downarrow \Rightarrow [T(\sigma), T(\sigma^\frown(1-i)) \downarrow \ \& \ T(\sigma^\frown 0), T(\sigma^\frown 1) \text{ split } T(\sigma)].$$

*(3) If T is a tree, we say σ is **on** T if $\sigma \in range(T)$ — written $\sigma \in T$.*
*And we say $A \subseteq \mathbb{N}$ is **on** T — or is a **branch** of T — if $A{\restriction}n$ is on T for infinitely many $n \in \mathbb{N}$.*

*(4) A tree T is **partial computable** if it is p.c. as a function $S \to S$ — and is **computable** if it is p.c. and total.*

(5) Let T, T^* be trees.
*We say T^* is a **subtree** of T if $range(T^*) \subseteq range(T)$, and write $T^* \subseteq T$.*

Notice that when we force with a string, it only puts a finitary constraint on the set we force. The idea is that a tree is a way of putting *infinitary* constraint on the sets on it — while still leaving plenty of freedom to choose particular branches satisfying further requirements.

I could now set up a forcing framework based on trees in place of strings, with subtrees playing a similar role to extensions of strings. This was first done by Gerald Sacks back in the 1960s. Loosely related to this is a large body of unpublished material of C.E.M. Yates on his *abundance theory*. Somehow the framework never became popular as a presentational tool. Nor were generic objects in this framework studied as interesting objects in themselves. In retrospect it was, and still is, easiest to get on and do what forcing with trees is needed in a natural and ad hoc way. And that is what we will do.

Here are some useful examples of trees:

EXERCISE 13.4.2 *Let Id: $2^{<\omega} \to 2^{<\omega}$ be the **identity tree** defined by $Id(\tau) = \tau$ for all $\tau \in 2^{<\omega}$.*
Verify that Id is a computable tree.

EXERCISE 13.4.3 *Let $T[\sigma]$ be defined by*

$$T[\sigma](\tau) = \begin{cases} T(\sigma^\frown \tau) & \text{if } T(\sigma^\frown \tau) \downarrow \\ \text{undefined} & \text{otherwise.} \end{cases}$$

*Show that $T[\sigma]$ is a subtree of T — called the **full subtree of T above** σ — such that $T[\sigma] \leq_T T$.*

Trees like these were first used by Clifford Spector in 1956 to construct a *minimal* Turing degree. It is thanks to a little paper of Shoenfield from 1966 that we have the elegant standard presentation seen here.

DEFINITION 13.4.4 *(1) Let $\mathcal{D}_r$ be a degree structure. We say $\mathbf{a} \in \mathcal{D}_r$ is **minimal** if it is minimal in $\mathcal{D}_r(> \mathbf{0}_r)$ — that is, if $\mathbf{a} > \mathbf{0}_r$, and the only $\mathbf{b} < \mathbf{a}$ is $\mathbf{b} = \mathbf{0}_r$. More generally:*

*(2) A set $\mathcal{C} \subseteq \mathcal{D}_r$ is called an **initial segment** of $\mathcal{D}_r$ if it is closed downward — that is, if for all $\mathbf{x} \leq \mathbf{a} \in \mathcal{D}_r$ we have $\mathbf{x} \in \mathcal{D}_r$.*

REMARK 13.4.5 Why are minimal Turing degrees so interesting?

At first it was because they were hard to find — because they were *not* there! As we saw earlier, there are no minimal c.e. degrees. And it is easy to see that the forcing and category methods we saw earlier are not going to help us find minimal degrees:

EXERCISE 13.4.6 *Show that the set of all minimal degrees is meager.* [**Hint:** Show that

$$B_i = \{\, X \oplus Y \subseteq \mathbb{N} \mid X \neq \Phi_i^Y \,\}$$

is arithmetical and contains every generic set. Apply Exercise 13.2.18.]

To add to this picture, Gerald Sacks showed that the set of all minimal degrees has measure zero.

Minimal degrees are interesting because they need new techniques.

A deeper and more enduring motivation is the way they lead us into the whole theory of intitial segments. We saw in Section 12.5 how important extensions of embeddings are in getting to grips with the elementary theory of the c.e. degrees. A minimal degree represents a very dramatic obstacle to extending embeddings. The theory of initial segments is concerned with characterising the extent of such obstacles.

By overcoming most other obstacles, and getting some very complicated initial segments, one can show that the Σ_2 theory of $\mathcal{D}$ is *undecidable*. Actually, embedding every bottomed finite distributive lattice suffices. An embedding as an initial segment is a stronger sort of embedding having what are called *saturation* properties we can only dream of in the c.e. degrees. It gives us a much stronger grip on the theory. You can even compute the degree of the theory, and do it for local substructures of $\mathcal{D}$, but we are getting ahead of ourselves now. □

This is how Shoenfield introduced the special trees needed for constructing a minimal degree:

DEFINITION 13.4.7   *Let Φ be a p.c. functional.*

(1) We say σ, τ Φ-split π on T if $\sigma, \tau \in T$ and there is some x for which $\Phi^\sigma(x), \Phi^\tau(x) \downarrow$ and are unequal.

(2) We say T is Φ-splitting if

$$\forall \sigma \left[T(\sigma^\frown 0), T(\sigma^\frown 1) \downarrow \implies T(\sigma^\frown 0), T(\sigma^\frown 1) \ \Phi\text{-split on } T \right].$$

We say T avoids Φ-splittings if no σ, τ Φ-split any π on T.

And this is how he captured Spector's core idea:

LEMMA 13.4.8 (The Computation Lemma)
Let A be on the p.c. tree T. Then:

(I) If T avoids Φ-splittings, then $\Phi^A \subseteq$ some p.c. function h — so if Φ^A is total, it is computable.

(II) If T is Φ-splitting, then $A \leq_T \Phi^A$.

PROOF   Let A, T be as described.

| To prove (I): | Look for the least $\langle \sigma, s \rangle$ for which $T(\sigma)[s] \downarrow$ and $\Phi^{T(\sigma)}(x)[s] \downarrow$.

Then define $h(x) = \Phi^{T(\sigma)}(x)[s]$. Then (I) follows from:

(i) h is p.c. — immediate.

(ii) If $\Phi^A(x) \downarrow$ then $h(x) \downarrow$ — true since A is on T. And

(iii) If $\Phi^A(x) \downarrow$ then $h(x) = \Phi^A(x)$.

But if this does not hold, there is some $T(\tau) \subset A$ such that $\Phi^{T(\tau)}(x) \downarrow \neq h(x) = \Phi^{T(\sigma)}(x)$ — giving the Φ-splitting $T(\sigma), T(\tau)$ on T, a contradiction.

| To prove (II): | We inductively work our way up the tree reading off longer and longer beginnings α_n of A, using oracle answers from Φ^A to verify that $T(\alpha_n) \subset \Phi^A$.

Start by defining $\alpha_0 = \emptyset$. Then assume we have computed $T(\alpha_n) \subset \Phi^A$.

Now compute the least $\langle x, s \rangle$ such that $\Phi^{T(\alpha_n^\frown 0)}(x)[s], \Phi^{T(\alpha_n^\frown 1)}(x)[s] \downarrow$ and are unequal — $\langle x, s \rangle$ must exist since A is on T and T is Φ-splitting.

Choose $\alpha_{n+1} = \alpha^\frown i$ with $\Phi^{T(\alpha_n^\frown i)}(x)[s] = \Phi^A(x)$ — there is only one such $i \leq 1$, of course, and we will have $\alpha_{n+1} \subset A$.

So $\alpha_n \leq_T \Phi^A$, and $A = \cup_{n \geq 0} \alpha_n$, giving $A \leq_T \Phi^A$. $\square$

You can see that you now have a powerful instrument for pushing Φ^A into **0** or $\deg(A)$ — if you can do that for every Φ you are well on the way to getting an A of minimal degree. All you need to do is work out how to get the splitting/nonsplitting trees, and find a way of putting them together. This is where the forcing comes in.

LEMMA 13.4.9 (The Tree Existence Lemma)

Let T be a tree. Then for each p.c. Φ there exists a T-computable $T^* \subseteq T$ such that either:

(I)′ T^* avoids Φ-splittings, or

(II)′ T^* is Φ-splitting.

PROOF Assume T is a given tree.

We get Case (I)′ if there exists a $T(\pi) \in T$ such that no σ, τ Φ-splits $T(\pi)$ on T. In this case we just define $T^* = T[\pi]$.

If there is no such $T(\pi)$ we get (II)′.

In this case we define $T^* = \mathrm{Spl}_\Phi(T)$, where we inductively define:

$$\mathrm{Spl}_\Phi(T)(\emptyset) = T(\emptyset)$$
$$\mathrm{Spl}_\Phi(T)(\pi{}^\frown 0), \mathrm{Spl}_\Phi(T)(\pi{}^\frown 1) = T(\sigma), T(\tau),$$

where $\langle s, \sigma, \tau \rangle$ is the least such that $T(\sigma), T(\tau)$ Φ-split $\mathrm{Spl}_\Phi(T)(\pi)$ at stage $s + 1$.

It is straightforward to verify now that:

(i) $\mathrm{Spl}_\Phi(T) \leq_T T$,

(ii) $\mathrm{Spl}_\Phi(T)$ is Φ-splitting, and

(iii) $\mathrm{Spl}_\Phi(T) \subseteq T$. ⬜

Notice that the way we choose our tree T^* in this proof looks very much like forcing with strings within the tree T, and you can formulate it like that as Lerman does in his book.

Anyway, we have done most of the work. So let us now prove:

THEOREM 13.4.10 (Spector, 1956)

There exist uncountably many minimal degrees.

PROOF Let $r \in 2^\omega$ denote a binary real. For each such r we define a unique set A_r satisfying each requirement of the form:

$$\mathcal{R}_\Phi : \quad \Phi^{A_r} \text{ total} \implies \Phi^{A_r} \text{ computable} \vee A_r \leq_T \Phi^{A_r},$$

where we assume $\mathcal{R}_\Phi$ is the ith requirement on a computable list. We identify Φ with its index i — and write $\Phi + 1$ for Φ_{i+1}, and $r \upharpoonright \Phi$ for $r \upharpoonright i$, etc.

The obvious strategy for $\mathcal{R}_\Phi$ is to choose A_r on a tree $T_{r \upharpoonright \Phi + 1}$ to which the Computation Lemma applies. So we actually have a tree of trees. To put all the strategies together, we need to nest the trees along each branch:

$$T_{r \upharpoonright 0} \supset \dots \supset T_{r \upharpoonright \Phi} \supset T_{r \upharpoonright \Phi + 1} \supset \dots$$

Here is how we do it:

$$T_{r \upharpoonright 0} = \text{Id}$$

$$T_{r \upharpoonright \Phi + 1} = \begin{cases} \text{Spl}_\Phi(T_{r \upharpoonright \Phi}[r(\Phi)]) & \text{if this tree is total} \\ T_{r \upharpoonright \Phi}[\pi], \ \pi \supseteq r(\Phi), \text{ no } T_{r \upharpoonright \Phi}(\sigma), T_{r \upharpoonright \Phi}(\tau) & \\ \quad \Phi\text{-splitting } T_{r \upharpoonright \Phi}(\pi) & \text{otherwise.} \end{cases}$$

Taking $A_r = \cup_{\Phi \geq 0} T_{r \upharpoonright \Phi}(\emptyset)$, we just need to observe:

(i) For each Φ, *either* $T_{r \upharpoonright \Phi + 1}$ avoids Φ-splittings, so that (I) of the Computation Lemma gives a total Φ^{A_r} computable — *or* $T_{\upharpoonright \Phi + 1}$ is Φ-splitting, so that (II) gives $A_r \leq_T \Phi^{A_r}$.

(ii) If $r \neq r'$ then $A_r \neq A_{r'}$.

It follows that there are $2^{\aleph_0}$ sets A_r such that $A_r \in \mathbf{0}$ or $\deg(A_r)$ is minimal. Hence — remembering that a countable union of countable sets is countable — there are uncountably many minimal degrees. □

Of course, if your set theory is good enough, you will see from this proof that there are actually $2^{\aleph_0}$ minimal degrees.

Can we extract some local minimal degrees? Yes, of course. But not as local as we would like.

To actually extract an A_r of minimal degree from the above proof, we would need an oracle for an r which makes $A_r \notin \mathbf{0}$, and an oracle to tell at step $\Phi + 1$ whether to choose a splitting or non-splitting subtree of $T_{r \upharpoonright \Phi}$.

Back in 1981, David Posner disposed of the need for the first of these oracles.

PROPOSITION 13.4.11 (Posner's Lemma)
Say for every p.c. Φ, A is on a p.c. tree T^* which either avoids Φ-splittings or is a Φ-splitting tree.
Then A cannot be computable.

PROOF Assume A *is* computable.

Define a p.c. Φ by

$$\Phi^\sigma = \begin{cases} \sigma & \text{if } \sigma \not\subset A \\ \uparrow & \text{if } \sigma \subset A. \end{cases}$$

Then no p.c. T^* with infinite branches avoids Φ-splittings. On the other hand, A cannot lie on any Φ-splitting T^*. □

So a computable r would do fine in the proof of Spector's theorem — *every* r gives an A_r of minimal degree.

COROLLARY 13.4.12 (of Spector's Theorem 13.4.10)
There exists a minimal degree $\leq \mathbf{0}''$.

PROOF By Posner's Lemma, if we take r computable, then Theorem 13.4.10 gives us an A_r of minimal degree — and this A_r is computable in an oracle which tells us whether $\mathrm{Spl}_\Phi(T_{r \restriction \Phi}[r(\Phi)])$ is total or not.

But for any computable T,

$$\mathrm{Spl}_\Phi(T) \text{ is total} \iff (\forall \pi \in T)(\exists \sigma, \tau) [\sigma, \tau \ \Phi\text{-split } \pi \text{ on } T]$$
$$\iff (\forall \pi \in T)(\exists \sigma, \tau \in T, x, s \geq 0) [\sigma, \tau \supset \pi$$
$$\& \ \Phi_s^\sigma(x), \Phi_s^\tau(x) \downarrow \text{ and are unequal.}]$$

Since $\mathbf{0}''$ is Π_2-complete, we are done. □

EXERCISE 13.4.13 Say $A \leq_T \emptyset'$, and for every p.c. Φ, A is on a p.c. tree T_Φ which either *avoids* Φ-splittings or *is* a Φ-splitting tree.

Show that A is the only infinite branch of some T_Φ.

[**Hint:** Define a p.c. Φ by

$$\Phi^\sigma = \begin{cases} \sigma & \text{if } \sigma \subseteq A^s, \text{ some } s \\ \uparrow & \text{otherwise.} \end{cases}$$

Show that if A is on a p.c. T^*, T^* cannot avoid Φ-splittings unless A is computable.]

You can see from this that total trees are not going to be any use in building minimal degrees more locally, below $\mathbf{0}'$ anyway. I would love to show you how to build minimal degrees below $\mathbf{0}'$ using a $\mathbf{0}'$ oracle, and even below a c.e. $ba > \mathbf{0}$ using full approximation — but no room unfortunately. Odifreddi is very good on tree arguments, by the way, if you feel adventurous.

This is a good place to notice that we can *relativise* our new kind of tree constructions.

DEFINITION 13.4.14 *(1)* **b** > **a** *is a* **minimal cover** *for* **a** *if it is minimal in* $\mathcal{D}(>\mathbf{a})$.

   (2) **b** > **a** *is a* **strong minimal cover** *for* **a** *if* $\mathcal{D}(<\mathbf{b}) = \mathcal{D}(\leq \mathbf{a})$.

By Example 13.2.21, the only degree whose minimal covers are all strong minimal covers is **0**.

Can you guess what we get when we relativise the minimal degree construction — minimal covers or strong minimal covers? Of course, relativised facts about $\mathcal{D}$ usually only give parallel facts about $\mathcal{D}(\geq \mathbf{a})$. And in this case it is easy to see:

EXERCISE 13.4.15 *Show that no CEA degree can be a strong minimal cover. Deduce that no degree $\geq \mathbf{0}'$ has a strong minimal cover.*

EXERCISE 13.4.16 (The Relativised Computation Lemma) *Let A be on the B–p.c. tree T. Show:*

   (I)   If T avoids Φ-splittings, then $\Phi^A \subseteq$ some B–p.c. function h, and

   (II) If T is Φ-splitting then $A \leq_T B \oplus \Phi^A$.

COROLLARY 13.4.17 (of Theorem 13.4.10)
Every Turing degree **b** *has uncountably many minimal covers.*

PROOF We just modify $T_{r\restriction 0} = \mathrm{Id}$ in the proof of Theorem 13.4.10 by relativising Id to a $B \in \mathbf{b}$:

$$\mathrm{Id}^B(\sigma) = \sigma \oplus B\restriction |\sigma|, \quad \text{for all } \sigma \in 2^{<\omega}.$$

Then the Tree Existence Lemma still works, from which $B \leq_T$ every infinite branch through Id^B, giving $B \leq_T$ each A_r. The result follows by Exercise 13.4.16. You can even relativise Posner's Lemma to get *every* $\deg(A_r)$ being a minimal cover of **b**.   ⬚

Strong minimal covers are not very well understood. There is a very old open question of Michael Yates asking if every minimal degree has a strong minimal cover. If the answer were "yes" we could build many useful intial segments bit by bit, instead of in one big complicated construction.

We can now go on to do all sorts of interesting things using trees. The next exercise is a bit of a disappointment, it has to be admitted. We fail to export our minimal Turing degrees into the enumeration degrees — although we do get some modest returns for trying.

EXERCISE 13.4.18 *Show that there exists a total function f which is not computable, but for which every partial function $\psi \leq_e g$ has a p.c. extension.*

Deduce that there exists an e-degree incomparable with all the 1-generic e-degrees.

[**Hint:** In the construction of a minimal Turing degree, replace the p.c. functionals Φ_i by e-operators Ψ_i.

Define: σ, τ *i-split* π *on* T if σ, τ are on T, $\pi \subset \sigma, \tau$, and $\Psi_i^\sigma, \Psi_i^\tau$ are incompatible — that is, there exist $\langle x, y \rangle \in \Psi_i^\sigma$ and $\langle x, z \rangle \in \Psi_i^\tau$ with $y \neq z$. Adapt the Computation Lemma to get a total f such that for each partial function Ψ_i^f either $f \leq_e \Psi_i^f$ or Ψ_i^f has a p.c. extension.]

Here is another obstacle to extensions of embeddings into $\mathcal{D}$ — arising in connection with *ascending sequences* like $\mathbf{0} < \mathbf{0}' < \mathbf{0}'' < \dots$.

When does an ascending sequence $\mathbf{a}_0 < \mathbf{a}_1 < \mathbf{a}_2 < \dots$ have a least upper bound in $\mathcal{D}$? Never! But we still get an obstacle:

DEFINITION 13.4.19 *(1) An upper bound $\mathbf{b}$ for the ascending sequence $\mathbf{a}_0 < \mathbf{a}_1 < \dots$ is a **minimal upper bound** if there is no upper bound $\mathbf{c} < \mathbf{b}$.*

*(2) An upper bound $\mathbf{b}$ for $\mathbf{a}_0 < \mathbf{a}_1 < \dots$ is **uniform** if there are $B \in \mathbf{b}$, $A_i \in \mathbf{a}_i$, each $i \geq 0$, such that $\oplus_{i \geq 0} A_i \leq_T B$.*

Which degrees are minimal upper bounds for some ascending sequence? There are many which are not — for instance, no minimal degree is. On the other hand:

EXERCISE 13.4.20 *Show that if $\mathbf{b}$ is CEA some $\mathbf{a} < \mathbf{b}$, then $\mathbf{b}$ is a minimal upper bound for an ascending sequence $\mathbf{a}_0 < \mathbf{a}_1 < \dots$. Deduce that $\mathbf{0}'$ is a minimal upper bound.*

[**Hint:** Take $\mathbf{a}_0 = \mathbf{a}$, and get $\mathbf{a}_{i+1} > \mathbf{a}_i$ with $\mathbf{a}_{i+1} < \mathbf{b}$ and $\mathbf{a}_{i+1} \not\leq \deg(\Phi_i^B)$ using the relativised Sacks Splitting Theorem.]

Uniform minimal upper bounds are a much harder proposition.

Which ascending sequences have minimal upper bounds? All of them! The exercise and the minimal cover construction are good preparation for the next result.

THEOREM 13.4.21

Every ascending sequence $\mathbf{b}_0 < \mathbf{b}_1 < \ldots$ *has uncountably many minimal upper bounds.*

PROOF   By now you have all the machinery you need in place.

You take representatives $B_0, B_1, \ldots$ of $\mathbf{b}_0, \mathbf{b}_1, \ldots$.

You build Φ-splitting/nonsplitting trees $T_{r \restriction 0} \supset \ldots \supset T_{r \restriction \Phi} \ldots$ as we did in Spector's Theorem 13.4.10.

You make $A_r \geq_T B_i$ by relativising $T_{r \restriction \Phi}$ to B_i prior to choosing $T_{r \restriction \Phi+1}$ — define

$$T^+_{r \restriction \Phi}(\sigma) = T_{r \restriction \Phi}(\sigma \oplus B_i \restriction |\sigma|),$$

and then replace $T_{r \restriction \Phi}$ by $T^+_{r \restriction \Phi}$ in the definition of $T_{r \restriction \Phi+1}$.

Then apply the Computation Lemma as in Corollary 13.4.17 to show that for each Φ^{A_r} total — *either* $\Phi^{A_r} \leq_T \oplus_{j \leq i} B_j$ — or $A_r \leq_T \Phi^{A_r} \oplus (\oplus_{j \leq i} B_j)$.

Hence, the degree of no Φ^{A_r} can be an upper bound for $\mathbf{b}_0 < \mathbf{b}_1 < \ldots$. And since each $\deg(A_r)$ *is* an upper bound, the theorem follows. $\square$

EXERCISE 13.4.22   *Show that if* $\mathbf{b}_0 < \mathbf{b}_1 < \ldots$ *is a* uniformly c.e. as-cending sequence — *that is,* $\mathbf{b}_0 < \mathbf{b}_1 < \ldots$ *has representatives* $B_0, B_1, \ldots$ *with* $\oplus_{i \geq 0} B_i$ *c.e.* — *then* $\mathbf{b}_0 < \mathbf{b}_1 < \ldots$ *has a minimal upper bound* $\mathbf{a} \leq \mathbf{0}^{(3)}$.

The minimal upper bound $\mathbf{a}$ cannot be $\mathbf{0}'$, by the way. Although you *can* choose $\mathbf{b}_0 < \mathbf{b}_1 < \ldots$ so that $\mathbf{a}$ can be c.e. — but that is another story.

Here is a good place to fill a gap in what we know about $\mathbf{\mathcal{D}}$. It does not need trees, but what you now know about working above an ascending sequence should bring the next exercise within reach.

DEFINITION 13.4.23 (Spector)   *We say* $\mathbf{a}, \mathbf{b}$ *form an* **exact pair** *for the ascending sequence* $\mathbf{c}_0 < \mathbf{c}_1 < \ldots$ *if* $\mathbf{a}, \mathbf{b}$ *are upper bounds for* $\mathbf{c}_0 < \mathbf{c}_1 < \ldots$, *and for all* $\mathbf{c} \leq \mathbf{a}$ *and* $\leq \mathbf{b}$, $\mathbf{c} \leq$ *some* $\mathbf{c}_i$.

EXERCISE 13.4.24   *Show that every ascending sequence has an exact pair.*

Deduce that $\mathbf{\mathcal{D}}$ *is not a lattice.*

[**Hint:** Build $B, C \geq_T$ every $C_i \in \mathbf{c}_i$ satisfying the requirements:

$$\mathcal{R}_{\Phi, \Theta} : \quad \Phi^B = \Theta^C \text{ total} \implies \Phi^B \leq_T C_{\langle \Phi, \Theta \rangle}.]$$

More generally you can replace ascending sequences by ideals $\mathcal{I} \subset \mathcal{D}$ and get a sort of *definability* result — if $\mathbf{a}, \mathbf{b}$ is an exact pair for $\mathcal{I}$, $\mathcal{I}$ is Turing definable from the *parameters* $\mathbf{a}, \mathbf{b}$ via:

$$\mathbf{x} \in \mathcal{I} \iff \mathbf{x} \leq \mathbf{a} \ \& \ \mathbf{x} \leq \mathbf{b}.$$

Exact pairs are very much the forerunners of today's powerful degree theoretic coding techniques, which have told us so much about the global structure of $\mathcal{D}$ and other degree structures.

There is no parallel theory of minimal degrees and minimal upper bounds for $\mathcal{D}_e$, by the way. But:

EXERCISE 13.4.25 *Show that every ascending sequence of e-degrees has an exact pair.*
Deduce that $\mathcal{D}_e$ is not a lattice.

By now I have probably given you the wrong impression of these tree techniques. There are many uses apart from embedding theorems. Let me give just one example.

You may remember from Theorem 12.2.19 and Exercise 12.2.20 that every degree $\geq$ some nonzero degree below $\mathbf{0}'$ is hyperimmune. Also, from Exercise 13.3.13 the set of all hyperimmune sets is comeager. Maybe all nonzero Turing degrees contain hyperimmune sets?

DEFINITION 13.4.26 *A degree $\mathbf{a}$ is **hyperimmune free** if it contains no hyperimmune set.*

EXAMPLE 13.4.27 *Show there exists a hyperimmune free degree $\mathbf{a} \leq \mathbf{0}''$.*

SOLUTION We build A to satisfy all requirements of the form:

$\mathcal{R}_\Phi :$ If Φ^A is the characteristic function of an infinite set, then A fails to avoid the c.e. array $\{D_{f(n)}\}_{n \geq 0}$,

where we build $\{D_{f(n)}\}_{n \geq 0}$ as needed.

The strategy for $\mathcal{R}_\Phi$ is to make A lie on a computable tree T_Φ such that for each X on T_Φ *either* Φ^X cannot be the characteristic function of an infinite set, *or* we build $\{D_{f_\Phi(n)}\}_{n\geq 0}$ in such a way that the $(n+1)^{\text{th}}$ level of T_Φ witnesses $\Phi^X \cap D_{f_\Phi(n)} \neq \emptyset$.

Define $T_0 = \text{Id}$.

For $T_{\Phi+1}$ ask: $\boxed{(\exists\sigma \supset \emptyset, n)(\forall\tau \supset \sigma, x > n, s \geq 0)\,[\,\Phi^{T_\Phi(\tau)}(x)[s] \neq 1\,]\,?}$

$\boxed{\text{Yes:}}$ Define $T_{\Phi+1} = T_\Phi[\sigma]$.

$\boxed{\text{No:}}$ Inductively define $T_{\Phi+1}(\emptyset) = T_\Phi(0)$.

Let $n \geq 0$ and assume $T_{\Phi+1}(\sigma)$ defined on T_Φ for $|\sigma| = n$, and $D_{f_\Phi(i)}$ defined for $i < n$.

For each σ with $|\sigma| = n$, choose a $\tau \supset \sigma$ such that there is some number $y_\sigma \in \Phi^{T_\Phi(\tau)}[|\tau|] - \cup_{i<n}D_{f_\Phi(i)}$, and define

$$T_{\Phi+1}(\sigma{}^\frown 0), T_{\Phi+1}(\sigma{}^\frown 1) = T_\Phi(\tau{}^\frown 0), T_\Phi(\tau{}^\frown 1).$$

Let $D_{f_\Phi(n)} = $ the set of all such y_σ, $|\sigma| = n$.
Finally take $A = \cup_{\Phi\geq 0}T_\Phi(\emptyset)$.

It is now straightforward to verify:

(i) $T_{\Phi+1}$ is a computable subtree of T_Φ for each Φ.

(ii) A can be built using an oracle for the Σ_2 question asked in building $T_{\Phi+1}$ — so $A \leq_T \emptyset''$.

(iii) If $\boxed{\text{Yes}}$ — then Φ^A trivially avoids every c.e. array.

If $\boxed{\text{No}}$ — then $\{D_{f_\Phi(i)}\}_{i\geq 0}$ is a c.e. array, and $A \cap D_{f_\Phi(i)} \neq \emptyset$ for each $i \geq 0$ — so again, Φ^A is not hyperimmune. □

Notice that the reason we cannot get hyperimmune free degrees $\leq \mathbf{0}'$ is that we really do need total trees to define the c.e. arrays in the above proof. With non-total trees you would only get weak c.e. arrays, giving a hyperhyperimmune free degree.

In Exercise 13.3.12 we found out that since every 1-generic set is immune, there exist lots of bi-immune sets — in fact, the set of bi-immune sets is comeager.

EXERCISE 13.4.28 *We say that* **a** *is* **bi-immune free** *if* **a** *contains no bi-immune set.*
Show that there exists a bi-immune free degree $\leq \mathbf{0}''$.

Chapter 14

Applications of Determinacy

Games are widely used for modelling conflict situations. Typically, they involve two or more decision makers — or *players* — with differing objectives but sharing the same resources and rules.

Game theory is all about discovering regularities governing outcomes — or end results — of games. *Zermelo's Theorem* from 1913 on chess as a determined game is credited with being game theory's first real theorem. Some very famous names have been associated with the subsequent development of the area. These include Emile Borel — who in the 1920s published four notes on the existence of strategies for winning games — and John von Neumann, whose *Minimax Theorem* appeared in 1928, and who co-authored with Oskar Morgenstern in 1944 the seminal *Theory of Games and Economic Behavior*. John Nash's 1994 Nobel Prize for game theoretic applications to economics highlighted how striking the real world applications can be.

Games can just as dramatically bring out regularities in mathematical structures. The 1999 appearance of Hugh Woodin's book on the *Axiom of Determinacy, Forcing Axioms and Its Nonstationary Ideal, Vol. 1* — all 915 pages of it — is a monument to the importance of strategies for winning games in set theory.

Once again we will indulge in a little profitable asset-stripping — taking just what is useful from the set theoretic framework and applying it to get new results in computability.

14.1 Gale–Stewart Games

We saw in Section 12.6 how local structure and information content came together in relation to classes of degrees — such as **PS** — which are *closed upwards*.

> **DEFINITION 14.1.1** *Let $\mathcal{C}$ be a degree structure. We say that $X \subseteq \mathcal{C}$ is* **closed upwards** *in $\mathcal{C}$ if $\mathbf{a} \in X$ and $\mathbf{a} \le \mathbf{b} \in \mathcal{C}$ imply $\mathbf{b} \in \mathcal{C}$.*

We have already seen that the complete degrees are closed upwards (in $\mathcal{D}$). Upward closure turns out to be one of the principal regularities of the Turing universe. And a particular sort of determined game — called a *Gale–Stewart game* — is a powerful tool for establishing upward closure in $\mathcal{D}$.

DEFINITION 14.1.2 *(1) Let $\mathcal{C} \subseteq 2^{\omega}$. The **Gale–Stewart game** $G_{\mathcal{C}}$ is played by two players, I and II, as follows:*

• Player I chooses an $i \in 2 = \{0, 1\}$, then Player II does, and so on, the two players alternating moves.

• Let A be the sequence $A(0), A(1), \dots$ constructed in a particular play of the game.

*• We say Player I **wins** if $A \in \mathcal{C}$, and Player II **wins** otherwise.*

*(2) We say that $g : 2^{<\omega} \to 2$ is a **strategy** for the game $G_{\mathcal{C}}$.*

*A player **follows** strategy g if whenever it is that players turn the number the player chooses is $g(\sigma)$ where σ is the sequence so far constructed.*

*We say g is a **winning strategy** for Player I if whenever I follows strategy g then I wins. Define g is a **winning strategy** for II similarly.*

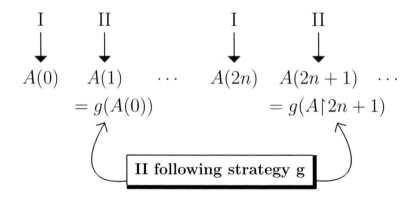

DEFINITION 14.1.3 *The game $G_{\mathcal{C}}$ is **determinate** whenever Player I or Player II has a winning strategy.*

The *Axiom of Determinacy (AD)* says:

> For all $\mathcal{C} \subseteq 2^{\omega}$, $G_{\mathcal{C}}$ is determinate.

As is well known, (AD) is a very powerful axiom but is inconsistent with another important and generally accepted axiom — the *Axiom of Choice (AC)*. Roughly speaking, (AC) says that if you have a set X of sets x then there exists a function f which *chooses* a member $f(x) \in x$ for each $x \in X$. A hard axiom to reject in favour of (AD)!

But all is not lost — the sort of games G_C we need to be determined here involve quite down-to-earth sets C. The following result will be more than enough.

THEOREM 14.1.4 (D.A. Martin, 1975)
If $C \subseteq 2^\omega$ is a Borel set then G_C is determinate — in particular, G_C is determinate for every arithmetical set C.

REMARK 14.1.5 (The Borel Hierarchy) We already know what an arithmetical $C \subseteq 2^\omega$ is — remember Definition 13.1.7 — but what is a *Borel* set? You do not need to know here, but I am going to try and tell you anyway! It is exciting mathematics, and if we are going to include Martin's celebrated result, we should state it fully and have some idea of what it means.

To avoid confusion, you should bear in mind that there are two approaches to building up hierarchies of sets of reals. One is topological and uses set theoretical operations to build complexity. The other is effective and uses quantification and computation. The former was developed by people like Borel, Baire, Lebesgue, Suslin, Luzin and Sierpinski during the first few decades of the last century — the golden age of classical *descriptive set theory*. The effective phase coincides with the early development of computability theory, so Stephen Kleene figures strongly, as does John Addison. The two approaches complement each other.

So, for instance, the classes at the nth level of the Borel hierarchy for 2^ω contain the Σ_n^0 and Π_n^0 classes of sets. Remember from Remark 13.1.10 that Σ_n^0, Π_n^0, $n \geq 0$, classified the arithmetical sets of sets according to their quantifier forms. These classes form an effective counterpart to the classical finite Borel hierarchy.

The Borel hierarchy for 2^ω is defined in Cantor space. There is a natural parallel between the inductive use of existential and universal quantifiers, and logical negation, used to get the arithmetical hierarchy, and the topological operations of forming countable unions and intersections, and taking complements, needed to get the Borel hierarchy. We just lose the effective content of the definition.

So in a nutshell, this is how we define the *Borel hierarchy*.

We start with the topologically simplest sets of binary reals and take

$$G = \text{ the set of all open subsets of } 2^\omega$$
$$F = \text{ the set of all closed subsets of } 2^\omega.$$

Next come:

$$G_\delta = \text{ the set of all countable intersections of open sets}$$
$$F_\sigma = \text{ the set of all countable unions of closed sets.}$$

And so on:

$$G_{\delta\sigma} = \text{ the set of all countable unions of sets in } G_\delta$$
$$F_{\sigma\delta} = \text{ the set of all countable intersections of sets in } F_\sigma.$$

Continuing like this, we get the whole family of Borel sets. So the Borel sets comprise the smallest class of sets of reals which contains all open sets and is closed under the formation of complements and countable unions. The effective counterpart is the *hyperarithmetic sets* — the hyperarithmetic hierarchy is built by transfinitely extending the arithmetical hierarchy in an effective way by iterating the jump.

Notice that at the bottom we get complete agreement between the Borel and arithmetical hierarchies for 2^ω:

The Σ_1^0 classes of sets $= G$, and the Π_1^0 classes of sets $= F$

where $\Sigma_0^0 = \Pi_0^0 = F \cap G = $ the *clopen* subsets of 2^ω.

Please do not think this hierarchy is just pure mathematics with little relationship to the real world. Quite down-to-earth areas of computer science involve the world of second order quantification, to which the Borel hierarchy gives a step-by-step approach.

Just to fill out the picture, let me connect us up to this world. If you take the images of Borel sets under continuous functions on 2^ω, you get the *analytic sets* — with effective counterpart the Σ_1^1 sets, got by applying an existential set quantifier to an arithmetical relation on sets. The complements of the analytic sets are the *coanalytic sets*. Continuing taking continuous images and complements, one constructs the *projective* sets. The nth level of this projective hierarchy corresponds to the effective version got by applying n alternating set quantifiers to an arithmetical relation on sets. Just to be confusing — sorry, this is standard terminology! — the effective counterpart to the projective hierarchy is called the *analytical hierarchy*, the classes at the nth level being contained in their projective partners and written Σ_n^1, Π_n^1.

Notice that these hierarchies, and the games that go with them, could just as well be based on Baire space ω^ω — our choice of 2^ω is with our degree theoretic applications in mind.   ▯

In preparation for the next section, we need to look more closely at how we assign a particular class $\mathcal{C} \subseteq 2^\omega$ to its natural level in the arithmetical hierarchy of sets of sets.

EXAMPLE 14.1.6 *Show that the set*

$$\mathcal{M} = \{A \subset \mathbb{N} \mid A \text{ is of minimal degree}\}$$

is a Π_4^0 class of sets.

SOLUTION Here is how we calculate the quantifier form of $\mathcal{M}$:

$$A \in \mathcal{M} \iff (\forall i)\,[\,\Phi_i^A \text{ not total } \vee \, \Phi_i^A \text{ computable } \vee A \leq_T \Phi_i^A\,]$$
$$\iff (\forall i)\,[\,(\exists x)(\forall s)\,\Phi_{i,s}^A \uparrow \, \vee (\exists j)(\forall x)(\exists s)\,\Phi_{i,s}^A(x) = \varphi_{j,s}(x)\, \vee$$
$$(\exists j)(\forall x)(\exists s)\,A(x) = \Phi_{j,s}(\Phi_{i,s}^A)\,]$$
$$\iff \forall(\exists\forall \vee \exists\forall\exists \vee \exists\forall\exists), \text{ etc.}$$

You can see that with a little manipulation of quantifiers we can organise this as a Π_4^0 relation of A. □

EXERCISE 14.1.7 *Show that the set*

$$\mathcal{MC} = \{A \subset \mathbb{N} \mid A \text{ is a minimal cover}\}$$

is a Σ_5^0 class of sets.

14.2 An Upper Cone of Minimal Covers

At first sight you might think that the arithmetical degrees — that is, the set of degrees of arithmetical sets of numbers — form a microcosm of the wider incomputable universe. And in many ways they do. But you would be mistaken to think that not much new happens when you get far enough up the structure. Borel determinacy will tell us that there is a sort of limiting process, but that its onset can be at very high levels of $\mathcal{D}$ indeed.

First, let us find out an interesting fact about the distribution of minimal covers in the arithmetical degrees. Notice that $\mathbf{0}'$ is not a minimal cover. If you did Exercise 12.4.8, you will have seen the same for $\mathbf{0}''$.

THEOREM 14.2.1 (Jockusch and Soare, 1970)
For no $n \geq 0$ is $\mathbf{0}^{(n)}$ a minimal cover.

PROOF   Let $\mathbf{0}^{(n+1)}$ be a minimal cover of $\mathbf{d} < \mathbf{0}^{(n+1)}$.

Let M be the greatest number $\leq n$ such that $\mathbf{0}^{(M)} \leq \mathbf{d}$.

So $\mathbf{0}^{(M+1)} \leq \mathbf{0}^{(n+1)}$, and $\mathbf{0}^{(M+1)}$ is c.e. in $\mathbf{0}^{(M)}$ — and so is c.e. in $\mathbf{d}$.

Hence, $\mathbf{d} < \mathbf{0}^{(M+1)} \cup \mathbf{d} \leq \mathbf{0}^{(n+1)}$ — "$<$" since $\mathbf{d} \not\geq \mathbf{0}^{(M+1)}$.

But since $\mathbf{0}^{(n+1)}$ is a minimal cover of $\mathbf{d}$, we have $\mathbf{0}^{(M+1)} \cup \mathbf{d} = \mathbf{0}^{(n+1)}$.

But $\mathbf{0}^{(M+1)} \cup \mathbf{d}$ is c.e. in $\mathbf{d}$.

So by the relativised Sacks Splitting Theorem, $\mathbf{0}^{(n+1)}$ splits over $\mathbf{d}$ — which contradicts $\mathbf{d}$ being minimally covered by $\mathbf{0}^{(n+1)}$.   ⬜

It follows, of course, that there is no upper cone of the arithmetical degrees which consists entirely of minimal covers. We will now see that things are different for $\mathcal{D}$ as a whole.

DEFINITION 14.2.2   *(1) We say $\mathcal{C} \subseteq 2^\omega$ is **degree invariant** if it is closed under $\equiv_T$ — that is if*

$$\forall A, B \subseteq \mathbb{N} \, [\, A \in \mathcal{C} \ \& \ B \equiv_T A \implies B \in \mathcal{C} \,].$$

*(2) $G_\mathcal{C}$ is a **degree game** if $\mathcal{C}$ is degree invariant.*

*(3) We say $\mathcal{C}$ **contains a cone** of degrees if $\cup\{\mathbf{a} \mid \mathbf{a} \geq \mathbf{d}\} \subseteq \mathcal{C}$ for some fixed $\mathbf{d} \in \mathcal{D}$.*

Here is the key application of Borel determinacy:

LEMMA 14.2.3
Let $G_\mathcal{C}$ be a determinate degree game.

If Player I has a winning strategy for $G_\mathcal{C}$ then $\mathcal{C}$ contains a cone of degrees — otherwise $2^\omega - \mathcal{C}$ contains a cone of degrees.

PROOF   Assume I has a winning strategy g. Let $\mathbf{d} = \deg(g)$.

We show: $\{\, X \in 2^\omega \mid \mathbf{d} \leq \deg(X) \,\} \subseteq \mathcal{C}$

Let $A \in \mathbf{a} \geq \mathbf{d}$.

Consider the play of $G_{\mathcal{C}}$ where I plays strategy g and II plays $A(n)$ at that players n^{th} move.

Say $B \in 2^{\omega}$ is constructed by this play. Then $A \leq_T B \leq_T A \oplus g$.

But $\mathbf{d} \leq \mathbf{a}$, giving $g \leq_T A$. So $A \leq_T B \leq_T A$ — that is, $A \equiv_T B$.

But since g is a winning strategy for I, we have $B \in \mathcal{C}$. So $A \in \mathcal{C}$ since $G_{\mathcal{C}}$ is a degree game — so $\{ X \in 2^{\omega} \mid \mathbf{d} \leq \deg(X) \} \subseteq \mathcal{C}$ as required.

The case where II has a winning strategy is similar. $\quad\square$

COROLLARY 14.2.4
There exists a cone $\mathcal{D}(\geq \mathbf{d})$ of minimal covers in $\mathcal{D}$.

PROOF Let $\mathcal{C} = \{ X \in 2^{\omega} \mid \deg(X) \text{ is a minimal cover} \}$.

Then $\mathcal{C}$ is Borel — Σ_5^0 in fact, by Exercise 14.1.7. And $\mathcal{C}$ is degree invariant by definition.

So by Martin's theorem, $G_{\mathcal{C}}$ is determinate. Then by Lemma 14.2.3, $\mathcal{C}$ or $2^{\omega} - \mathcal{C}$ contains a cone.

But $2^{\omega} - \mathcal{C}$ does not contain a cone, since by Corollary 13.4.17 there exists a minimal cover above any given degree.

So $\mathcal{C}$ contains a cone. $\quad\square$

Notice what different results we get using determined games than we do using forcing — generic degrees were best at avoiding cones!

What the methods have in common though is weak effective content which takes digging out. You might be a little dissatisfied with not knowing where the base degree $\mathbf{d}$ of this cone of minimal covers lies in $\mathcal{D}$. However, another really quite beautiful fact is that $\mathbf{d}$ can be about as close to home as Theorem 14.2.1 will allow — on closer examination, one finds that $\mathbf{0}^{(\omega)}$ is the base of a cone of minimal covers.

I say "about as close" since $\mathbf{0}^{(\omega)}$ is not actually a minimal upper bound for the arithmetical degrees, but is the least element of a cone of double jumps of such a minimal upper bounds. The fact that $\mathbf{d}$ cannot be arithmetical — every arithmetical degree is, by definition, below some $\mathbf{0}^{(n)}$ — gives us a difference between the theories of $\mathcal{D}$ and $\mathcal{D}(\text{arith})$ (= the set of all arithmetical degrees).

Hopefully you remember from Remark 10.6.8 how we defined the elementary — that is first order — theory $\text{Th}(\mathcal{D})$ of $\mathcal{D}$. We can, of course, use the language of $\text{Th}(\mathcal{D})$ to make *elementary statements* about any other degree structure

$\boldsymbol{\mathcal{D}}_r$ — we say $\boldsymbol{\mathcal{D}}$ and $\boldsymbol{\mathcal{D}}_r$ *interpret the same language.* The elementary theory Th($\boldsymbol{\mathcal{D}}_r$) of $\boldsymbol{\mathcal{D}}_r$ will be the set of those wfs of Th($\boldsymbol{\mathcal{D}}_r$) true in $\boldsymbol{\mathcal{D}}_r$.

DEFINITION 14.2.5 *We say that two degree structures $\boldsymbol{\mathcal{D}}_r$, $\boldsymbol{\mathcal{D}}_R$ are* **elementarily equivalent** — *written $\boldsymbol{\mathcal{D}}_r \equiv \boldsymbol{\mathcal{D}}_R$ — if $Th(\boldsymbol{\mathcal{D}}_r) = Th(\boldsymbol{\mathcal{D}}_R)$.*

Notice that two isomorphic structures which interpret the same first order language must be elementarily equivalent. Anyway:

COROLLARY 14.2.6
The degree structures $\boldsymbol{\mathcal{D}}$ and $\boldsymbol{\mathcal{D}}$(arith) are not elementarily equivalent.

PROOF The sentence φ given by

$$(\exists \mathbf{x})(\forall \mathbf{y} \geq \mathbf{x})[\mathbf{a} \text{ is a minimal cover}]$$

can be expressed as an elementary wf — where $\boldsymbol{\mathcal{D}} \models \varphi$ and $\boldsymbol{\mathcal{D}}$(arith) $\not\models \varphi$, by Theorem 14.2.1. ▯

EXERCISE 14.2.7 *Show that $\boldsymbol{\mathcal{D}}(\leq \mathbf{0}')$ is not elementarily equivalent to $\boldsymbol{\mathcal{E}}$.*

EXERCISE 14.2.8 *Show that $\boldsymbol{\mathcal{E}}$ is not elementarily equivalent to $\mathbf{D}_2$.*

EXERCISE 14.2.9 *Show that $\boldsymbol{\mathcal{D}}_e$ and $\boldsymbol{\mathcal{P}}$ (= the set of all partial degrees) are elementarily equivalent.*

14.3 Borel and Projective Determinacy, and the Global Theory of $\boldsymbol{\mathcal{D}}$

What happens to games $G_\mathcal{C}$ when $\mathcal{C}$ is only projective instead of Borel?

We cannot prove them determinate. Instead we get a powerful set theoretic alternative to full (AD):

THE AXIOM OF PROJECTIVE DETERMINACY (PD)

If $C \subseteq 2^\omega$ is a projective set then G_C is determinate — in particular, G_C is determinate for every analytical set C.

One use of (PD) is to sort out the theory of the projective hierarchy, something which was beyond those early developers of the classical theory.

I will give just one application of (PD) to the theory of the Turing degrees.

THEOREM 14.3.1 (Martin, 1968)

Assuming projective determinacy, there is a cone $\mathcal{D}(\geq \mathbf{d})$ of bases of elementarily equivalent cones of $\mathcal{D}$.

PROOF   Given a φ of $\mathrm{Th}(\mathcal{D})$, take

$$C = \{\, A \mid \mathcal{D}(\geq \deg(A)) \models \varphi \,\}.$$

You can translate $\mathcal{D}(\geq \deg(A)) \models \varphi$ into an analytic statement about A based on φ, including the lower bound $\deg(A)$, replacing degrees $\mathbf{x}$ by $\deg(X)$, quantifiers $(\exists \mathbf{x})$ by $(\exists X)$, etc.

Then since C is analytic, C is projective. So (PD) means the degree game G_C is determinate. This means either C or $2^\omega - C$ contains a cone of degrees $\mathcal{D}(\geq \mathbf{d}_\varphi)$.

If we take an upper bound $\mathbf{d}$ for the countable set of all such degrees $\mathbf{d}_\varphi$, we get the required upper cone.   ⬜

EXERCISE 14.3.2   *Show that there exists a comeager set of degrees which are bases of elementarily equivalent cones of $\mathcal{D}$.*

Theorem 14.3.1 is a mathematically big statement. It is important as an example of how we can get a grip on a *global* assertion about $\mathcal{D}$. The algorithmic content and computational complexity of our Universe make the Turing model a natural source of answers to basic questions. The global theory of $\mathcal{D}$ is concerned with overall features of the incomputable universe. Although these may not obviously impact on what we experience locally, we do live inside global structures, and answers to global questions can ultimately change the way we see the world.

Most of the global questions — not just about $\mathcal{D}$, but also $\mathcal{D}_e$ and the Medvedev lattice — go back to Hartley Rogers' 1967 paper, based on his talk to the 1965 Tenth Logic Colloquium in Leicester, England. As Rogers states in his introduction to the paper "I shall be more concerned with presenting certain open problems in recursive function theory than with giving new results," going on to say that these problems "are easily stated and appear to be of central significance in the foundations of recursive function theory."

It is a remarkable fact that these "easily stated" questions — some of which became better known via his book — have guided the direction of research in the area ever since.

What sort of things did he ask? Most were about definability and invariance of various structures — such as that of $\mathbf{D}$ in $\mathbf{D}_e$ — and relations such as "c.e. in" on those structures. Another concerned *homogeneity* of the structure of $\mathbf{D}$. Rogers made the ultimate "big statement" on the extent of relativisation in $\mathbf{D}$:

THE HOMOGENEITY CONJECTURE

For every $\mathbf{a} \in \mathbf{D}$ *we have* $\mathbf{D} \cong \mathbf{D}(\geq \mathbf{a})$.

There is also a "strong" version of the conjecture where we include a symbol for the jump operator in the language for $\mathbf{D}$.

At the time Theorem 14.3.1 above must have seemed an interesting but very feeble response to Rogers' conjecture — the base of the cone not $\mathbf{0}$, elementary equivalence and not isomorphism, and not even an outright theorem — needing (PD) with not even Borel determinacy known yet. Today it is still about the best we can do. Homogeneity fails very extensively.

Much is now known that Rogers could have hardly dreamt of. If you would like to know more, Odifreddi volumes I and II are an encyclopaedic and readable source.

Chapter 15

The Computability of Theories

Is what we know what we compute? If so, everything we have seen so far says we do not know very much — our place in the Universe is indeed a lowly one! No wonder human affairs are in such a mess.

On the other hand, the experience of AI tells us something different. Pure computing machines turn out to be very bad at thinking as we do. Only when we ask our machines to engage with the real world in an adaptive non-convergent way do we see even the most basic human skills replicated. Our great aptitude is copying, that is, computing at best like an oracle Turing machine, and copying something that is not itself purely mechanical. Our minds seem to reflect — to be a microcosm of — the apparent complexity of the Universe.

But knowledge we can communicate — the stuff of science in particular — is language based and certainly has a very high algorithmic content. It must be captured in theories. And even if knowledge in the widest sense may not be axiomatic, the theories we use do have reductive structure.

In this chapter we look at the computability of information content which is related to theories in various ways.

15.1 Feferman's Theorem

History nowadays gives proper attention to the social context from which leaders and events emerge. The little people, not visible as individuals in the large picture, are certainly "naturally occurring", and can be thought of as "defining" the historical features we recognise. Or again — we may find a drifting cloud to be momentarily full of interest, but soon gone. It is no less naturally occurring than the geographical landmarks shown on the hiker's map we carry.

We should keep such images in mind in our search for natural examples of incomputable sets. As we saw earlier, diophantine sets have a good claim to be mathematics "existing in or caused by nature" — remember (page 99) our definition of *natural* taken from the *The New Oxford Dictionary of English*. And if we are talking about the mathematics of information, so do axiomatisable theories, or *formal systems* as they are sometimes more obscurely called. Mathematicians give special status to only a very few of these, which all reside in very select neighbourhoods — **0** or **0**′. But our next theorem should reinforce your picture of naturally occurring information content dispersed throughout the c.e. Turing degrees.

As you may remember from Exercise 9.3.9, the theory of pure monadic predicate calculus — that is, PC with no function of constant symbols, and restricted to predicates of just one argument — is decidable. We can use this to get:

THEOREM 15.1.1 (Feferman, 1957)
Every c.e. degree **a** *contains an axiomatisable first order theory.*

PROOF   Let A c.e. $\in$ **a** be given. We define our theory $\mathcal{T}$ as follows:

(1) Take pure monadic predicate calculus with the unary predicate symbols $P_0, P_1, \ldots$.

(2) Add proper axioms of the form $(\forall x)P_n(x)$ with $n \in A$.

Then $\boxed{A \leq_T \mathcal{T}}$ — since $n \in A \iff \vdash_{\mathcal{T}} (\forall x)P_n(x)$.

To see this, notice that: (i) If $n \in A$ then $(\forall x)P_n(x)$ is an axiom of $\mathcal{T}$.

And: (ii) If $n \notin A$ then $\nvdash_{\mathcal{T}} (\forall x)P_n(x)$ — since there is clearly a model of $\mathcal{T}$ in which $(\forall x)P_n(x)$ is not true, given that P_n does not appear in any proper axiom of $\mathcal{T}$.

Also $\boxed{\mathcal{T} \leq_T A}$ — Let φ be a sentence of $\mathcal{T}$. Then:

$$\vdash_{\mathcal{T}} \varphi \iff \varphi \text{ is true in every model of } \mathcal{T}.$$

But just as in showing pure monadic predicate calculus decidable in Exercise 9.3.9, we need only consider a finite number of models, this number being dependent on the number of predicate symbols appearing in φ. And since each such model can be selected with help from an oracle for A, we can decide whether or not $\vdash_{\mathcal{T}} \varphi$ computably in A.

To finish off the proof, we notice that if A is c.e., then so are the proper axioms of $\mathcal{T}$. And hence $T_{\mathcal{T}}$ is c.e.

Then by Craig's result from Exercise 8.1.4, $\mathcal{T}$ is computably axiomatisable.

∎

You would probably be even more impressed if we could find a *finitely* axiomatisable theory in any given c.e. **a**. Well, we can — there is a theorem of Hanf from 1962 which tells us just that.

EXERCISE 15.1.2 *Show that there is a first order theory $\mathcal{T}$ of any given Turing degree* **a**.

15.2 Truth versus Provability

We saw that PA was creative, so contained in $\mathbf{0}'$. We also saw that $\mathrm{Th}(\mathfrak{N})$, the theory of true first order arithmetic, is not even axiomatisable.

Can we also locate the degree of $\mathrm{Th}(\mathfrak{N})$ exactly? This would give us a better idea of *how much* of arithmetic our axiomatic theories do capture. Well, it turns out that the theorems of PA hardly scrape the surface of true arithmetic.

THEOREM 15.2.1
 The degree of $\mathrm{Th}(\mathfrak{N})$ is $\mathbf{0}^{(\omega)}$.

PROOF   We start by noticing we can take $(\oplus_{n\geq 0}\emptyset^{(n)}) \in \mathbf{0}^{(\omega)}$.

For $\boxed{(\oplus_{n\geq 0}\emptyset^{(n)}) \leq_T \mathrm{Th}(\mathfrak{N})}$ — we observe that each $\emptyset^{(n)}$ is defined in the language of $\mathrm{Th}(\mathfrak{N})$. So we can uniformly decide $m \in \emptyset^{(n)}$ by deciding it in $\mathrm{Th}(\mathfrak{N})$.

The converse $\boxed{\mathrm{Th}(\mathfrak{N}) \leq_T (\oplus_{n\geq 0}\emptyset^{(n)})}$ is just a little more technical.

For each sentence φ of $\mathrm{Th}(\mathfrak{N})$ we need a uniform way of computably reducing it to some $\emptyset^{(n)}$.

What we do know is that $\emptyset^{(n+1)}$ is Σ_{n+1} complete. And that given an index for a set A c.e. in $\emptyset^{(n)}$, we can uniformly compute A from $\emptyset^{(n+1)}$.

We know from logic that we can uniformly put φ in what is called *prenex normal form* — that is, in the form $(Q_1 x_1)\ldots(Q_k x_k)\psi$, where each $Q_i x_i$ is an existential or universal quantifier, and ψ has no quantifiers. We can then form a new expression $(\exists y)(Q_1 x_1)\ldots(Q_k x_k)[y = y \,\&\, z = z \,\&\, \psi]$ where y, z are new variables not appearing in φ.

The new expression presents φ as a Σ_{n+1} set A for some $n \geq 0$, and allows us to find — uniformly again — an index for A as a set c.e. in $\emptyset^{(n)}$. We can then compute A uniformly from $\emptyset^{(n+1)}$.

Now notice that $A = \mathbb{N}$ if φ is true — and $= \emptyset$ if φ is false. So φ is true $\iff 0 \in A \leq_T \emptyset^{(n+1)}$.

The result follows. $\Box$

REMARK 15.2.2 Just as PA turns out to be computably equivalent to a whole range of incomputable sets, all of them inhabiting $\mathbf{0}'$ — so $\mathrm{Th}(\mathfrak{N})$ is equivalent to various other less well-understood theories.

For instance, it is easy to see that first order statements about the lattice of c.e. sets, or about the c.e. Turing degrees, or the Turing degrees below $\mathbf{0}'$, can be expressed as statements in $\mathcal{L}_{\mathrm{PA}}$. So the corresponding theories — $\mathrm{Th}(\mathcal{E})$, $\mathrm{Th}(\boldsymbol{\mathcal{E}})$ or $\mathrm{Th}(\mathcal{D}(\leq \mathbf{0}'))$ — are computable from $\mathrm{Th}(\mathfrak{N})$.

What is very remarkable and completely nontrivial is that these theories are all as complicated as they can be — sufficiently so to enable us to code arithmetic into those structures so as to compute $\mathrm{Th}(\mathfrak{N})$ from any one of $\mathrm{Th}(\mathcal{E})$, $\mathrm{Th}(\boldsymbol{\mathcal{E}})$ or $\mathrm{Th}(\mathcal{D}(\leq \mathbf{0}'))$. So all these theories have degree $\mathbf{0}^{(\omega)}$.

What about $\mathrm{Th}(\mathcal{D})$? Obviously you can still reduce statements of $\mathrm{Th}(\mathcal{D})$ to statements in arithmetic, but to do that you need to quantify over sets. So $\mathrm{Th}(\mathcal{D})$ is part of the *second order* theory $\mathrm{Th}^2(\mathfrak{N})$ of arithmetic. Again it is sufficiently complicated for the two theories to be computably equivalent. ☐

15.3 Complete Extensions of Peano Arithmetic and Π_1^0-Classes

Theorem 15.2.1 says that complete and perfect information on anything of which arithmetic is a part is not merely impossible to compute. It is so computably remote from us, it is right over the horizon.

What do we do in real life when we can be sure of some things and not of others?

When what is knowable is clear and the task in hand is well defined, we stick to what we know. But sometimes the extent of what is sure is not clear. Even in science, a theory may turn out to be useful but dramatically wrong. But there may be global benefits from working with complete but possibly imperfect information. Scientific theories can provide a working framework for decades, even centuries, before the secure core is transcended. While a system of law based on what is later seen as bad social assumptions can deliver great societal benefits.

Mathematically, Gödel's Theorem tells us that in a sufficiently strong language, complete information cannot be computably generated. But are there complete extensions of PA, or other useful theories, which are closer to home? We can find out more by looking at the exact relationship between the computably enumerable sets and particular Π_1^0 classes.

For the purposes of the following definition, a *computable tree* is a computable ideal of $\omega^{<\omega}$ (with the ordering $\subseteq$ on strings). So a computable tree $T \subseteq \omega^{<\omega}$ is computable, and satisfies $\sigma \subseteq \tau \in T \implies \sigma \in T$.

Such trees are not much different to those of Definition 13.4.1. They are essentially the identity tree on $\omega^{<\omega}$ with terminating branches — where T is now computable as a set of strings, rather than as a function over the strings.

DEFINITION 15.3.1 $\mathcal{A} \neq \emptyset$ *is a* Π_1^0 *class of functions if and only if there exists a computable tree T such that*

$$f \in \mathcal{A} \iff \forall x \in \omega \, [f \restriction x \in T].$$

That is, a Π_1^0 *class $\mathcal{A}$ is the set of infinite branches of a computable tree T, where we write $\mathcal{A} = [T]$.*

Notice that the notation fits perfectly with our previous description — see Remark 13.1.10 — of level 1 of the arithmetical hierarchy of sets of sets and functions.

EXAMPLE 15.3.2 (Jockusch) *Define the class DNC of all **diagonally non-computable** functions to be those functions f for which*

$$(\forall i)[\, f(i) \neq \varphi_i(i)\,].$$

Show that DNC is a Π_1^0 *class of functions.*

SOLUTION Define a computable tree T by

$$\sigma \in T \iff (\forall s < |\sigma|)\,[\varphi_{i,s}(i) \downarrow \iff \sigma(i) \neq \varphi_{i,s}(i)].$$

| Show DNC $\subseteq [T]$: | Say $f \in$ DNC. |

Then $(\forall i)[f(i) \neq \varphi_i(i)$ — so that for every $\sigma \subset f$ and $s \geq 0$ we have $\sigma(i) \neq \varphi_{i,s}(i)$ — giving $f \in [T]$.

| Show $[T] \subseteq$ DNC: | Assume $f \in [T]$, and $i \geq 0$. |

Then for every $z \geq 0$ we have $f \restriction z \in T$. So

$$\varphi_{i,s}(i) \downarrow \ \& \ f \restriction s \in T \implies f(i) \neq \varphi_{i,s}(i).$$

So $f(i) \neq \varphi_i(i)$, giving $f \in$ DNC. $\qquad\qquad$ ◻

Where do complete extensions of PA arithmetic come in here?

To make the connection, we need to consider Π_1^0 classes of sets. This means looking at 0–1 valued Π_1^0 classes got by restricting our trees T to 0–1 valued strings. This is a special case of *computably bounded* Π_1^0 classes — where $[T]$ is computably bounded if we know computably exactly how many successors each $\sigma \in T$ has. Obviously every 0–1 valued Π_1^0 class is computably bounded.

We can look at a theory as the set of Gödel numbers of its theorems.

THEOREM 15.3.3 (Shoenfield, 1960)
The set of all complete extensions of a consistently axiomatised first order theory is a Π_1^0 class of sets.

PROOF Write ψ_i for the sentence of the consistently axiomatised first order theory $\mathcal{T}$ with Gödel number i. Let $\{T_{\mathcal{T}}^s\}_{s \geq 0}$ be a c.e. approximating sequence for the set $T_{\mathcal{T}}$ of Gödel numbers of theorems of $\mathcal{T}$. We assume as usual that $i \in T_{\mathcal{T}}^s \Rightarrow i < s$.

Write $\vdash_{\mathcal{T}}^s \psi_i$ for $i \in T_{\mathcal{T}}^s$ — intuitively, "ψ is provable in $\mathcal{T}$ in $\leq s$ steps."

Define the required computable binary tree T to be all $\sigma \in 2^{<\omega}$ satisfying the following conditions:

(1) If $\vdash_{\mathcal{T}}^{|\sigma|} \psi_i$ then $\sigma(i) = 1$
   (so every $A \in [T]$ codes an extension of $\mathcal{T}$)

(2) If $\vdash_{\mathcal{T}}^s (\psi_i \Rightarrow \psi_j)$ and $\sigma(i) = 1$ then $\sigma(j) = 1$, and

(3) If $\psi_k = \psi_i \& \psi_j$ with $k < |\sigma|$, and $\sigma(i) = 1$ and $\sigma(j) = 1$, then $\sigma(k) = 1$
   (then (1), (2) and (3) say that every $A \in [T]$ codes a theory)

(4) If $\sigma(i) = 1$ and $\psi_j = \neg \psi_i$ with $j < |\sigma|$, then $\sigma(j) = 0$
   (so each $A \in [T]$ codes a *consistent* theory)

(5) If $\psi_j = \neg \psi_i$ with $j < |\sigma|$, then either $\sigma(i) = 1$ or $\sigma(j) = 1$
   (so each $A \in [T]$ codes a *complete* theory)   ⬚

It immediately follows of course that:

COROLLARY 15.3.4
The set of all complete extensions of first order Peano arithmetic is a Π_1^0 class of sets.

EXERCISE 15.3.5 *Show that the set of all consistent extensions of a consistently axiomatised first order theory is a Π_1^0 class of sets.*

This next result is a sort of abstract version of Theorem 15.3.3.

EXERCISE 15.3.6   *We say that a set S **separates** two disjoint sets $A, B \subseteq \mathbb{N}$ if $A \subseteq S \subseteq \overline{B}$.*

Show that the set of all sets separating two disjoint c.e. sets is a Π_1^0 class.

Of course, by Exercise 9.2.13, this Π_1^0 class may have no computable members.

When *does* a Π_1^0 class have computable members?

EXERCISE 15.3.7   *Let $T \subseteq \omega^\omega$ be a computable tree. Show that if $[T]$ is computably bounded and $[T] = \{f\}$ then f is computable.*

EXERCISE 15.3.8   *Let $T \subseteq 2^{<\omega}$ be a computable tree. We say $A \in [T]$ is **isolated** in T if there is a $\sigma \in T$ such that $N_\sigma \cap [T] = \{A\}$.*

Show that each isolated $A \in [T]$ is computable.

Deduce that if a Π_1^0 class of sets has no computable members, it must have $2^{\aleph_0}$ members.

EXERCISE 15.3.9   *Show that every Π_1^0 class of sets without computable members is meager in the Cantor topology for 2^ω.*

The next exercise shows that there exist infinite Π_1^0 classes of sets with no computable member. In fact:

EXERCISE 15.3.10   *Show that there exists an infinite computable tree with no c.e. $A \in [T]$.*

[**Hint:** Show there exists a Π_1^0 class of bi-immune sets. To define T, wait for a large enough $x, y \in W_{i,s}$, then freeze all branches above any σ with $|\sigma| > s$ and $\sigma(x) = 1$ or $\sigma(y) = 0$.]

What we are interested in, of course, are the *degrees* of complete extensions of PA. Are there any such extensions closer to home than in $\mathbf{0}^{(\omega)}$?

> **DEFINITION 15.3.11**   *We write **PA** for the class of all PA degrees — that is, the set of degrees of all complete extensions of first order Peano arithmetic.*

We could equivalently take **PA** to be the set of degrees of all *consistent* extensions of PA.

EXERCISE 15.3.12   *Let $T \subseteq \omega^\omega$ be an infinite computable tree, with $[T]$ computably bounded. Show that $[T]$ has a member of c.e. degree.*

[**Hint:** Let $A =$ the leftmost infinite path through T. Show that $A \equiv_T$ the c.e. set of strings on T to the left of A.]

It immediately follows that there is a complete extension of PA of c.e. degree.

EXERCISE 15.3.13   *Let $T \subseteq \omega^\omega$ be a computable tree. We say $[T]$ is* **bounded** *if T is finitely branching.*

Show that every non-empty bounded Π_1^0 class has a member of Σ_2^0 degree.

[**Hint:** Use an oracle for $\emptyset'$ to show your chosen path is of Σ_2^0 degree.]

If we are only interested in the degrees of members of Π_1^0 classes, we can reduce computably bounded classes to 0–1 valued ones:

EXERCISE 15.3.14   *Let $[T]$ be a computably bounded Π_1^0 class. Show that there is a 0–1 valued Π_1^0 class $[T^*]$ such that the sets of degrees of $[T]$ and $[T^*]$ are the same.*

[**Hint:** Define a function Bin: $T \to 2^{<\omega}$ by

$$\text{Bin}(\emptyset) = \emptyset,$$
$$\text{Bin}(\tau^\frown n) = \text{Bin}(\tau)^\frown 0^n {}^\frown 1,$$

for each successor $\tau^\frown n$ of τ on T. Check that $T^* = \{\sigma \mid \sigma \subseteq \text{Bin}(\tau), \tau \in T\}$ has the right properties.]

REMARK 15.3.15   There are close connections between **PA** and two other very interesting degree classes. Say we define the *fixed point free* degrees **FPF** to be the degrees containing *fixed point free* functions $f \in \omega^\omega$ for which $(\forall i)(W_i \neq W_{f(i)})$. Then

$$\boxed{\textbf{FPF} = \textbf{DNC}}$$

and

$$\boxed{\textbf{PA} = \text{the degrees of } 0\text{–}1 \text{ valued DNC functions}}$$

It turns out that both **PA** and **FPF** are closed upwards. The inclusion **PA** $\subset$ **FPF** is proper, even on degrees below $\mathbf{0'}$ — in fact, there is a jump inversion theorem for degrees in **FPF** $-$ **PA**.   ⬜

EXERCISE 15.3.16 *Show that if g is diagonally non-computable, there exists a fixed point free $f \leq_T g$.*

[**Hint:** Let ψ p.c. be such that $W_i \neq \emptyset \Rightarrow \psi(i) \in W_i$. Let $\varphi_{q(i)} = \lambda y[\psi(i)]$. Take $W_{f(i)} = \{g(q(i))\}$ for all i.]

It is time to put Exercise 15.3.12 in a more general context.

15.4   The Low Basis Theorem

The beauty of the approach via Π_1^0 classes is the way it brings together so many apparently different examples within a concise and approachable framework. We can now work out the relationship of **PA** to other important degree classes in the general context of Π_1^0 classes of sets.

DEFINITION 15.4.1 *(1) We say that $\mathcal{B} \subseteq 2^\omega$ is a **basis** for Π_1^0 classes of sets if every such Π_1^0 class has a member in $\mathcal{B}$.*

*(2) A set of degrees is a **basis** for Π_1^0 classes if the union of the set of degrees is.*

So Exercise 15.3.12 says that $\mathcal{E}$ is a basis for 0–1 valued Π_1^0 classes. And Exercise 15.3.10 tells us that **0** is not such a basis, a result going back to Kleene in 1952.

Here is a result telling us how consistent extensions of PA relate to Π_1^0 classes:

THEOREM 15.4.2 (The Scott Basis Theorem, 1962)
The set $\mathcal{D}(\leq \mathbf{a})$ for any given $\mathbf{a} \in \mathbf{PA}$ is a basis for 0–1 valued Π_1^0 classes.

PROOF   This result holds because we can prove enough in first order Peano arithmetic to be able to trace an infinite path through any given infinite computable tree.

Let $[T]$ be a Π_1^0 class of sets, and let $\mathcal{T}$ be a consistent extension of PA. We inductively define our infinite branch on T computably in $T_{\mathcal{T}}$:

Define $\sigma_0 = \emptyset$.

Assume σ_n already defined, with $\sigma_n \subset$ some $A \in [T]$.

Let $\tau \succ_m \sigma$ say that "τ has an extension of length m on T, but σ does not" — so $\tau \succ_m \sigma$ can be written as a computable statement in the language of PA. Then let

$$\sigma_{n+1} = \begin{cases} \sigma_n\hspace{-2pt}\frown\hspace{-2pt}1 & \text{if } \vdash_T (\exists m)[\sigma_n\hspace{-2pt}\frown\hspace{-2pt}1 \succ_m \sigma_n\hspace{-2pt}\frown\hspace{-2pt}0] \\ \sigma_n\hspace{-2pt}\frown\hspace{-2pt}0 & \text{otherwise.} \end{cases}$$

So we can compute σ_{n+1} uniformly using T.

Now say we define $\sigma_{n+1} = \sigma_n\hspace{-2pt}\frown\hspace{-2pt}0$.

Then $(\exists m)[\sigma_n\hspace{-2pt}\frown\hspace{-2pt}1 \succ_m \sigma_n\hspace{-2pt}\frown\hspace{-2pt}0]$ is not true — otherwise, since every Σ_1 relation is semi-representable in PA by the Semi-Representability Theorem 8.1.10, it would be provable in PA and hence in T.

So by the inductive assumption, there must be an $A \in [T]$ with $\sigma_n\hspace{-2pt}\frown\hspace{-2pt}0 \subset A$.

On the other hand, assume we define $\sigma_{n+1} = \sigma_n\hspace{-2pt}\frown\hspace{-2pt}1$.

Then we cannot have $(\exists m)[\sigma_n\hspace{-2pt}\frown\hspace{-2pt}0 \succ_m \sigma_n\hspace{-2pt}\frown\hspace{-2pt}1]$ being true, since then it would be provable in PA and hence in T, contradicting the consistency of T.

So this time there is an $A \in [T]$ with $\sigma_n\hspace{-2pt}\frown\hspace{-2pt}1 \subset A$.

It follows that $A = \cup_{n \geq 0} \sigma_n \leq_T T_T$ is the required member of $[T]$. $\quad\Box$

It is a remarkable fact due to Robert Solovay that there is a converse to this result — if $\mathcal{D}(\leq \mathbf{a})$ is a basis for 0–1 valued Π_1^0 classes then $\mathbf{a}$ is a PA degree.

Here is our most useful basis, proved by Jockusch and Soare around 1972. You often come across applications of it in the most unexpected places.

THEOREM 15.4.3 (The Low Basis Theorem)
The low degrees form a basis for Π_1^0 classes of sets.

PROOF This involves forcing with Π_1^0 classes.

Assume a given Π_1^0 class $[T]$ of sets. We define a sequence $T_0 \supset T_1 \supset \ldots$ of infinite computable subtrees of T.

The idea is that we build $T_0 \supset T_1 \supset \ldots$ computably in $\emptyset'$, in such a way that we decide whether $i \in A'$ or not in the same way for *every* $A \in [T_{i+1}]$ in defining T_{i+1}. Then if we take $A \in \cap_{i \geq 0}[T_i]$, we will be able to compute A' from the construction computably in $\emptyset'$.

Here is our inductive definition of the trees: $T_0 = T$.

Let $X_i = \{\sigma \in T_i \mid i \notin W_{i,|\sigma|}^\sigma\}$, a computable set. Define:

$$T_{i+1} = \begin{cases} X_i & \text{if } X_i \text{ is infinite,} \\ T_i & \text{otherwise.} \end{cases}$$

Then $T_0 \subset T_1 \subset \ldots$ has the properties we need — T_{i+1} is a computable subtree of T_i. And "Is X_i infinite?" is a Δ_2 question — I will leave you to check that — and so answerable using an oracle for $\emptyset'$.

So we have an $A \in \cap_{i \geq 0}[T_i] \subseteq [T]$ — where $A' \leq_T \emptyset'$. Since $i \in A' \iff$ we chose $T_{i+1} = T_i$ through the second part of the definition of T_{i+1}.   ⬜

EXERCISE 15.4.4   *Say why the above proof does not work if we take $X_i = \{\tau \subseteq \sigma \in T_i \mid i \in W_{i,|\sigma|}^{\sigma}\}$.*

For us the most interesting application is:

COROLLARY 15.4.5

There exists a complete extension of Peano arithmetic of low degree.

PROOF   Use the Low Basis Theorem with the Π_1^0 class of all complete extensions of PA.   ⬜

Can this degree be c.e.? Or of minimal degree? The answer is "No" on both counts — as we will see from the exercise which follows, and the next section.

EXERCISE 15.4.6   *Find a Π_1^0 class of sets none of whose members are computable or of minimal degree.*

Hence, show that no consistent extension of PA can be of minimal degree.
[**Hint:** Define a computable binary tree T such that

$$A \oplus B \in [T] \iff (\forall i)[B(i) \neq \Phi_i^A(i)] \ \& \ (\forall i)[A(i) \neq \Phi_i^B(i)].]$$

EXERCISE 15.4.7   *Show that the hyperimmune free degrees below $0^{(2)}$ form a basis for 0–1 valued Π_1^0 classes.*
[**Hint:** Force with Π_1^0 classes based on Example 13.4.27.]

15.5   Arslanov's Completeness Criterion

The next result from around 1981 has become a classic, with immediate consequences for the PA degrees.

THEOREM 15.5.1 (Arslanov Completeness Criterion)
The only fixed point free c.e. degree is $\mathbf{0}'$. In particular, $\mathbf{PA} \cap \mathcal{E} = \{\mathbf{0}'\}$.

PROOF Let A be c.e. We will show that A is complete if and only if there is a function $f \leq_T A$ such that for every i, $W_i \neq W_{f(i)}$.

($\Rightarrow$) Computably in $\emptyset'$ we can find uniformly an $f(i)$ such that

$$W_{f(i)} = \begin{cases} \{0\} & \text{if } 0 \notin W_i, \\ \emptyset & \text{otherwise.} \end{cases}$$

Then $W_i(0) \neq W_{f(i)}(0)$.

($\Leftarrow$) Let $f = \Theta^A$ be a fixed point free function.

The idea is that we can use the Fixed Point Theorem on the computable approximations to f. And then these approximations must be very clever — $\emptyset'$ clever — if their fixed point is not to carry over to f itself. Here is the proper argument.

Given any p.c. function ψ, consider $W_{\Theta^A(i)[\psi(x)]}$. By the Fixed Point Theorem with parameters — see Exercise 4.4.2 — there is a computable h such that

$$W_{h(x)} = W_{\Theta^A(h(x))[\psi(x)]}.$$

Now if $A \upharpoonright \theta(h(x))[\psi(x)] = A \upharpoonright \theta(h(x))$ then

$$W_{h(x)} = W_{\Theta^A(h(x))} = W_{f(h(x))},$$

so that $h(x)$ is a fixed point of f.

This means that the A-computable function $C_A(\theta(h(x))$ dominates every p.c. function $\psi(x)$. So by Exercise 12.2.21 we have $\emptyset' \leq_T A$.

For the last part of the theorem, we use $\mathbf{PA} \subset \mathbf{FPF}$ to see that the only possible c.e. member of PA is $\mathbf{0}'$. The result follows since by Exercise 15.3.12 every non-empty Π_1^0 class of sets has a member of c.e. degree. $\square$

Notice that we do seem to need the assumption that A is c.e. in the second part of this proof. And the next exercise seems to confirm that.

EXERCISE 15.5.2 *Show that there exist incomplete fixed point free degrees below $\mathbf{0}'$.*

[**Hint:** Use $\mathbf{FPF} = \mathbf{DNC}$ and Example 15.3.2.]

However — Arslanov's completeness criterion for c.e. degrees was extended in 1989 by Jockusch, Lerman, Soare and Solovay to all finite levels of the n-c.e. (and even n-CEA) hierarchy. So, for instance, there are no fixed point free d.c.e. degrees other than $\mathbf{0}'$.

You can think of Arslanov's Theorem as saying that truth in mathematics has little to do with c.e. sets or axiomatic theories. As you can guess, there are implications for non-monotonic reasoning — that is, reasoning which allows for the correcting of mistakes. You should go to Cenzer and Remmel's article on "Π_1^0 Classes in Mathematics" in volume 2 of the *Handbook of Recursive Mathematics* to find out more.

15.6 A Priority-Free Solution to Post's Problem

Our next result is a bit of a curiosity, which has caused quite a lot of interest. For a while it gave some comfort to those who did not like priority arguments, and who looked forward to a world where you could get away with knowing nothing about them.

In proving it, Antonín Kučera was able to use the Low Basis Theorem to get a low 0–1 valued DNC function, and hence a low $\mathbf{a} \in \mathbf{FPF}$. Since Jockusch and Soare's proof only uses a $\mathbf{0}'$ oracle, Kučera can apply his theorem to obtain a priority free solution to Post's problem.

> **THEOREM 15.6.1 (Kučera 1986)**
> *Every fixed point free — and in particular PA — degree below $\mathbf{0}'$ has a nonzero c.e. predecessor.*

PROOF Let $\{f^s\}_{s \geq 0}$ be a Δ_2-approximating sequence for f — where $f(x) = \text{Lim}_s\, f^s(x)$ for each x.

We will define a simple set A as in Theorem 6.2.3, with $A \leq_T f$. Once again, as in the proof of Arslanov's Theorem, we use the fact that f is fixed point free to make the approximations f^s to f — which *will* have fixed points — cleverly avoid communicating their fixed points to f. This time we will use the cleverness to compel f to Δ_2 permit the necessary A changes.

As in Theorem 6.2.3 we build A c.e. satisfying the requirements:

$$\mathcal{N}_i: \qquad |A \cap \{0, 1, \ldots, 2i\}| \leq i$$
$$\mathcal{P}_i: \qquad \text{If } W_i \text{ is infinite, then } W_i \cap A \neq \emptyset.$$

The construction will be relative to a given number y. At stage s of the construction we will computably define a number $s(i, y) = s$. At the end of the construction we will consider $W_{f^{s(i,y)}(y)}$, and use the Fixed Point Theorem to choose a computable *function* $y(i)$ such that

$$W_{y(i)} = W_{f^{s(i,y(i))}(y(i))}.$$

But having described what we are going to do, why not make things clearer, and describe the construction with $y = y(i)$ *already chosen*?! This is a common device, well worth remembering.

So, writing $s(i) = s(i, y(i))$, we assume given y with $W_{y(i)} = W_{f^{s(i)}(y(i))}$, where, if $s(i) \uparrow$, we have $W_{y(i)} = \emptyset$, of course.

The Construction:

(1) For each as yet unsatisfied $\mathcal{P}_i$, wait for a stage s at which there is a number $x \in W_{i,s}$ with $x > 2i$ — and with $x > y(i)$, and $f^t(y(i)) = f^s(y(i))$ for each $t \in [x, s]$.

(2) If such an x appears, choose the x for the least such i.
Define $s(i) = s$, and throw x into A. $\mathcal{P}_i$ is *satisfied* at all stages $t > s$.

The verification that A is simple goes just as for Theorem 6.2.3. We just need to notice that if W_i is infinite, there will certainly be an $x \in W_i$ such that $x > y(i)$ and $f^t(y(i)) = f(y(i))$ for all $t \geq x$.

| Show $A \leq_T f$ | How do we compute $A(x)$ from f?

First notice that x can only be thrown into A on behalf of some $\mathcal{P}_i$ with $2i < x$. Then:

(1) For each such i compute $y(i)$, and ask: " Is $f^x(y(i)) = f(y(i))$?"

(2) If "Yes" — then x can never be thrown into A on behalf of $\mathcal{P}_i$. Since otherwise, when $x \in A^s - A^{s-1}$, we have $s(i) = s$ and $f^{s(i)}(y(i)) = f^x(y(i))$, giving

$$W_{y(i)} = W_{f^{s(i)}(y(i))} = W_{f(y(i))}$$

— contradicting f being fixed point free.

(3) For each i for which we get "No", we look for a stage $s > x$ at which $f^s(y(i)) \neq f^x(y(i))$.
Let $s^* > x$ be the largest of such stages ($= x$ if no such stages exist).
Then $A(x) = A^{s^*}(x)$. $\square$

So you now see how to get an incomplete c.e. degree without using a priority argument. Neither the Low Basis Theorem nor the simple set construction involves any injury.

You should see Kučera's 1989 paper for a range of ingenious applications to other results previously needing finite (and even infinite) injury priority constructions. There are only a limited number of such applications though, and their proofs get increasingly tortuous. I am afraid you will still need to learn about priority!

A good question though: Are the priority free proofs any easier? One measure of complexity of proofs — remember our mention of reverse mathematics in Remark 8.2.9 — is how much induction is needed to carry them out. In this case both Kučera's proof and the standard priority one need Σ_1 induction — that is, the induction axiom of PA for Σ_1 wfs.

Another intriguing non-standard construction is due to Jockusch and Simpson from 1980. They first construct a 0–1 valued Π_1^0 class $\mathcal{P}$ with no computable members such that any two branches determine a minimal pair of truth-table degrees. Then — observing that every member of this $\mathcal{P}$ which is of hyperimmune free degree is also of minimal Turing degree — they get from Exercise 15.4.7 an alternative construction of a minimal degree.

15.7 Randomness

The concept of randomness is a very significant one for computer scientists. Its mathematical theory took off with Martin-Löf's notion of algorithmic randomness around 1966. Since then randomness has been pursued in many different ways, most notably by Andrei Kolmogorov, Gregory Chaitin, Ray Solomonoff and Robert Solovay.

I will just outline the basics of the pure theory of randomness, and tell you how this relates to what we have been doing in this chapter so far.

Unlike incomputability, there is no clear-cut definition of randomness. Here is the natural statistical definition of a random real Martin-Löf came up with, one of a number of related definitions. Let μ be the Lebesgue measure we met in Section 13.2.

DEFINITION 15.7.1 *(1) An $f \in 2^\omega$ is **Martin-Löf random** — or **1-random** — if for every c.e. sequence $\{A_i\}_{i \geq 0}$ of Σ_1^0 sets $A_i \subseteq 2^\omega$ such that $\mu(A_i) \leq 2^{-i}$, we do not have $f \in$ each A_i.*

(2) More generally, we say $\mathcal{A} \subset 2^\omega$ has Σ_1^0-measure zero if $\mathcal{A} \subseteq \cap_{i \geq 0} A_i$ for some such $\{A_i\}_{i \geq 0}$.

Another way of phrasing this definition is in terms of a *Martin-Löf test T* which is a c.e. set of pairs of the form (i, σ) of numbers i and strings $\sigma \in 2^{<\omega}$. Then the sets A_i arise via

$$A_i = \bigcup_{(i,\sigma) \in T} N_\sigma.$$

We say f *fails* the test T if $f \in \cap_{i \geq 0} A_i$, and otherwise *passes* the test T. The

1-random reals are the ones which pass every Martin-Löf test.

Intuitively, random reals are very hard to approximate. f is random if it is not in any constructively defined null subset of the measure space. The practical effects of randomness are very much like those of incomputability — random objects are indeed very hard to computably predict. But you should not confuse these two overlapping but distinct ideas. What makes for randomness is not just incomputability, but *concealed* information content — sometimes very high information content, but inaccessible to the point of worthlessness in any practical sense. For this reason notions of randomness have important consequences for coding theory.

The coding connection can be made clearer via Chaitin's definition of a random real, which is also very persuasive and is equivalent to Martin-Löf's definition. Chaitin says f should be random if it is *algorithmically incompressible* — that is, if the "size" of any program for describing $f \restriction x$ is not significantly smaller than x. I would recommend Chaitin's book on *Algorithmic Information Theory* as an entertaining introduction to his ideas on randomness and its wide-ranging consequences.

I should also mention that since $\mu(2^\omega) = 1$, μ is called a *probability measure*. It turns out that if $\mathcal{A}$ is a sufficiently constructive property of reals which is true with probability 1, then every random real *must* have property $\mathcal{A}$.

Here is a basic result of Martin-Löf from 1970.

THEOREM 15.7.2

There is a universal Martin-Löf test T with $\mathcal{U}_i = \bigcup_{(i,\sigma) \in T} N_\sigma$ each i, where

(1) $\mathcal{U}_0 \supseteq \mathcal{U}_1 \supseteq \dots$.

(2) $\forall i \, [\, \mu(\mathcal{U}_i) < 2^{-i} \,]$.

(3) $\mathcal{A} \subseteq \bigcap_{i \geq 0} \mathcal{U}_i$ for every $\mathcal{A} \subset 2^\omega$ of Σ_1^0-measure zero.

PROOF Let T_i be a standard listing of c.e. tests.

For each $n > i + 1$ enumerate into T all (i, σ) with $(n, \sigma) \in T_n^s$, some $s \geq 0$, with

$$\mu\left(\bigcup_{(n,\sigma) \in T_n^s} N_\sigma \right) < 2^{-n}.$$

(1) follows immediately from the definition of T.

For (2) — By Exercise 13.2.31 part (5), we have

$$\mu(\mathcal{U}_i) = \mu\left(\bigcup_{(i,\sigma) \in T} N_\sigma \right) < 2^{-(i+2)} + 2^{-(i+3)} + \dots < 2^{-i}.$$

For (3) — Say T_n is a Martin-Löf test. Then $\cup_{(n,\sigma)\in T_n} N_\sigma \subseteq \mathcal{U}_i$ for each $i < n - 1$.

But by the Padding Lemma there are infinitely many indices m of T_n giving

$$\bigcup_{(m,\sigma)\in T_n} N_\sigma = \bigcup_{(m,\sigma)\in T_m} N_\sigma \subseteq \mathcal{U}_m.$$

Hence if $\mathcal{A}$ fails test T_n, it also fails test T. □

EXERCISE 15.7.3 *Let $\mathcal{P}_0, \mathcal{P}_1, \ldots$ be the complements of the classes $\mathcal{U}_0, \mathcal{U}_1, \ldots$ in the above theorem.*

Show that $\mathcal{P}_0, \mathcal{P}_1, \ldots$ are Π_1^0 classes of reals with

$$\mathcal{P}_0 \subseteq \mathcal{P}_1 \subseteq \ldots$$

We now get immediately from Theorem 15.7.2:

COROLLARY 15.7.4

(1) *For each $i \geq 0$ the Π_1^0 class $\mathcal{P}_i$ contains only 1-random reals.*

(2) *The class of all random reals has measure one.*

What can we say about the degrees of 1-random reals?

The following result and proof is due to Antonín Kučera, as is the corollary which follows. As usual, we say a degree is 1-random if it contains a 1-random set. I will write $\deg(\mathcal{P}_i)$ for the set of all Turing degrees of members of $\mathcal{P}_i$, **RAND** for the set of all 1-random degrees.

LEMMA 15.7.5

*For each $i \geq 0$, $\deg(\mathcal{P}_i) = $ **RAND**.*

PROOF Let T be the universal test from Theorem 15.7.2.

Define a sequence $T^{(0)}, T^{(1)}, \ldots$ by

$$T^{(0)} = T$$
$$T^{(k+1)} = \{(i, \tau {}^\frown \sigma) \mid (i, \tau) \in T^{(k)} \ \& \ (i, \sigma) \in T\},$$

and let

$$\mathcal{U}_i^{(k)} = \left(\bigcup_{(i,\sigma)\in T^{(k)}} N_\sigma \right).$$

It is easy to see that for each i, $\mathcal{U}_i^{(0)}, \mathcal{U}_i^{(1)}, \ldots$ gives a new test, where for each i, k we have

$$\mu(\mathcal{U}_i^{(k)}) < 2^{-i(k+1)}. \qquad (15.1)$$

Now given i, any given random real f must pass this test. Let k be the least number for which $f \notin \mathcal{U}_i^{(k)}$.

If $k = 0$, then $f \in \mathcal{P}_i$.

If $k > 0$, then $f \in \mathcal{U}_i^{(k-1)} - \mathcal{U}_i^{(k)}$.

So there is some $(i, \sigma) \in T^{(k-1)}$ such that $f \in N_\sigma$, and for any $(i, \tau) \in T$ we have $\sigma^\frown \tau \not\subseteq f$.

Let $g \in 2^\omega$ with $\sigma^\frown g = f$. Then $g \equiv_T f$, and $g \notin \mathcal{U}_i$.

Hence $g \in \mathcal{U}_i$ — giving $\deg(f) \in \deg(\mathcal{P}_i)$.

The reverse inclusion comes from Corollary 15.7.4. $\qquad\qquad$ ⬜

Even better — we now see that the 1-random degrees have a similar sort of relationship to the Π_1^0 classes of nonzero measure, as do the generic degrees to the arithmetical comeager sets of degrees. And — in case you were wondering — that is part of the reason why they fit into this chapter.

COROLLARY 15.7.6

If $\mathcal{A}$ is a Π_1^0 class for which $\mu(\mathcal{A}) > 0$, then **RAND** $\subseteq deg(\mathcal{A})$.

PROOF Let $\mathcal{A}$ be a Π_1^0 class with $\mu(\mathcal{A}) > 0$.

As in the previous proof, we can define a sequence $\overline{\mathcal{A}}^{(0)}, \overline{\mathcal{A}}^{(1)}, \ldots$. Since $\mu(\mathcal{A}) > 0$, we get a corresponding Martin-Löf test, using a modified Equation 15.1 for $\mathcal{A}^{(k)}$.

Since every 1-random f passes this test, we get $\deg(f) \in \deg(\mathcal{A})$, in the same way as before. $\qquad\qquad$ ⬜

And just as we found 1-generics could be used to get us into all sorts of comeager sets of degrees, so 1-random sets give us members of particular degree classes of nonzero measure. For instance, using Sacks' technique for showing the minimal degrees have zero measure, we can get an infinite independent set of degrees below any 1-random degree. So the Σ_1 theory of $\mathcal{D}(\leq \mathbf{a})$ is decidable for $\mathbf{a}$ 1-random.

Randomness without Measure

Just as computability gave us genericity without forcing in the form of Jockusch's definition of 1-generic — so computability theory gives us randomness without measure.

Here is an equivalent way of describing the 1-random reals. It is essentially Bob Solovay's characterisation of randomness, as simplified by Rod Downey.

DEFINITION 15.7.7 *(1) A **Solovay test** T is a c.e. set of binary strings with $\sum_{\sigma \in T} 2^{-|\sigma|} < \infty$.*

*(2) A set A **fails** a Solovay test T if for infinitely many $\sigma \in T$ we have $\sigma \subset A$ — and otherwise A **passes** the test T.*

(3) The set A is 1-random if it passes every Solovay test.

A Solovay test is a c.e. sequence of guesses about beginnings of A. The condition $\sum_{\sigma \in T} 2^{-|\sigma|} < \infty$ makes sure the guesses are bold enough — the lengths of the strings in T must increase fast enough to ensure the infinite sum converges. T cannot play safe. Random sets are chaotic enough to eventually escape the predictions made by T. *Eventually* is the key word here. We cannot ask f to completely avoid any sufficiently thin Σ_1^0 class of functions. The above definition is the next best thing.

EXERCISE 15.7.8 *Let T be a Solovay test. Show that the set of binary reals which fail the test T is a Π_2^0 class.*

Of course, there is no such thing as complete randomness. If f avoids one kind of class, it necessarily gets captured by the members of another — a 1-random f is captured by all sufficiently fat Σ_2^0 classes of reals.

The following example, due to George Barmpalias, illustrates the usefulness of Definition 15.7.7. It is known that 1-random sets satisfy various immunity properties. However:

EXAMPLE 15.7.9 *Show that no 1-random set is hyperimmune.*

PROOF Assume that A is random and hyperimmune. Consider the c.e. array $D_{f(0)} = \{1\}$, $D_{f(1)} = \{2, 3\}$, $D_{f(2)} = \{4, 5, 6, 7\}$, ... defined by

$$D_{f(0)} = \{1\}$$
$$D_{f(n+1)} = \{\max D_{f(n)} + 1, \ldots, \max D_{f(n)} + 2 \cdot |D_{f(n)}|\}.$$

Then $D_{f(n)} = 2^n = 2 \cdot |D_{f(n-1)}|$, and

$$\min D_{f(n)} = \max D_{f(n-1)} + 1 = \sum_{i \leq n-1} 2^i + 1 = 2^n.$$

Since A is hyperimmune, there are infinitely many n with $D_{f(n)} \cap A = \emptyset$. We will define a Solovay test T failed by A.

At stage s, enumerate in T all the strings of length 2^{s+1} whose last $|D_{f(s)}| = 2^s$ digits are 0 — that is, 2^{2^s} strings of length $2 \cdot 2^s = 2^{s+1}$.

To show that T is a Solovay test, let $(\sigma_{si})_{i < 2^{2^s}}$ be a one–one enumeration of the strings enumerated in T at stage s. We have

$$T = \bigcup_{s \in \mathbb{N}} \{\sigma_{si} \mid i < 2^{2^s}\} \quad \text{and}$$

$$\mu(\sigma_{si}) = 2^{-|\sigma_{si}|} = 2^{-2 \cdot 2^s},$$

and so

$$\sum_{\sigma \in T} 2^{-|\sigma|} = \sum_{s \in \mathbb{N}} \sum_{i < 2^{2^s}} 2^{-|\sigma_{si}|} = \sum_{s \in \mathbb{N}} 2^{-2 \cdot 2^s} \cdot 2^{2^s} = \sum_{s \in \mathbb{N}} 2^{-2^s} < \sum_{s \in \mathbb{N}} 2^{-s} < \infty.$$

Consider an increasing sequence $(s_i)_{i \in \mathbb{N}}$ of stages such that $D_{f(s_i)} \cap A = \emptyset$. This means that at stage s_i one of the strings enumerated in T is a beginning of A. In fact, considering A as a binary sequence, in the positions $r \in D_{f(s_i)}$ there are 0's; so that the last $|D_{f(s_i)}| = 2^{s_i}$ digits agree with $A \restriction 2^{s_i+1}$. And for the first 2^{s_i} digits we have considered all cases.

Since all strings enumerated are distinct and (s_i) is infinite, A fails the test T and so cannot be random. $\qquad \Box$

The 1-random degrees turn out to have a quite intimate relationship with the Π_1^0 classes, and their special companions, the complete extensions of Peano arithmetic. Here is another connection with earlier parts of this chapter.

PROPOSITION 15.7.10

Every 1-random degree is fixed point free — that is, **RAND** $\subset$ **FPF**.

There is a nice 1985 paper by Kučera which gives this proof, along with many other interesting results and comments.

EXERCISE 15.7.11 *Show that* $\mathbf{0}'$ *is the only c.e. degree of a 1-random real number.*

Given that there are 1-random members of $\mathbf{0}'$, we can ask if there are any *natural* 1-random members. Well, if we look at the halting problem for the universal Turing machine U in a rather different way, we get one. This was Chaitin's idea:

Think of any string $p \in 2^\omega$ as a program for U, which you run by inputting it to U in the usual way. A program p may or may not halt — but you can compute the probability that for any set of instructions, a universal Turing machine will halt in the form of the binary real:

$$\Omega = \sum_{p \text{ halts}} 2^{-|p|}$$

Then guess what? Ω turns out to be 1-random! And although it is very different from the special c.e. members of $\mathbf{0}'$ we found earlier, it is in $\mathbf{0}'$. There has been a lot of interest in Chaitin's Ω, for obvious reasons — it is not every day we come across a new real number. We will come across this number again in Section 16.4.

EXERCISE 15.7.12 *Show that there exist 1-random degrees $< \mathbf{0}'$.*

EXERCISE 15.7.13 *Show that there exists a 1-random degree below any degree of a complete extension of Peano arithmetic.*

EXERCISE 15.7.14 *Show that the only c.e. degree which bounds a 1-random degree is $\mathbf{0}'$.*
Deduce that no properly n-c.e. degree bounds a 1-random degree.

Kučera has shown that although every complete degree is 1-random, the 1-random degrees are not closed upwards.

If you were wondering what 2-random — or even n-random — might be, the answer is simple.

If you just replace c.e. Martin-Löf or Solovay tests with Σ_n tests in defining random, you get the notion of *n-random*. Alternatively, you can relativise the definition be replacing Σ_1 with Σ_1^A, to get *1-random relative to A*. Of course, $(n+1)$-random is the same as 1-random relative to $\emptyset^{(n)}$.

A bizarre discovery of Kučera and Terwijn is that there exist incomputable c.e. sets A for which the set of degrees 1-random relative to A is just **RAND**.

REMARK 15.7.15 This chapter has become rather technical. Let me try and draw out some sort of general picture from the growing complexities.

We started off trying to extend our familiar incomplete theories. In doing this we were willing to sacrifice global validity in the interests of completeness. An important discovery was the underlying relevance of Π_1^0 classes. And we found that within the constraints of the Π_1^0 class we could indeed find a lot of Turing local infrastructure — but the locality came at a cost. Below $\mathbf{0}'$ there was a pull away from the familiar and preferred context of the c.e. degrees.

We found a close association with unpredictability in the form of fixed point free and diagonally non-computable functions.

Finally — in some ways the most disturbing — was the emergence of randomness. The full cost of our sought-for Turing locality and theoretical completeness appeared to be hidden information content and the link to chaotic phenomena.

At the start of this chapter I talked about scientific theories as attempts at completing the uncompletable. Maybe we should now view them as mere approximations to one big nonaxiomatic and ultimately incorrect theory! The lessons are far from clear, though. What is certain is that theoretical development and an engagement with the real world will continue to throw up new ideas and, in the course of time, a better understanding of the complexities around us.   □

Chapter 16

Computability and Structure

Sometimes we look at real situations and struggle to come up with usable descriptions of them. In the mathematical context, that is what much of this book has been about. In a fairly abstract sense, of course.

But it can also happen that we make plans and theorise, and then have to manage reality to fit the design. This is where *models* come in. In this chapter I will be concerned with the computability of the structures on which our theories are based and which, in turn, arise from those theories.

16.1   Computable Models

Gödel's Completeness Theorem tells us that every consistent first order theory has a model. So far we have not looked much further than models of Peano arithmetic. But as you can imagine, the variety of models, of classes of models, and their interactions between theories and other ways of describing structures is huge. And we can examine this entire area for effective content if we want to.

What can I tell you in a few pages that you actually need to know? To start with, some basic notions are pretty essential.

To talk about computability of a model — for us, a structure $\mathcal{M}$ with a domain M and a countable set of relations and functions over M — we need to assume M is *countable*. This is reasonable, given the downward Löwenheim–Skolem Theorem 3.1.15.

Even more — we need to assume a *presentation* of $\mathcal{M}$, whereby the members of M have labels we can compute over. Usually we will implicitly assume $M = \mathbb{N}$, with the relations and functions and constants of $\mathcal{M}$ number-theoretic. To talk about $\mathcal{M}$ we need the first order language $\mathcal{L}_\mathcal{M}$ which has function and predicate symbols corresponding to the functions and relations of $\mathcal{M}$, together with individual constants $\mathbf{a}$ corresponding to the members $a \in M$. We write $\mathrm{Th}(\mathcal{M}, a \in M)$ for the set of all sentences of $\mathcal{L}_\mathcal{M}$ which are true in $\mathcal{M}$.

DEFINITION 16.1.1   *(1) A model $\mathcal{M}$ is* **computable** *if its domain M is computable and its relations and functions are uniformly computable.*

(2) A model is **decidable** *if M is computable, and $Th(\mathcal{M}, a \in M)$ is computable.*

To take account of the underlying presentation we occasionally say $\mathcal{M}$ is *computably presented* instead of just computable. And we sometimes say $\mathcal{M}$ is *computably presentable* to mean it is *isomorphic* to a computable — that is, computably presented — model. Similarly we might sometimes say, a little more pedantically, that $\mathcal{M}$ is *decidable* to mean it is isomorphic to a *decided* model.

EXERCISE 16.1.2   *Show that every decidable model $\mathcal{M}$ is computably presentable.*

The converse does not hold. The standard model $\mathfrak{N}$ of arithmetic is certainly computable — all the relations and functions are recursive. But no model of true PA can be decidable, since that would give the decidability of the first order theory $Th(\mathfrak{N})$ of true arithmetic, contradicting Exercise 9.1.3.

Notice that $\mathfrak{N}$ is what we call a *prime* model of PA, where:

DEFINITION 16.1.3   *(1) We say a model $\mathcal{M}^*$ is a* **submodel** *of $\mathcal{M}$ — written $\mathcal{M}^* \subseteq \mathcal{M}$ — if $M^* \subseteq M$, and the relations and functions of $\mathcal{M}^*$ are just those of $\mathcal{M}$ restricted to M^*.*

(2) We say a model $\mathcal{M}$ of a theory $\mathcal{T}$ is a **prime model** *if every model of $\mathcal{T}$ has a submodel isomorphic to $\mathcal{M}$.*

EXERCISE 16.1.4   *Show that if a theory $\mathcal{T}$ has two prime models then they are isomorphic to each other.*

EXERCISE 16.1.5   *Let $\mathcal{T}$ be the theory of a densely ordered set with least and greatest elements.*

Show that $\mathbb{Q}[0,1]$ — set of all rationals between 0 and 1 — is a prime model of $\mathcal{T}$. Show also that every countable model of $\mathcal{T}$ is decidable.

[**Hint:** Use Exercise 9.3.10.]

$\mathfrak{N}$ is special in yet another sense — it turns out to be the *only* computable model of PA, up to isomorphism.

THEOREM 16.1.6 (Tennenbaum, 1959)
No nonstandard model $\mathcal{M}$ of PA can be computable.

PROOF Let $\mathcal{M}$ be a nonstandard model of PA, and for $a \in M$, define

$$\text{Div}_a = \{\, k \in \mathbb{N} \mid \mathcal{M} \models p_k \,|\, a \,\}.$$

We can decide whether or not $\mathcal{M} \models p_k \,|\, a$ by searching through the atomic sentences of $\mathcal{M}$ for one which says $a = d \times p_k + r$ for some $r < p_k$. So all we have to do is prove:

There exists an $a \in M$ with Div_a not computable.

To show this, we choose — using Exercise 9.2.13 — computably inseparable c.e. sets A and B. Let $\varphi(y, s)$ be the wf of PA which says

$$(\forall x < s)\,[\,(x \in A^s \Rightarrow p_x \,|\, y) \;\&\; (x \in B^s \Rightarrow p_x \,\nmid\, y)\,],$$

A^s and B^s standard c.e. approximations to A and B.

But $(\exists y)\varphi(y, s)$ holds in $\mathcal{M}$ for all finite members s of M. So the induction axiom (N7) of PA says that the set of s satisfying $(\exists y)\varphi(y, s)$ must be unbounded in $\mathcal{M}$, giving an infinite $s \in M$ for which $(\exists y)\varphi(y, s)$ holds.

But then if $\mathcal{M} \models \varphi(a, s)$ for some $a \in M$, we have that Div_a separates A and B, and so Div_a cannot be computable. $\qquad\qquad$ ▯

Actually, with a little sleight-of-hand, you can get incomputable *standard* models of PA.

But let us first say what we mean by the *degree* of a model $\mathcal{M}$, where we assume the domain $M = \mathbb{N}$. We want it to be the l.u.b. of the degree of M and the degree of its relations and functions. A good way of doing this is to use the *atomic diagram* of $\mathcal{M}$, which puts together all the basic information about $\mathcal{M}$ in one set of wfs. You will remember we already came across the notion of a diagram of a structure in our Section 13.1 on forcing.

DEFINITION 16.1.7 *(1) The **atomic diagram** $D_{\mathcal{M}}$ of $\mathcal{M}$ is the set of all atomic wfs of $\mathcal{L}_{\mathcal{M}}$, and their negations, true in $\mathcal{M}$.*

*(2) The **degree** $\deg(\mathcal{M})$ of a model $\mathcal{M}$ is defined to be the degree of $D_{\mathcal{M}}$.*

EXERCISE 16.1.8   *Show that M is computable if and only if $deg(M) = 0$.*

EXERCISE 16.1.9   *Show that for every c.e. degree $\mathbf{a}$ there is an axiomatisable first order theory T with a model M of degree $\mathbf{a}$.*
[**Hint:** Use the proof of Feferman's Theorem 15.1.1.]

Here are the incomputable standard models I promised. There are standard models of PA of all Turing degrees.

EXERCISE 16.1.10   *Given $A \subseteq \mathbb{N}$, show that there is a model A of PA such that $A \cong \mathfrak{N}$ and $deg(A) = deg(A)$.*
[**Hint:** Turn A into a standard model of PA by replacing every $n \in \mathbb{N}$ in the atomic diagram $D_{\mathfrak{N}}$ of $\mathfrak{N}$ by $p_A(n)$, where p_A is the principal function of A, as in Definition 12.2.13.]

Let us finish this brief introduction to computable models with a closer look at the general question:

> *What is the relationship between the degree of a theory T and the degrees of its models?*

If you have seen a proof of Gödel's Completeness Theorem, it is almost certain that it will have been the 1949 proof of Leon Henkin. This has the advantage that it gives you a way of building a model of a consistent first order theory T constructively — or at least computably in the set of theorems of T. As a result we find that a consistent theory has a model computable in T. And in particular:

THEOREM 16.1.11
A decidable consistent theory has a decidable model.

If you understood the proof of Theorem 15.2.1, you might be brave enough to try this:

EXERCISE 16.1.12   *Given a model M, show that $Th(M) \leq_T (D_M)^{(\omega)}$.*
Deduce that any consistent axiomatisable theory has a model $\leq_T \emptyset^{(\omega)}$.

Of course, this is not a very good result — PA has a computable model, whereas $Th(\mathfrak{N}) \equiv_T \emptyset^{(\omega)}$! We can use our results on bases for Π^0_1 classes to get some strikingly strong bounds on the degrees of models. For instance:

THEOREM 16.1.13

If T is a consistent axiomatisable theory, then T has a countable model of degree $< \mathbf{0}'$.

PROOF By Theorem 15.3.3 the set of complete extensions of T is a Π_1^0 class. By the Low Basis Theorem 15.4.3, this class has a member T^+ of low degree, and hence of degree $< \mathbf{0}'$.
So T has a model $\leq_T T^+ \leq_T \emptyset'$. $\qquad\qquad\square$

It follows of course that there are nonstandard models of PA which have degree below $\mathbf{0}'$, and can even be low.

There are all sorts of interesting results concerning the connections between the degrees of nonstandard models and those of their theories. For instance Harrington has found a nonstandard model $\mathcal{M}$ of PA which has degree $\leq \mathbf{0}'$, and $\mathrm{Th}(\mathcal{M})$ has degree $\mathbf{0}^{(\omega)}$.

What about the degrees of isomorphic copies of a given nonstandard model?

DEFINITION 16.1.14 *Given a model $\mathcal{M}$, define*

$$DI(\mathcal{M}) = \{\, \deg(\mathcal{A}) \mid \mathcal{A} \cong \mathcal{M} \,\}.$$

EXERCISE 16.1.15 *Show that for every degree $\mathbf{a}$ there is model $\mathcal{M}$ such that $DI(\mathcal{M}) = \{\mathbf{a}\}$.*

[**Hint:** Take $A \in \mathbf{a}$ and let $\mathcal{M} = \langle \mathbb{N}, R_i, i \in \omega \rangle$ where for each i

$$R_i = \begin{cases} \mathbb{N} & \text{if } i \in A, \\ \emptyset & \text{if } i \notin A. \end{cases}]$$

EXERCISE 16.1.16 *Let $\mathcal{M}$ be a linear ordering on $\mathbb{N}$. Show that $DI(\mathcal{M})$ is closed upwards.*

[**Hint:** Take $A \in \mathbf{a} > \deg(\mathcal{M})$. Let π be the permutation which only interchanges $2n$ and $2n+1$ when $n \in A$. Define $\mathcal{A}$ to be the structure for which π is an isomorphism $\mathcal{M} \to \mathcal{A}$.]

As you saw from Exercise 16.1.15, sometimes $DI(\mathcal{M})$ has a least element. A less trivial example comes from the theory of groups — for every degree $\mathbf{a}$ there exists a countable abelian group $\mathcal{M}$ for which $DI(\mathcal{M}) = \mathcal{D}(\geq \mathbf{a})$.

But the only countable linear orderings for which this happens are the computably presentable ones. And — see Julia Knight's article in Volume 1 of the *Handbook of Recursive Mathematics* — if $\boldsymbol{M}$ is a nonstandard model of PA, then $DI(\boldsymbol{M})$ does not have a least element.

If this very brief look at computable models has left you looking for more, I would recommend the survey by Valentina Harizanov — in the same volume as the article by Julia Knight I just mentioned. But to read that you will have to learn a lot more model theory!

16.2   Computability and Mathematical Structures

Computable model theory can involve looking at complicated theories, and even cooking up unnatural theories with strange models. But everyday mathematics is populated with just a few basic kinds of structure, describable via simple axioms.

Is there anything to say about the constructive content of such simple structures? Yes — you cannot get much more basic than linear orderings, and even here we have a rich and interesting theory. I will focus on orderings in this section, but you can approach structures such as Boolean algebras, groups, rings, or fields with a similarly constructive outlook.

I will rely heavily on linear orderings of the form $\mathbf{A} = \langle \mathbb{N}, \leq_A \rangle$. And again, in keeping with the last section:

DEFINITION 16.2.1   *(1) We say* $\mathbf{A} = \langle \mathbb{N}, \leq_A \rangle$ *is* **computably presented** — *or just* **computable** *when there is no ambiguity* — *if* $\leq_A$ *is a computable relation.*

(2) And $\mathbf{A}$ *is* **computably presentable** *if it is isomorphic to a computably presented ordering.*

I often extend this terminology and say, for example, that $\mathbf{A}$ is Σ_1 *presented* if $\leq_A$ is a c.e. relation.

It is sometimes helpful to present an ordering $\mathbf{A}$ in the familiar context of the rationals $\mathbb{Q}$. A *computable subordering* of $\mathbb{Q}$ is a computable subset of $\mathbb{Q}$ with its usual ordering $\leq$.

We will see (Exercises 16.2.9, 16.2.10, 16.2.11) that suborderings of $\mathbb{Q}$ are more easily *related* to each other — while presented orderings of $\mathbb{N}$ can be *built* more easily to bring out *differences* between notions (see Theorem 16.2.12 and

Exercise 16.2.19). If we do some translating between the two settings, we get the best of all worlds.

Here is an effective version of an old result of Cantor. Having become used to computing over $\mathbb{Q}$ and other countable domains, it is easy to talk about *computable* functions $\mathbb{N} \to \mathbb{Q}$, for example.

PROPOSITION 16.2.2

  If **A** *is computable, then* **A** *is computably isomorphic to a computable subordering of* $\mathbb{Q}$.

PROOF   Take a computable list $r_0, r_1, \ldots$ of $\mathbb{Q}$. Write $i =_A j$ if $i \leq_A j$ and $j \leq_A i$.

I will build the computable isomorphism $\theta : \mathbb{N} \to \text{range}(\theta)$ by stages.

Stage s

Assume $\theta(i) = r_{k_i}$ defined for all $i < s$, with $i \leq_A j \iff \theta(i) \leq \theta(j)$.

If $s =_A i$, any $i < s$, define $\theta(s) = \theta(i)$.

Otherwise look for any closest neighbour(s) $i <_A s <_A j$ of s amongst $i, j < s$. Choose a rational r_{k_s} with $k_s > s$ with appropriate neighbour(s) $\theta(i) < r_{k_s} < \theta(j)$.

Define $\theta(s) = r_{k_s}$.

$\theta : \mathbb{N} \to \text{range}(\theta)$ is clearly an isomorphism by the way I defined it. And range(θ) is computable since k is computable and $k_s > s$ for all s. □

The converse also holds:

EXERCISE 16.2.3   *Show that every computable subordering of* $\mathbb{Q}$ *is computably isomorphic to a computably presented ordering* **A** *of* $\mathbb{N}$.

In fact:

EXERCISE 16.2.4   *Show that every* Π_n *subordering of* $\mathbb{Q}$ *is isomorphic to a* Σ_n *presented ordering* **A** *of* $\mathbb{N}$ — *and conversely.*

[**Hint:** Prove it for $n = 1$ and then relativise.]

Also:

EXERCISE 16.2.5   *Show that if* **A** *is* Δ_n *presented, then* **A** *is* Δ_n-*isomorphic to a* Δ_n *subordering of* $\mathbb{Q}$.

[**Hint:** Relativise the proposition.]

How do things change in the proof of the proposition if $\mathbf{A}$ is only Σ_1 and not computable? What is the difference between a computable ordering and a Σ_1 presented ordering?

The main problem in the above proof is that we can no longer recognise if $s =_A$ some $i < s$ at stage s — for each i we can decide whether $s \leq_A i$ or $i \leq_A s$ by enumerating $\leq_A$. But say we find $s \leq_A i$ — we cannot know computably if $i \leq_A s$ ever gets enumerated into $\leq_A$. Anyway, here is the correctly modified result.

THEOREM 16.2.6

If $\mathbf{A}$ is a Σ_1 presented linear ordering, then $\mathbf{A}$ is Δ_2-isomorphic to a Π_1 subordering of $\mathbb{Q}$.

PROOF   We start off just as before building θ. But if we find we made a mistake — we mapped some $i <_A s <_A j$ to $\theta(i) < \theta(s) < \theta(j)$, only to find $s =_A j$, say — then we have to *change* $\theta(s)$ to $= \theta(j)$, and *throw out* from range(θ) the old value of $\theta(s)$, never to be chosen again.

Let $\{=_A^s\}_{s \geq 0}$ be a c.e. approximating sequence for $=_A$. We define a Δ_2-approximating sequence for θ in such a way that range(θ) $\in \Pi_1$.

Stage 0

Define $\theta^0 = \emptyset$.

Stage $s+1$

Assume $\theta^s(i)$ defined for all $i <$ some $m < s$. All values are carried over to stage $s+1$ unless changed below.

Case I — There is some least i such that $\theta^s(i) \downarrow$, $\theta^s(j) \downarrow$ some j, with $\theta^s(i) \neq \theta^s(j)$ — but $i =_A^s j$.
  Then for each such j, redefine $\theta^{s+1}(j) = \theta^{s+1}(i) = \theta^s(i)$.

Case II — Not Case I. Let i be the least number such that $\theta^s(i) \uparrow$.
  If $i =_A^s j$, any $j < m$, define $\theta^s(i) = \theta^s(j)$.
  Otherwise look for any closest neighbour(s) $j <_A^s i <_A^s j'$ of i among the numbers $j, j' < m$.
  Choose a fresh rational r with appropriate neighbour(s) $\theta^s(j) < r < \theta^s(j')$. Define $\theta^{s+1}(i) = r$.

Then $\theta : \mathbb{N} \to$ range(θ) is clearly a Δ_2-isomorphism by the way I defined it — $\mathrm{Lim}_s \, \theta^s(i)$ exists for all i since $\theta^s(i)$ eventually gets defined and only changes value to some $\theta^s(j)$ with $j < i$. And range(θ) is Π_1 since we always choose a fresh rational for a new value of θ. $\qquad\square$

You should now be ready to try:

EXERCISE 16.2.7 *If* **A** *is* Π_1 *presented, then* **A** *is* Δ_2-*isomorphic to a* Σ_1 *subordering of* $\mathbb{Q}$.

Relativising to $\emptyset^{(n-1)}$ you should be able to generalise what we have now proved:

EXERCISE 16.2.8 *Show:*

(1) If **A** *is a* Σ_n *presented linear ordering, then* **A** *is* Δ_{n+1}-*isomorphic to a* Π_n *subordering of* $\mathbb{Q}$.

(2) If **A** *is* Π_n *presented linear ordering, then* **A** *is* Δ_{n+1}-*isomorphic to a* Σ_n *subordering of* $\mathbb{Q}$.

What is so useful about these results is that we can now look at linear orderings via suborderings of $\subseteq \mathbb{Q}$. And having shifted the complexity of our orderings from the ordering relation to the domain of the ordering, we get some surprising comparisons.

Here are some more nice properties, all got by relativising and variants of the proof of Proposition 16.2.2.

EXERCISE 16.2.9 *Show that any* Σ_1 *ordering* $\subseteq \mathbb{Q}$ *is isomorphic to a computable ordering* $\subseteq \mathbb{Q}$.

Deduce that any Σ_n *ordering* $\subseteq \mathbb{Q}$ *is isomorphic to a* Δ_n *ordering* $\subseteq \mathbb{Q}$.

[**Hint:** Replace the Σ_1 set of rationals with an isomorphic computable set.]

EXERCISE 16.2.10 *Show that any* Δ_2 *ordering* $\subseteq \mathbb{Q}$ *is isomorphic to a* Π_1 *ordering* $\subseteq \mathbb{Q}$.

Deduce that any Δ_{n+1} *subordering of* $\mathbb{Q}$ *is isomorphic to a* Π_n *subordering of* $\mathbb{Q}$.

[**Hint:** Given a Δ_2-approximating sequence for an $A \subseteq \mathbb{Q}$, replace it by a Π_1-approximating sequence for a set isomorphic to A.]

We have now built up in steps to the following quite impressive result, which Lawrence Feiner put into his thesis in 1967.

EXERCISE 16.2.11 *Show that any* Σ_{n+1} *subordering of* $\mathbb{Q}$ *is isomorphic to a* Π_n *subordering of* $\mathbb{Q}$.

We now see that the arithmetical hierarchy of suborderings of $\mathbb{Q}$ — up to isomorphism — simplifies to:

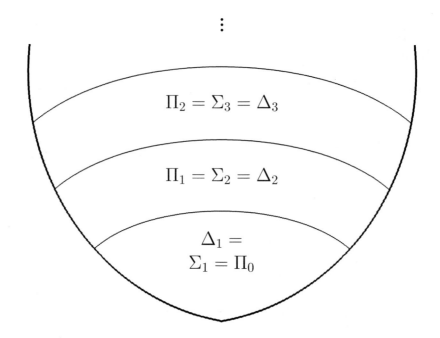

Does this hierarchy collapse any further? Corollary 16.2.18 below tells us the answer is "No." Or at least it will when we relativise it to higher levels. First let us prove the parallel result for presentations.

THEOREM 16.2.12

There is a Σ_1-presented linear ordering not isomorphic to a computably presented linear ordering.

PROOF   Here is the plan of the proof:

(1) Choose any set $X \in \Delta_2$ such that $\Sigma_3^X \nsubseteq \Sigma_3$ — for example, we can take $X \in \mathbf{0}'$, so that $\emptyset^{(4)} \in \Sigma_3^X - \Sigma_3$.

(2) Here is the tricky part: Construct an X-computably presented ordering $\mathbf{A}$ — and define a special property — or *invariant* — B of the *ordering* (not just the presentation) which is always a Σ_3 relation for any computably presentable $\mathbf{A}$ — but which for the $\mathbf{A}$ we build is Σ_3^X complete.

(3) Then $\mathbf{A}$ is Δ_2 presented — so by Exercise 16.2.20 — is isomorphic to a Σ_1 presented linear ordering.

(4) And $\mathbf{A}$ is not isomorphic to a computably presentable ordering — since

$B(\mathbf{A}) \notin \Sigma_3$.

All that remains to do is to say what $B(\mathbf{A})$ is, and build $\mathbf{A}$ as described in part (2) of the plan. On the way we will learn how to work with Σ_3 sets.

Defining $B(\mathbf{A})$

Given $\mathbf{A}$, write just $\leq$ for $\leq_A$.

(a) We say (a, b) is a *successivity* if $a < b$ and for all c not $a < c < b$ in $\mathbf{A}$.

(b) $a_1 < a_2 < \cdots < a_n$ is an *n-block of* $\mathbf{A}$ if every (a_i, a_{i+1}) is a successivity, but no (x, a_1) or (a_n, y) is a successivity.

(c) Define the *block relation* — or *n-block relation* — $B(n)$ on $\mathbf{A}$ by

$$B(n) \iff \text{defn} \ \mathbf{A} \text{ contains an } n\text{-block}$$

Then $n \in B(\mathbf{A}) \iff$ defn $B(n)$ holds.

Notice that $B(\mathbf{A})$ depends only on the ordering itself — we only use the degree of the presentation to locate $B(\mathbf{A})$ in the arithmetical hierarchy.

> **LEMMA 16.2.13**
> If $\mathbf{A}$ is X-computably presented, then $B(\mathbf{A}) \in \Sigma_3^X$.

PROOF Just write formally:

$$n \in B(\mathbf{A}) \iff (\exists a_1, \ldots, a_n)(\forall y)(\exists x)$$

$$\left[a_1 < \cdots < a_n \ \& \bigwedge_{1 \leq i \leq n-1} \neg(a_i < y < a_{i+1}) \right.$$

$$\left. \& \ (y < a_1 \Rightarrow y < x < a_1) \ \& \ (a_n < y \Rightarrow a_n < x < y) \right]. \quad \Box$$

So given X, we are going to build an $\mathbf{A}$ which is X-computable with $B(\mathbf{A})$ Σ_3^X complete. To do this we take $C = X^{(3)}$, and make $C \leq_T B(\mathbf{A})$.

Working with Σ_3 sets

At this point your main worry will be the thought of dealing with a Σ_3 set. With good reason — three quantifiers is the limit of what we commonly deal with in mathematics — for example, in defining convergence or continuity. Remembering that $C \in \Sigma_3$ if

$$n \in C \iff (\exists x) D(x, n) \text{ with } D \in \Pi_2,$$

let us define:

DEFINITION 16.2.14 *(1) A uniformly computable sequence $\{D^s\}_{s \geq 0}$ of finite sets is a Π_2-approximating sequence to $D \subseteq \mathbb{N}$ if for all n*

$$n \in D \iff (\forall t)(\exists s > t)\,[\,n \in D^s\,].$$

(2) A uniformly computable sequence $\{\widetilde{C}^s\}_{s \geq 0}$ of finite sets is a Σ_3-approximating sequence for $C \subseteq \mathbb{N}$ if for all n

$$n \in C \iff (\exists x)(\forall t)(\exists s > t)\,[\,\langle n, x \rangle \in \widetilde{C}^s\,].$$

EXERCISE 16.2.15 *Show every $D \in \Pi_2$ has a Π_2-approximating sequence.*

[**Hint:** Use Exercise 10.5.15.]

EXERCISE 16.2.16 *Show every $C \in \Sigma_3$ has a Σ_3-approximating sequence.*

[**Hint:** Given $C = \{n \mid (\exists x)\, D(x, n)\}$ with $D \in \Pi_2$, use Exercise 16.2.15 on the graph of D.]

I have gone to the trouble to present Σ_3 sets this way, so that in the future you will have a way of visualising an approximation process:

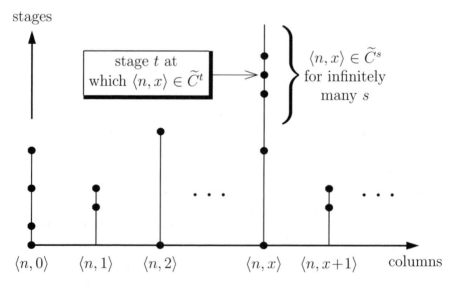

The above diagram shows how to visualise n entering C via the xth column.

The principles established here will stand you in good stead for working with Π_3 and Δ_3 sets.

EXERCISE 16.2.17 *Draw pictures approximating (a) $n \in C \in \Pi_3$, and (b) $n \in C \in \Delta_3$. What is the difference between the two approximating pictures?*

Anyway, we are now in good shape to describe the main part of the proof.

Building the ordering A

Let C be Σ_3^X complete, and let $\{\widetilde{C}^s\}_{s \geq 0}$ be a Σ_3^X-approximating sequence for C.

We build **A** X-computably presentable such that for all $n > 0$

$$\boxed{(n-1) \in C \iff B(n)} \quad - (\maltese)$$

Let ζ denote the ordering $\omega^* + \omega$, and n denote the ordering $\overbrace{1 + 1 + \cdots + 1}^{n \text{ times}}$ of $n - 1$ successivities.

We start with the ordering $\omega.\zeta = \zeta + \zeta + \dots$.

For each $n \geq 1$, we modify the $(n-1)^{\text{th}}$ copy of ζ — call it ζ_{n-1} — to ensure $(\maltese)$.

To do this, we start off by organising ζ_{n-1} into $\omega^* + n + n + \dots$.

Now — let $s_0, s_1, \dots$ be the (possibly finite) list of stages s at which $\langle n - 1, x \rangle \in \widetilde{C}^s$. Let α be the ordering $s_0, s_1, \dots$ and α^* be $\dots, s_1, s_0$.

Then replace the xth n of ζ_{n-1} with $\alpha + n + \alpha^*$. So ζ_{n-1} becomes:

- $\zeta + n + \zeta + n + \zeta + \dots$ if α is infinite,

- $\zeta + n + \zeta + \cdots + n + \zeta$ if α is finite but non-empty,

- ζ if α is empty.

From this we get $(\maltese)$, as required — and the theorem follows. $\quad\square$

We now get:

COROLLARY 16.2.18
There is a Π_1 linear ordering $\subseteq \mathbb{Q}$ not isomorphic to a computable linear ordering $\subseteq \mathbb{Q}$.

PROOF By Theorem 16.2.12 we can build a linear ordering **A** which is Σ_1 presented, and so by Exercise 16.2.10 is isomorphic to a Π_1 subordering of $\mathbb{Q}$ — and is such that **A** is not isomorphic to a computably presentable ordering.

Then **A** is not isomorphic to a computable subordering of $\mathbb{Q}$, by Exercise 16.2.3.   ⬜

A straightforward relativisation gives Feiner's full result:

EXERCISE 16.2.19 (Feiner, 1967)   *Show that for all $n \geq 0$ there is a Π_{n+1} linear ordering $\subseteq \mathbb{Q}$ not isomorphic to a Π_n linear ordering $\subseteq \mathbb{Q}$.*

EXERCISE 16.2.20   *Show that any Π_{n+1} presentable linear ordering is isomorphic to a Σ_n presented linear ordering.*

Deduce that the arithmetical hierarchy of isomorphism types of presentations of linear orderings takes the form:

$$\Delta_1 \subsetneq \Sigma_1 \subsetneq \Sigma_2 \subsetneq \cdots$$

[**Hint:** Use Exercises 16.2.11, 16.2.8, 16.2.4, and 16.2.19.]

The following diagram summarises all the equivalences we have proved between classes of linear orderings at the nth level:

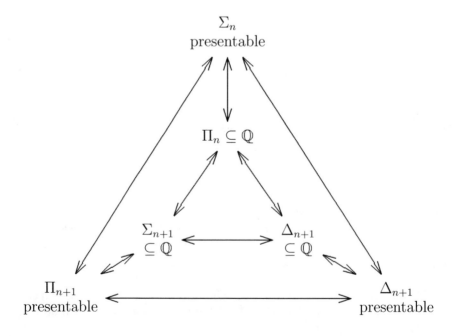

REMARK 16.2.21 There is a remarkable application of this result to the global theory of $\mathcal{D}$. The relativised theory of initial segments of D tells us that we can embed linear orderings of increasingly higher degree as initial segments of $\mathcal{D}[0^{(n)}, 0^{(n+2)}]$. Since Feiner's hierarchy does not collapse, we get increasingly *complicated* segments embedded, which refutes the strong version of Rogers' Homogeneity Conjecture for $\mathcal{D}$ — which was discussed at the end of Chapter 14.

Nowadays there are simpler proofs of nonhomogeneity, but not prettier. What all the proofs have in common is the coding of information into the structure of $\mathcal{D}$. Feiner did it by coding the information into his linear orderings, and then into $\mathcal{D}$ via segments of $\mathcal{D}$. ☐

You can look at other mathematical structures constructively. For instance Feiner was able to get similar results to those above for Boolean algebras — see Downey's 1997 survey article on *On presentations of algebraic structures*.

REMARK 16.2.22 When you go to structures such as groups and rings with a rich classical theory, there is more scope for finding constructive counterparts to classical theorems and concepts. The results become less about computability theoretic notions and connect up more with what "real" mathematicians do. For instance, there is the well-known result of M.O. Rabin that the algebraic closure of a computable field is computable. Computable algebra, and the wider theory of computable structures, has had a valuable clarifying role for mathematics.

The origins of computable algebra are usually traced back to a paper of Fröhlich and Shepherdson from 1956. Since then there has been a huge amount of very diverse work done. Much deep and important work originates with A.I. Mal'cev and the Novosibirsk school associated with Yu. L. Ershov, S.S. Goncharov and others. In the U.S.A. Anil Nerode has influenced a rather different approach. For a very thorough introduction to the background I would recommend Viggo Stoltenberg-Hansen and John Tucker's article on "Computable Rings and Fields" in the *Handbook of Computability Theory*. ☐

Before leaving computable linear orderings, we should look at one of the most important and useful of all such orderings.

Computable ordinals and Kleene's $\mathcal{O}$

A special kind of linear ordering is an *ordinal*. Ordinals are invaluable for notating inductive processes of all sorts — and if we want constructive content to these processes, it is vital to use computable ordinals.

Let us start off with an overview of computable ordinals. First I will remind you of some basic properties of ordinals. I will take a set-theoretically naïve point of view, and assume you already have some set theoretical sophistication, or just do not care that much.

Ordinal numbers

> **DEFINITION 16.2.23**   *Let* $\mathbf{A} = \langle A, \leq_A \rangle$ *be a linear ordering.*
>
> *We say that* $<_A$ *is* **wellordering** *if* $\mathbf{A}$ *contains no infinite descending sequence* $\cdots <_A a_2 <_A a_1 <_A a_0$ — *in which case we say that* $\mathbf{A}$ *is a* **wellordered set**.

Equivalently — using the Axiom of Choice — $\mathbf{A}$ is wellordered if every subset of A has a least element.

EXERCISE 16.2.24   *Show that every subset of* $\mathbb{N}$, *with the usual ordering, is a wellordered set — but that the set* $\mathbb{Z}$ *of all integers is not wellordered.*

The *ordinal numbers* are special wellordered sets constructed inside set theory. Conventionally, we get the *finite ordinals* $0, 1, 2, \ldots$ as

$$0 = \emptyset, \; 1 = \{\emptyset\}, \; 2 = \{\emptyset, \{\emptyset\}\}, \; 3 = \{0, 1, 2\}, \; \text{etc.},$$

and write $\omega = \{0, 1, \ldots\}$ for the first infinite ordinal.

The ordering relation $<$ on the ordinals is just the membership relation $\in$. You get any ordinal number as the set of all its predecessors — so, for example, you get $\omega + 1 = \omega \cup \{\omega\} = \{x \mid x \in \omega + 1\}$.

Ordinals like 3 or $\omega + 1$ of the form $\alpha + 1 = \alpha \cup \{\alpha\}$ are called *successor* ordinals; the others (apart from 0) — such as ω — are *limit* ordinals. You can do arithmetic on the ordinals, although the details look different at higher levels than they do on ω.

The ordinals are wellordered by $\in$ (although set theory does not allow us to take the set of *all* ordinals). The idea is that given any wellordered set, it will be isomorphic to some ordinal. So the ordinals act as unique representatives of the order types of wellordered sets.

All our ordinals will be countable, but there are a lot of those!

Computable ordinals

Looked at from above, the computable ordinals are easy to describe:

> **DEFINITION 16.2.25**   *A countable ordinal* α *is* **computable** *if it is finite or is isomorphic to a computable wellordering* $\mathbf{A} = \langle A, \leq_A \rangle$.

It is not hard to see that the computable ordinals form a countable initial segment of the countable ordinals.

EXERCISE 16.2.26 *Show that if the ordinal β is $<$ a computable ordinal α, then β is a computable ordinal.*

EXERCISE 16.2.27 *Show that there is a countable ordinal which is not computable.*

[**Hint:** Take the least ordinal greater than all the computable ordinals — this least noncomputable ordinal is called ω_1^{CK} after its 1937 discoverers Church and Kleene.]

But the above definition does not give us a good grip on the computable ordinals and in practice, we need a more useful *characterisation*. We *either* need to be able to recognise a given computable linear ordering as being wellordered (some hope!) — *or* we need some more constructive way of building the computable ordinals up, starting with the finite ones. The latter approach — from below — turns out to be the only practical one.

Kleene's $\mathcal{O}$

What Kleene did was build up a numbering — or set of *ordinal notations* — of the computable ordinals from below, using computable operations on notations corresponding to the set theoretic operations used to generate new ordinals.

Let us first look at an alternative way of visualising the computable ordinals. Remember from Section 15.3 the definition of a *computable tree* as a computable ideal of $\omega^{<\omega}$.

DEFINITION 16.2.28 *A tree $T \subset \omega^{<\omega}$ is **wellfounded** if it has no infinite branches.*

This time you will see that our trees need to grow downwards. Wellfoundedness is a notion from set theory meaning there are no infinite descending sequences of members of sets. I will now start laying out ordinals on wellfounded trees, so that the trees give a sort of graphic picture of how the ordinal is built up from smaller ordinals.

Let T be wellfounded. We assign the ordinal $\mathrm{Ord}_T(\sigma)$ to each $\sigma \in T$ as follows:

$$\mathrm{Ord}_T(\sigma) = \begin{cases} \mathrm{lub}\{\, 1 + \mathrm{Ord}_T(\sigma^\frown n) \,\} & \text{if } (\exists n)[\,\sigma^\frown n \in T\,] \\ 1 & \text{otherwise.} \end{cases}$$

Then:

> The *ordinal of* $T \neq \emptyset$ is $\mathrm{Ord}(T) = \mathrm{Ord}_T(\emptyset)$ — and $\mathrm{Ord}(\emptyset) = 0$.

So for example the following tree represents $\omega + 1$:

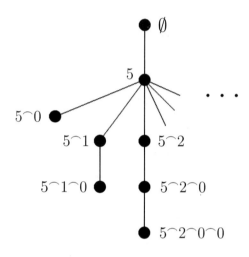

The important thing about ordinals of course is that you can do induction on them. You can inductively verify that every countable ordinal is the ordinal of some wellfounded tree. Say you have verified this for every β less than some countable ordinal α. If $\alpha = \beta + 1$, with $\beta = \mathrm{Ord}(T)$, then $\alpha = \mathrm{Ord}(T_0)$ where $T_0 = \{0 ^\frown \sigma \mid \sigma \in T\}$ — that is, get T_0 by adding a single extra branching above the root $\emptyset$ of T.

If we limit ourselves to taking the ordinals of *computable* wellfounded trees, we get exactly the computable ordinals. Clearly all we want is to get a single wellfounded tree on which to computably represent all the computable ordinals. Unfortunately, we cannot do such a thing.

What Kleene did was the next best thing. He took the intuition behind our wellfounded trees, and built up from below what seems like one big tree of *notations* — that is, numbers — for *constructive* ordinals. But the notations will not be unique. As you move up the ordinals, the trees/notations corresponding to different ordinals proliferate.

The set of all notations for constructive notations we call $\mathcal{O}$. Along with the notations for ordinals comes an ordering $<_{\mathcal{O}}$ on the notations. $<_{\mathcal{O}}$ is not linear due to an ordinal having more than one notation.

DEFINITION 16.2.29 (The Constructive Ordinals)   *We inductively define $\mathcal{O}$ and $<_\mathcal{O}$. We start by giving the ordinal 0 notation 1.*

Assume all ordinals $< \alpha$ have been assigned notations, and $<_\mathcal{O}$ has been defined on these notations.

(1) Say $\alpha = \beta + 1$, and β has notation x.

Then α gets notation 2^x and we add $z <_\mathcal{O} 2^x$ to $<_\mathcal{O}$ for each z such that $z = x$ or $z <_\mathcal{O} x$ already.

(3) Say α is a limit ordinal, and $\{\varphi_e\}_{e \geq 0}$ is a list of notations for ordinals with limit α, and $(\forall n)[\varphi_e(n) <_\mathcal{O} \varphi_e(n+1)]$ already.

Then give α notation 3.5^e, and add $z <_\mathcal{O} 3.5^e$ to $<_\mathcal{O}$ for all z for which $z <_\mathcal{O} \varphi_e(n)$ already, some $n \geq 0$.

*We call the set of ordinals notated by $\mathcal{O}$ the **constructive ordinals**, and write $|x|_\mathcal{O}$ for the ordinal notated by x.*

It is straightforward to verify that $<_\mathcal{O}$ is a wellfounded partial ordering. Also notice that if a notation $x \in \mathcal{O}$ is given, then $<_\mathcal{O}$ is linear on $\{y \mid y <_\mathcal{O} x\}$.

Of course, there are lots of different ways of defining the details of the notations. What is important is whatever way we do it, so long as we follow the principles behind the above definition, we get:

THEOREM 16.2.30

The set of constructive ordinals is exactly the set of all computable ordinals.

You will have to go back to Rogers' book to find a detailed proof of this fact. It is not hard to prove, but too tedious and technical to include here.

One of the most useful applications of $\mathcal{O}$ is in notating the levels of the *hyperarithmetic hierarchy*, got by iterating the Turing jump through the computable ordinals. We do this by using the notations to build a transfinite hierarchy similar to how we got the arithmetical hierarchy containing the sets computable from some $\emptyset^{(n)}$. At limit points we take constructive joins in a similar way to how we got $\emptyset^{(\omega)}$. A justly celebrated theorem of Kleene says that this hierarchy takes us all the way up through the Δ^1_1 sets of numbers. The very brave might like to look for more in Gerald Sacks' book on *Higher Recursion Theory*.

But I have already digressed too far, and it is time to return to more homely mathematics.

16.3   Effective Ramsey Theory

There are interesting infinitary versions of familiar topics in combinatorics. Many of these were first looked at by the famous Hungarian mathematician Paul Erdős. One of the most interesting and technically challenging of these is that to do with *Ramsey theory*. This is named after the innovative and versatile mathematician Frank Plumpton Ramsey, who died in 1930 at the early age of 26, leaving behind a small number of remarkably influential publications on mathematics, philosophy and economics — not to mention a younger brother Michael who went on to become Archbishop of Canterbury, leader of the Anglican Church.

Unfortunately, even the basic definitions of Ramsey theory are a little abstract and hard to digest. So let us start with a simple example from the finite case. If you know some graph theory, you might have seen this before.

EXAMPLE 16.3.1   *Assume that any two people either are friends or are not friends.*

   Show that in any group of six people, there are three people each of whom are friends with the other two — or there are three people none of whom is friends with either of the other two.

SOLUTION   We can visualise this by representing the six people by the vertices of a graph. We join each pair of vertices by an edge, coloured either red or green. If two vertices represent people who are friends, we join them with a red edge — otherwise the edge joining them must be green.

Now consider just one particular vertex v, say. v is incident with five edges, so is incident with a set of exactly three edges of the same colour — say red. Now if any pair of the three vertices so joined to v is joined by a red edge, we get a triangle of red edges — giving three friends. But if each pair of the three vertices joined to v by red edges is joined by a green edge, we obviously get a green triangle — three non-friends.   ▯

Taking a very broad view, Ramsey theory is about the way apparently disordered structures contain orderly substructures. You only need a group of six people to be sure it contains three mutual friends or three mutual strangers. It is an example of how complete disorder is impossible. Computability theory aims to tell you how well hidden the order is.

Anyway, let us get the definitions we need out of the way. We will relate them to the example a bit later.

DEFINITION 16.3.2 *(1) Let $A \subseteq \mathbb{N}$. We write*

$$[A]^n = \text{the class of all } n\text{-element subsets of } A$$
$$= \{X \subseteq A \mid |X| = n\}.$$

(2) If X is some set, we say $P \subseteq 2^X$ is a **partition** *of X if*

$$X = \bigcup \{Y \mid Y \in P\} \text{ and } (\forall Y, Z \in P)[Y \neq Z \Rightarrow Y \cap Z = \emptyset).$$

(3) Let P be a finite partition of $[\mathbb{N}]^n$ — say $P = \{C_1, \ldots, C_p\}$.

We say an infinite $A \subseteq \mathbb{N}$ is a **homogeneous set** *for P if $[A]^n \subseteq C_i$ for some i between 1 and p. We write*

$$H(P) = \{\, A \text{ infinite} \subseteq \mathbb{N} \mid [A]^n \subseteq C_i, \, 1 \leq i \leq p \,\}$$
$$= \text{the set of all homogeneous sets for } P.$$

Taking the cue from our introductory example, we sometimes call a general partition $P = \{C_1, \ldots, C_p\}$ of $[\mathbb{N}]^n$ a *p-colouring* of $[\mathbb{N}]^n$.

The classic result is:

RAMSEY's THEOREM (1930)

Every finite partition P of $[\mathbb{N}]^n$ has a homogeneous set.

The ingenious proof of this is very nonconstructive (I will give you a basic version of it a little later). So there are some obvious questions. We say $P = \{C_1, \ldots, C_p\}$ is *computable* if each C_i is computable. Then:

QUESTIONS

(1) If P is computable, does there exist a computable homogeneous set for P? And more generally:

(2) What are the Turing degrees of sets $A \in H(P)$? Or the jumps of such degrees?

We will concentrate on the most important case — the one on which the general theory is largely based — that is, $n = p = 2$:

DEFINITION 16.3.3   *We say P is a **basic** partition if P is a partition of $[\mathbb{N}]^2$ into just two classes.*

You can now start thinking in terms of graphs and connect the definitions with the above Example 16.3.1. In the example we represented the 2-element subsets of the 6 people by edges. The partition was got by the colouring of edges red or green. The homogeneous set consisted of the vertices of the monochromatic triangle we found. This way of looking at homogeneous sets for basic partitions will come in useful later on.

Our first result will show that we can find basic partitions with very constrained homogeneous sets. We will use these constraints later to get a negative answer to question (1) above.

THEOREM 16.3.4

If $A \leq_T \emptyset'$, then there exists a computable basic partition P such that for all $B \subseteq \mathbb{N}$ we have

$$B \in H(P) \implies B \subseteq A \text{ or } B \subseteq \overline{A}.$$

PROOF   Since $A \leq_T \emptyset'$, we can choose an i with $A = \Phi_i^K$.
Here is an ingenious definition of a suitable partition:

$$C_0 = \{\, \{n, s\} \mid n < s \ \& \ \Phi_i^K(n)[s] \downarrow = 1 \,\}$$
$$C_1 = [\mathbb{N}]^2 - C_0$$
$$P = \{\, C_0, C_1 \,\}.$$

From the definition P is clearly computable.

To show P has the other property, we assume $B \in H(P)$ and show $B \subseteq A$ or $B \subseteq \overline{A}$.

Case I. $\boxed{\text{Say } [B]^2 \subseteq C_0}$ — then we can show $B \subseteq A$.

Assume $n \in B$.

Since $B \in H(P)$, we have for all $s > n$ that $\{n, s\} \in [B]^2 \Rightarrow \{n, s\} \in C_0$. So by definition of C_0 we have

$$(\forall s \in B)[\, n < s \Rightarrow \Phi_i^K(n)[s] \downarrow = 1\,].$$

It is easy to deduce now that we must have $A(n) = \Phi_i^K(n) = 1$. Here are the details.

Let $\Phi_i^K(n)$ have use z, with $\Phi_i^{K \restriction z}(n) \downarrow = \delta$, say. Since B is infinite, there is an $s^* \in B$ such that $s^* > n$ and $K^{s^*} \restriction z = K \restriction z$. Then

$$1 = \Phi_i^K(n)[s^*] \downarrow = \Phi_i^K(n) = \delta,$$

giving $n \in A$.

Case II. $\boxed{\text{Say } [B]^2 \subseteq C_1}$ — then if $n \in B$ we have

$$(\forall s > n)[\, \{n, s\} \in [B]^2 \Rightarrow \{n, s\} \in C_1\,],$$

and we can argue just as in Case I to get $n \in \overline{A}$.

So $B \subseteq \overline{A}$. $\quad\square$

It is now easy to get:

COROLLARY 16.3.5 (Specker)
There exists a computable basic partition P with no computable homogeneous set.

PROOF Using Exercise 13.3.12, take A now to be a bi-immune set $\leq_T \emptyset'$. Then by Theorem 16.3.4 we have

$$B \in H(P) \Rightarrow B \subseteq A \ \text{ or } \ B \subseteq \overline{A}.$$

Since A is bi-immune, B is not c.e. — so certainly not computable. $\quad\square$

This last result is slightly surprising, until we look more closely at how we get a homogeneous set. So as to extract what constructive content the classical proof does have, we will prove:

THEOREM 16.3.6 (Ramsey, 1930)
If P is any basic partition, then $H(P)$ is not empty.

PROOF Let P be a basic partition — and write $P = \{P, \overline{P}\}$.

We now bring in our graph colouring analogy:

Colour $\{x, y\}$ **red** if $\{x, y\} \in P$.

Colour $\{x, y\}$ **green** if $\{x, y\} \notin P$.

Aim: To colour each member of an infinite $A \subseteq \mathbb{N}$ either red or green — in such a way that $(\forall \text{ red } x, y \in A) \ \{x, y\}$ is red, *and* $(\forall \text{ green } x, y \in A) \ \{x, y\}$ is green.

The Construction of A

Define inductively $a_0, a_1, \ldots$, giving each a_i its own colour — red or green — as we go.

Terminology: A number n is k-**acceptable** if for each $i < k$, we have $a_i < n$, and $\{a_i, n\}$, a_i have the same colour.

Overall requirement: If $i < k$ then a_k is k-acceptable.

Outcome (if the requirement is satisfied): Say we define

$$R = \{\, a_i \mid a_i \text{ is red }\}$$
$$G = \{\, a_i \mid a_i \text{ is green }\}.$$

Then if R or G is infinite, it is a homogeneous set for P.

Stage k

Assume a_i defined and coloured for all $i < k$, *and* there exist infinitely many k-acceptable numbers. Write $A_k =$ the set of k-acceptable numbers.

Computably define $a_k =$ the least $n \in A_k$.

Colour a_k *red* if $\{a_k, n\}$ is red for infinitely many $n \in A_k$ — and *green* otherwise.

Then with this colouring of a_k, it immediately follows that there exist infinitely many $(k + 1)$-acceptable numbers.

And hence the inductive hypothesis of stage k of the construction persists at stage $k + 1$, and the construction of our homogeneous R or G never breaks down.   ⬚

We now dig out the constructive content of this proof.

COROLLARY 16.3.7 *(of the proof of Theorem 16.3.6)*
If P is a computable basic partition, then there is a homogeneous set A for P with $A \leq_T \emptyset''$.

PROOF We just notice that at stage k of the construction of the a_i's to colour a_k, we just need to ask:

Is $\{a_k, n\}$ red for infinitely many nunbers $n \in A_k$?

Since P is computable, A_k is computable, and so c.e. — and we can find its index.

So the construction can be carried out with help from an oracle for Inf — which by Exercise 10.5.18 is a Π_2-complete set of indices — and so $\leq_T \emptyset''$. ⬚

Notice that although the oracle used here was Π_2, we cannot deduce that the homogeneous set it computed for us is also Π_2 — it is only computable in $\emptyset''$, and so is Δ_3. To get the stronger result we have to work a lot harder, and use a finite injury priority argument relative to an oracle for $\emptyset'$.

THEOREM 16.3.8 (Jockusch, 1972)
If P is a computable partition of $[\mathbb{N}]^2$ into $p \geq 2$ classes, then $H(P)$ contains a Π_2 set.

PROOF   I will just outline the proof for P basic — that is, for $p = 2$.

The idea: We carry out the construction as before, but instead of asking if a certain c.e. set is infinite, we *assume* it is (so avoiding the $\emptyset''$-computable oracle). If we assumed correctly, there is no problem, and the construction proceeds as before. If the assumption is *wrong*, we eventually find this out at a later stage when the construction breaks down, with help of an oracle for $\emptyset'$, and then go back and correct out mistake. We make sure that R and G are Π_1 in our oracle for $\emptyset'$, and so are Π_2.

The construction of A

At stage s we define some approximations

$$a_0^s, a_1^s, \ldots, a_{k(s)}^s$$

to the sequence $a_0, a_1, \ldots$ we want. $k(s)$ will tend to increase with s, but occasionally will fall back to a k for which a_k has been wrongly chosen at an earlier stage.

We initially colour each a_i red, until we find we must change to green.

Terminology: n is k-acceptable at stage s if for each $i < k$, a_i^s is defined and has the same colour at stage s as $\{a_i^s, n\}$, a_i and $a_i^s < n$.

Definition of the numbers a_k^{s+1} at stage $s+1$:

Case I. There exists an $n > s$ never having been chosen as an a_i, which is $k(s)$-acceptable at stage s.

Define $a_{k(s)+1}^{s+1} = a_{k(s+1)}^{s+1}$ the least such n, and colour n red.
All $a_i^{s+1} = a_i^s$, each $i \leq k(s)$.

Case II. Otherwise.

Let $j(s)$ be the largest j such that there is an $n > s$ never previously chosen as an a_i, and which is j-acceptable at stage s. [Notice that $j(s) < k(s)$ since Case I does not apply.]

Change the colour of $a_{j(s)}^s$ — and throw away every a_i^s with $i > j(s)$.

Define $k(s+1) = j(s)$ and $a_i^{s+1} = a_i^s$ for each $i \leq j(s)$.

Note: The only oracle question asked in this construction at stage $s + 1$ — in Case I — is whether there is an n as described. That is, we ask if a given c.e. set is non-empty — so we need oracle K_1 which is c.e. and so $\leq_T \emptyset'$ by Exercise 5.2.15.

I will leave you to verify:

EXERCISE 16.3.9   *Show:*

(1) For each k, we only change the value of a_k finitely often. And once a_k is fixed, we only change the colour of a_k from red to green.

(2) If $i < j$, $\{a_i, a_j\}$ has the same colour as the eventual colour of a_i.

(3) $A = \{a_0, a_1, \dots\}$, and $R =$ the set of red numbers in A, are each $\in \Pi_2$. And if R is finite then $G = A - R$ is also Π_2.

It will follow as before that either R or G gives the required homogeneous set, this time $\in \Pi_2$.   ⬜

REMARK 16.3.10   In this construction of a Π_2 homogeneous set for a computable basic P, we define at stage s numbers $a_0^s, a_1^s, \dots$. At each stage $> s$, if we redefine $a_i \neq a_i^s$, then we throw a_i^s into the Σ_1 (in $\emptyset'$) set $\overline{A}$.

It is interesting to notice that we can frame this proof as another *moving marker* argument, as used by Richard Friedberg to construct his maximal set. We use an infinite list of markers

$$\Gamma_0, \Gamma_1, \dots, \Gamma_i, \dots$$

where the *current position* of Γ_i at stage s is a_i^s.

Then $\text{Lim}_s\, a_i^s = a_i \iff$ we eventually stop moving Γ_i, and Γ_i comes to rest on a_i. And we obtain $A = \{a_0, a_1, \dots\}$ as the final positions of the markers $\Gamma_0, \Gamma_1, \dots$.   ⬜

EXERCISE 16.3.11   *Complete the proof of Theorem 16.3.8 for the case $p > 2$.*

[**Hint:** Work with p colours, one for each of the p elements of the partition P. Change colours successively as before, but now possibly $p - 1$ times. Check that colours eventually stabilise as before. Take the first of the p sets of coloured numbers to be finite as the required Π_2 homogeneous set.]

What about the full theorem of Ramsey on colourings of $[\mathbb{N}]^n$ — how computable is the homogeneous set then? To get the homogeneous set for colourings of $[\mathbb{N}]^n$ you need to do an induction on n. Each inductive step pushes your bound on the degree of the homogeneous set one level up the arithmetical hierarchy. So, for instance:

If $P = \{C_1, \ldots, C_k\}$ is a computable partition of $[\mathbb{N}]^n$, then there is a Π_n homogeneous set for P.

EXERCISE 16.3.12 *Let P_n, $n \geq 2$, be a computable 2-colouring of $[\mathbb{N}]^n$.*

Show that there is a computable 2-colouring P_{n+1} of $[\mathbb{N}]^{n+1}$ such that $H(P_{n+1}) \subseteq H(P_n)$.

[**Hint:** Take the 2-colouring

$$C_0 = \{\, \{m_1, \ldots, m_n, s\} \mid (\forall i, 1 \leq i \leq n)\,[m_i < s \ \& \ \Phi_i^K(m_i)[s] \downarrow = 1]\,\},$$
$$C_1 = [\mathbb{N}]^{n+1} - C_0.$$

Follow the proof of Theorem 16.3.4, with the role of n in $\{n, s\}$ taken over by that of $m_1, \ldots, m_n$ in $\{m_1, \ldots, m_n, s\}$.]

EXERCISE 16.3.13 *If $n \geq 2$, show there exists a computable 2-colouring of $[\mathbb{N}]^n$ for which there is no homogeneous set computable in $\emptyset^{n-1}$ and, hence, no Σ_n homogeneous set.*

[**Hint:** Use induction on n with $n = 2$ taken care of by Corollary 16.3.5. Assume the result for n, and relativise to $\emptyset'$. Apply Exercise 16.3.12.]

Actually we can do better:

THEOREM 16.3.14
For all $n > 1$, there is a 2-colouring of $[\mathbb{N}]^{n+1}$ for which every homogeneous set is of degree $\geq \mathbf{0}^{(n-1)}$.

Let us skip the proof, but use this to get an exquisitely unexpected consequence — both theorem and corollary thanks to Jockusch again. We can use Theorem 16.3.14 to show that the degrees of cohesive sets — and those for all the immunity properties in Section 12.2 — are closed upwards!

Let us start with a couple of simple definitions, and a basic proposition.

DEFINITION 16.3.15 *(1) We say $B \subseteq \mathbb{N}$ is **rich** if every degree above that of B contains a subset of B.*

(2) If P is a 2-colouring of $[\mathbb{N}]^k$, we write $H^+(P)$ for the set of all sets B such that $B - D \in H(P)$ for some finite D.

PROPOSITION 16.3.16

Let P be a 2-colouring of $[\mathbb{N}]^k$, and $B \subseteq \mathbb{N}$ be infinite. Then if B is not rich, we have $B \in H^+(P)$.

PROOF We take an infinite $B \notin H^+(P)$, and prove it is rich.

To do this, assume $B \leq_T A$ and build $C \subseteq B$ with $A \equiv_T C$.

Idea: Use $B \notin H^+(P)$ and the computability of P to code A into a set C made up from sets $C_i \in [\mathbb{N}]^k$.

The construction of the sets C_i

Stage 0. Define $C_0 = \{0\}$.

Stage $i + 1$. Assume we already have C_i together with a computable listing of the finite sets.

Choose C_{i+1} to be the least finite set ensuring:

(1) $C_{i+1} \in [B]^k$,

(2) $\max(C_i) < \min(C_{i+1})$, and

(3) $i \in A \iff C_{i+1} \in P$.

$\boxed{\text{Prove } C_{i+1} \text{ exists}}$ — First notice that $B - \{j \mid j \leq \max(C_i)\}$ is infinite and not in $H(P)$. This means there are sets C_{i+1}^1, C_{i+1}^2 satisfying (1) and (2), one in P and one not. So we can choose the one satisfying (3) to be C_{i+1}.

It is now easy to check $C = \bigcup_{i \geq 0} C_i$ has the required properties. □

The next two exercises complete the proof of the upward closure result.

EXERCISE 16.3.17 *Show that if B is infinite but not rich, then every arithmetical set is B-computable.*

[**Hint:** Use Theorem 16.3.14 and Exercise 16.3.16.]

It follows, of course, that all infinite arithmetical sets are rich. Are *all* infinite sets rich? No — never underestimate the strangeness of the non-computable universe. There exist infinite sets with *no* subsets of strictly higher degree.

EXERCISE 16.3.18 *Show that if the property $\mathcal{P}$ is hereditary under inclusion — remember Definition 12.2.3 — and there is some arithmetical set with*

property $\mathcal{P}$, then the class of all degrees of sets satisfying $\mathcal{P}$ is closed upwards.
[**Hint:** Take **a** containing a $\mathcal{P}$-set B. Find a rich $\mathcal{P}$-set $C \leq_T B$. To do this, consider two cases — either B is rich or B is not rich, and use Exercise 16.3.17.]

What a nice theorem. It applies to cohesive, hyperhyperimmune, hyperimmune, immune, and some others we have not mentioned — since all of these immunity properties have arithmetical representatives, and are hereditary under inclusion.

This section owes much to Carl Jockusch's beautiful little paper "Ramsey's theorem and recursion theory" from 1972, which I would very much recommend to you for further reading. Real gluttons for information might brave William Gasarch's massive "A survey of recursive combinatorics" in Volume 2 of the 1998 *Handbook of Recursive Mathematics*.

Much of the interest in Ramsey's Theorem, by the way, is because it has an interesting proof. It provides a real playground for *reverse mathematicians* — those who concern themselves with classifying everyday mathematics according to how much induction, etc., is needed to develop it. Gasarch has a nice summary of recent attempts to tell just *how* difficult it is to prove Ramsey's Theorem in various guises.

16.4 Computability in Analysis

Professor A has just finished his seminar on computable combinatorics. The first question is from a very senior and notoriously combative analyst sitting in the front row — where else would a man of such self-importance sit: " Computable structures? But *real* mathematics is done on *real* numbers! What relevance can Turing machines have to that?"

Professor A — not quite sure where to start — is interrupted from the back by a young computer scientist: "What do mathematicians know about what's real? You can assign a *finite* number to the total number of atomic particles in the Universe. And our most powerful computers will only ever handle rational numbers — and ones with not many decimal places at that."

As I was slipping quietly out of the door, I just caught Professor A's mild tones: "Of course, science without calculus is unthinkable — but computable approximations ... ".

Poor Professor A — fancy having to try and reconcile such extreme views. And that is what we will be trying to do, mathematically, in this short section. All I will attempt is a brief introduction to how people have dealt with computability of sets of reals and of functions over them. For a fuller treat-

More Advanced Topics

ment the current standard references are the books of Marian Pour-El and Ian Richards *Computability in Analysis and Physics*, and of Klaus Weihrauch on *Computable Analysis*.

Of course the notion of a "computable real number" goes right back to Turing. The standard definitions of computable real numbers and functions are based on formulations of Grzegorczyk and Lacombe from around 1955. A computable real is one which can be computably approximated by rationals in a predictable way — or equivalently, one for which we can program a Turing machine to write down its decimal expansion. As usual, reals will be presented in binary notation.

DEFINITION 16.4.1 *(1) We say the real number r is* **computable** *if there is a computable $\varphi : \mathbb{N} \to \mathbb{Q}$ such that $|r - \varphi(n)| \leq 2^{-n}$ for all $n \geq 0$. In which case we say φ* **binary converges** *to r.*

(2) If $\varphi : \mathbb{N} \to \mathbb{Q}$ is computable and $Lim_n \varphi(n) = r$ we say that φ is a **computable presentation** *of r.*

We have already been dealing in a more schematic way with $[0, 1] \subset \mathbb{R}$ when we discussed category, measure and randomness. Our new definition fits perfectly well with earlier notions of a computable real:

EXERCISE 16.4.2 *Let r be a real, and define $f(n) =$ the $(n + 1)^{\text{th}}$ digit in the binary decimal for r. Show that r is computable if and only if $f \in \mathbf{0}$.*

EXERCISE 16.4.3 *Show that every rational number is computable.*

It is easy to see that familiar numbers like $\sqrt{2}$ and π are computable. One way is to use suitable Taylor expansions and use partial sums and the remainder term to get the right convergence property. Much of everyday mathematics is computable.

There are lots of equivalent ways of defining the computable reals. For instance, we can do it using Dedekind cuts:

EXAMPLE 16.4.4 *Show that a real r is computable if and only if the set $L(r) = \{q \in \mathbb{Q} \mid q \leq r\}$ is computable.*

SOLUTION $(\Rightarrow)$ Say r is computable.

If r is rational, we can easily tell whether a given rational $q \leq r$ or not.

Otherwise just look for an $n > 0$ such that

$$|\varphi(n) - r| \leq 2^{-n} \quad \text{and} \quad |\varphi(n) - q| > 2^{-n}.$$

Then $q \leq r$ if and only if $q \leq \varphi(n)$.

($\Leftarrow$) Say $L(r) = \{q \in \mathbb{Q} \mid q < r\}$ is computable. Define $\varphi(n)$ as follows: Compute rationals q_0, q_1 such that $q_0 < r \leq q_1$ and $|q_1 - q_0| < 2^{-n}$. Take $\varphi(n) = q_0$. $\qquad$ $\square$

It does not make any difference how fast we ask $\varphi(n)$ to converge:

EXERCISE 16.4.5 *Show that r is computable if and only if there is a computable presentation φ of r and a computable $t : \mathbb{N} \to \mathbb{N}$ for which $Lim_n t(n) = \infty$ and $|\varphi(n) - r| \leq \dfrac{1}{t(n)}$ for all n.*

EXERCISE 16.4.6 *Show that r is computable if and only if there is a computable presentation φ of r and a computable $M : \mathbb{N} \to \mathbb{N}$ for which $|\varphi(i) - r| \leq 2^{-n}$ for all $i \geq M(n)$.*

However, we do need to know computably *how* fast $\varphi(n)$ converges. This is best expressed in terms of the modulus of convergence of Cauchy sequences.

DEFINITION 16.4.7 *Let $\{q_i\}_{i \geq 0}$ be a sequence for which $|q_i - q_j| \leq 2^{-n}$ for all $i, j \geq M(n)$. Then M is called the* **modulus of convergence** *of $\{q_i\}_{i \geq 0}$.*

We can assume the modulus M is a nondecreasing function:

EXERCISE 16.4.8 *Show that if M is a modulus of convergence for $\{q_i\}_{i \geq 0}$, then so is $M'(n) = \max\{M(k) \mid k \leq n\}$.*

We now find a role for M in ensuring reals are computable.

EXERCISE 16.4.9 *Assume that $Lim_i q_i = r$. Show that if $|r - q_i| \leq 2^{-n}$ for all $i \geq M'(n)$, with M' computable, then $\{q_i\}_{i \geq 0}$ has a computable modulus of convergence M.*

We now show that:

THEOREM 16.4.10

Let $\{q_i\}_{i \geq 0}$ be a computable sequence of rationals with computable modulus of convergence $M : \mathbb{N} \to \mathbb{N}$. Then $r = Lim_i \, q_i$ is a computable real number.

PROOF   Assume that $|q_i - q_j| \leq 2^{-n}$ for $i \geq M'(n)$, where M' is computable.

Define $M(n) = M'(n)$ and $\varphi(i) = q_i$, each $n, i \geq 0$. Then M, φ are computable. And since $Lim_i \, q_i = r$, we have $\inf_i q_i \leq r \leq \sup_i q_i$.

So $\sup\{|r - q_i| \text{ for } i \geq M(n)\} \leq \sup\{|q_i - q_j| \text{ for } i, j \geq M(n)\} \leq 2^{-n}$. Hence $|r - q_i| \leq 2^{-n}$ for all $i \geq M(n)$.

The result follows from Exercise 16.4.6.   ☐

You can see that the role of the computable modulus in Theorem 16.4.10 is a very necessary one:

EXERCISE 16.4.11   *Show that there exists a real r which is not computable, but which has a computable presentation $\varphi : \mathbb{N} \to \mathbb{Q}$ such that $Lim_n \, \varphi(n) = r$.*
[**Hint:** Take $r = 0.K(0)K(1)\ldots$ with $K = \{x \in \mathbb{N} \mid x \in W_x\}$, and define $\varphi(n) = 0.K^s(0)K^s(1)\ldots K^s(n)$. Show that the sequence $\{\varphi(n)\}_{n \geq 0}$ cannot have a computable modulus of convergence.]

EXERCISE 16.4.12   *For any $A \subseteq \mathbb{N}$ define*

$$r_A = \sum_{i \in A} 2^{-i}.$$

Show that r_A is computable if and only if A is computable.

What sort of reals do we get if we relax the computability condition on the modulus? If we do this in a controlled way, we get a natural notion of a c.e. real.

DEFINITION 16.4.13   *We say that r is (left) **computably enumerable** if it is the limit of a computable increasing sequence of reals.*
  *We can define r being **right c.e.** similarly.*

EXERCISE 16.4.14   *Show that r is computably enumerable if and only if $-r$ is right c.e.*

EXERCISE 16.4.15 *Show that r is computable if and only if r is both left and right c.e.*

We have already seen a c.e. real that is not computable in Section 15.7. You will remember that Chaitin's Ω was the probability that a universal Turing machine U halted on any program $p \in 2^{<\omega}$. It was defined to be

$$\Omega_U = \sum_{p \in PROG_U} 2^{-|p|},$$

where $PROG_U$ is the set of programs p on which U halts. Given a c.e. sequence $\{PROG_U^s\}_{s \geq 0}$ of approximations to $PROG_U$, you can get Ω_U as the limit of the computing ascending sequence of rationals $\{\Omega_U^s\}_{s \geq 0}$ where

$$\Omega_U^s =_{\text{defn}} \sum_{p \in PROG_U^s} 2^{-|p|}.$$

How do we know that the series $\sum 2^{-|p|}$ for Ω_U converges?

It is usual — following Chaitin — to make U *self-delimiting* in the sense that we only allow $p \in PROG_U$ if U halts on p *having actually used the whole program p*. Precisely defined, this ensures that the set of $p \in 2^{<\omega}$ on which U halts is *prefix-free*.

DEFINITION 16.4.16 *A set $A \subset 2^{<\omega}$ is **prefix-free** if for every $\sigma, \tau \in A$ if $\sigma \subseteq \tau$ then $\sigma = \tau$.*

EXERCISE 16.4.17 *Show that for every infinite prefix-free c.e. set $A \subset 2^{<\omega}$, the sum $\sum_{\sigma \in A} 2^{-|\sigma|}$ converges to a c.e. real number ≤ 1.*

There is a strong converse proved by Calude, Hertling, Khoussainov, and Wang in 2001.

PROPOSITION 16.4.18
If $r \in (0,1]$ is c.e., there is an infinite prefix-free computable set $A \subset 2^{<\omega}$ with $r = \sum_{\sigma \in A} 2^{-|\sigma|}$.

PROOF Let $\{q_i\}_{i \geq 0}$ be a computable increasing sequence of rationals with limit r, where we can assume $0 < q_i < r \leq 1$ for all i. We can use this sequence to build a nondecreasing computable sequence $\{n_i\}_{i \geq 0}$ of positive integers and an increasing computable sequence $\{k_i\}_{i \geq 0}$ of nonnegative integers such that for all j

$$\sum_{i \leq k_j} 2^{-n_i} < q_j < 2^{-j} + \sum_{i \leq k_j} 2^{-n_i}.$$

Then $\sum_{i \geq 0} 2^{-n_i}$. We can now computably choose an infinite prefix-free set $A = \{\sigma_i \in 2^{<\omega} \mid i \geq 0\}$ such that $n_i = |\sigma_i|$ for each i. To do this, you can inductively choose each σ_i leftmost on $2^{<\omega}$ such that $\{\sigma_0, \ldots, \sigma_i\}$ is prefix-free — where $\sum_{j \leq i} 2^{-|\sigma_j|} < 1$ guarantees the inductive process can continue. The set A is computable since the strings σ_i are nondecreasing in length.

Then $r = \sum_{\sigma \in A} 2^{-|\sigma|}$ as required.   □

EXERCISE 16.4.19   *Given a binary* $r = 0.r_0 r_1 \ldots$, *let* $X_r = \{i \mid r_i = 1\}$.

Show that if X_r *is c.e. then* r *is a c.e. real. Give an example of a c.e. real* r *such that* X_r *is not c.e.*

It turns out that Ω_U has a very special relationship to the set of all c.e. reals. In a very natural sense all c.e. reals are *reducible* to Ω_U — and there is a whole class of Ω-*like* reals which are reducible to each other, and have a similar relationship to the c.e. reals.

The following definition is due to Solovay and Chaitin from around the mid 1970s:

DEFINITION 16.4.20   *Let* $r, r' \in \mathbb{R}$. *We say* r **dominates** r' — *written* $r \geq_{dom} r'$ — *if there is a p.c. function* $f : \mathbb{Q}^{<r} \to \mathbb{Q}$ *and a* $c > 0$ *such that for all* $q \in \mathbb{Q}^{<r}$ *we have* $c(r - q) \geq r' - f(q)$.

Intuitively, r dominates r' if any rational approximation to r from below gives a rational approximation to r' from below.

EXERCISE 16.4.21   *Let* r, r' *be c.e. reals. Show that* r *dominates* r' *if and only if there are computable increasing sequences* $\{q_i\}_{i \geq 0}$, $\{q'_i\}_{i \geq 0}$ *of rationals converging to* r, r', *and a* $c > 0$, *such that* $c(r - q_n) \geq r' - q'_n$ *for all* $n \geq 0$.

EXERCISE 16.4.22   *Show that* $\geq_{dom}$ *is reflexive and transitive.*

EXERCISE 16.4.23   *Let* r, r', r_0 *be c.e. reals. Show that:*
    (a) $r + r'$ *is a c.e. real,*
    (b) $r + r' \geq_{dom} r$ *and* $r + r' \geq_{dom} r'$,
    (c) If $r_0 \geq_{dom} r$ *and* $r_0 \geq_{dom} r'$, *then* $r_0 \geq_{dom} r + r'$.

EXERCISE 16.4.24   *Write* $r \equiv_{dom} r'$ *if* $r \geq_{dom} r'$ *and* $r' \geq_{dom} r$, $[r] = \{r' \in \mathbb{R} \mid r' \equiv_{dom} r\}$, *and* $\mathbb{R}_{c.e.} = \{[r] \mid r$ *is a c.e. real*$\}$. *Let* $\leq$ *on* $\mathbb{R}_{c.e.}$ *be the ordering induced by* $\leq_{dom}$.

Show that $\langle \mathbb{R}_{c.e.}, \leq \rangle$ *is an upper semilattice with least element = the set of all computable reals.*

DEFINITION 16.4.25 (Chaitin, 1977)   *We say c.e. real is* Ω-like *if it dominates all c.e. reals.*

This definition is equivalent to a 1975 definition of Solovay:

EXERCISE 16.4.26   *We say that a computable increasing sequence* $\{q_i\}_{i \geq 0}$ *of rationals with* $r = Lim_i\, q_i$ *is* **universal** *if for every other computable increasing sequence* $\{q_i'\}_{i \geq 0}$ *of rationals with* $r' = Lim_i\, q_i'$ *we have a number* $c > 0$ *such that for all* n $c.(r - q_n) \geq r' - q_n'$.

Show that if r *is the limit of a universal computable increasing sequence of rationals, then* r *is* Ω-like.

We are now ready to describe the Ω-like reals and their remarkable role amongst the c.e. reals in general.

THEOREM 16.4.27

   (1) If $0 < r < 1$ *then* r *is* Ω-like *if and only if there is some universal self-delimiting Turing machine* U *such that* $r = \Omega_U$.

   (2) The structure $\langle \mathbb{R}_{c.e.}, \leq \rangle$ *has a greatest element consisting of all* Ω-like *reals.*

   (3) Every Ω-like *real is random — and hence* $0.\chi_K$ *is not* Ω-like, *since* χ_K *is not random.*

So the theorem says that any Chaitin Ω number is a c.e. random real, with the useful property that if you have a good approximation to it, you can get a good approximation to *any* c.e. real. No other kind of c.e. real — including χ_K — has this property.

For more detail and further results you should go to the very readable 2001 paper of Calude, Hertling, Khoussainov, and Wang.

Returning to the main development — we can now use Definition 16.4.1 to define the computable functions on the reals. Notice that we can ask our p.c.

functionals Φ to compute over functions $\varphi : \mathbb{N} \to \mathbb{Q}$ in the natural way (or more formally using a computable coding of $\mathbb{Q}$).

DEFINITION 16.4.28   *Let $f : \mathbb{R} \to \mathbb{R}$. We say f is **computable** if there is a p.c. functional Θ such that for every φ which binary converges to a real r we have that Θ^{φ} binary converges to $f(r)$ — that is, there exists a $\psi : \mathbb{N} \to \mathbb{Q}$ which binary converges to $f(r)$ where $\psi(n) = \Theta^{\varphi}(n)$ for each n.*

REMARK 16.4.29   At this point I should mention that we have already seen a parallel development of many of the basic ideas when we previously set up Cantor space — for instance, we already have a good idea of what a *continuous* function over the reals is from Definition 13.2.9. And in Proposition 13.2.24 we showed that the p.c. functionals are continuous. We could now in our much less schematic way define a more hands-on version of continuity, and verify that the computable functions over the reals are all continuous. The details are not pretty!

We could then go on to develop a constructive analogue of much of classical analysis. And we do not have to stop there. The theory takes in complex analysis, Banach spaces, and other very general areas with a basis in classical analysis. Unfortunately, the uneasy mix of notions from computability theory and analysis makes the detail not just complicated, but often open to alternative formulations. This particularly hinders the search for answers to some very basic questions concerning — once again! — the existence of natural examples of incomputable sets. Sets of reals this time, of course.

There is a very striking result of Marian Pour-El and Ian Richards concerning solutions of differential equations. Of course, differential equations are the basis for theoretical descriptions of much of the material universe. Their result concerns a rather innocent looking differential equation with initial condition, not obviously different from the sort of equations thrown up daily by working scientists:

$$y'(x) = f(x, y(x)), \qquad y(0) = 0. \tag{16.1}$$

Say the function f is a computable function defined on $[0,1] \times [-1,1]$. Does the equation have a computable solution y?

THEOREM 16.4.30 (Pour-El and Richards, 1979)
There is a computable function $f : [0,1] \times [-1,1] \to \mathbb{R}$ such that the Equation 16.1 has uncountably many solutions on $[0.1]$, but does not have a computable solution y on $[0, \delta]$ for any $\delta > 0$.

On the other hand, for a different choice of f the equation has a unique computable solution on $[0, 1]$.

The question of whether or not the Mandelbrot set is computable is a particularly tantalising one. When Roger Penrose discussed the problem in his book *The Emperor's New Mind* it achieved much more attention than is usually given mathematical problems! Even if people did not quite understand the origins of the set, they were familiar with the beautiful computer generated pictures and could grasp the association of the Mandelbrot set with chaos in nature.

In fact, its definition as a set of complex numbers is quite simple:

DEFINITION 16.4.31 *If $c \in \mathbb{C}$, let $f_c : \mathbb{C} \to \mathbb{C}$ be given by $f_c(z) = z^2 + c$.*

Then the **Mandelbrot set** M *is defined by*

$$M = \{\, c \in \mathbb{C} \mid (\forall n)(\, |f_c^n(0)| \leq 2\,) \,\}.$$

But how can a picture on a computer screen involve incomputability? Well, without going into detail, I think you can see that the *complement* of M is c.e., but not obviously so M itself. This is the basis of the computer screen approximations. But does it amount to full computability? There have been various proposed answers to the question, but nothing people in the know are willing to believe.

Just one final remark. Stephen Kleene spent much of his final years developing a theory of computability for sets of functions, and functions upon those sets, and so on. His framework — the basis for what Gerald Sacks calls *E-recursiveness* in his book — very clearly comes out of the *theory* of computability rather than its practice. It is not widely understood and currently neglected as a research topic. But maybe in the course of time Kleene's theory of computability for objects of higher finite type might be just what is needed to clear up some confusion. □

16.5 Computability and Incomputability in Science

In the seventeenth century science became well and truly about computability, thanks to Newton and his contemporaries. Whether or not people were happy with this, the world of ideas was moulded by the successes of the scientific outlook. In the twentieth century science started to deal with incomputability. The 1920s and '30s threw up quantum uncertainty and nonlocality, and the discovery of incomputability itself. As the century went on, scientific reductionism came under attack in all sorts of areas, at the same time as people became more aware of nonlinearity in nature.

You can find a more detailed description of the various crises effecting science in my article with George Odifreddi on "Incomputability in Nature". There are many areas in which the connection between small scale mechanisms and large scale structure is not really understood. What is the mathematics behind the emergence of subatomic structure, where there is not much more to build on than the algorithmic implications of causality itself? How can computability theory help us avoid arbitrary basic assumptions in explaining the way the Universe is? In our article we argued that matter is information, and its organised forms *are* algorithmic content. And that for us this means looking at computability over the reals.

We argued that in this context mathematical definability is the missing mechanism linking the local and the global. I said something about this at the start of Chapter 12. You might remember the metaphor of the rushing stream and its surface patterns — the parallel with the emergence of form in chaotic environments. This image is relevant to very many puzzling aspects of the material universe.

Anyway, a whole swathe of fundamental issues raise questions about the algorithmic content of the world we live in. Computability theorists cannot lose, you have to admit. Things which can be done computably are best done with some knowledge of algorithmic structures and complexity of programs. Even at the most practical level, a little theory can often smooth the way in surprising ways. And if computability breaks down — great! — computability theorists are on hand to give us reasons and analyses of how to deal with incomputability. And that has been what this book is about. I have tried to give you some idea of the wide range of approaches taken by people working in the area.

I have tried to give some idea of how all this rather detailed body of knowledge relates to the "big picture". I thought of concluding with much more about the relevance of the Turing universe to science and the humanities. But there are no "theorems" to state regarding the balance of computability and incomputability in the material universe, let alone "proofs" of the effects of this balance on human affairs. Nor are there likely to be.

What there is is a large body of very beautiful mathematics, which for some people has wider significance. There is no collective consensus to report on. Some very famous scientists have tried to anticipate the emerging big picture, not always adding to their reputations! This book is an invitation to join the formative forces at work. To become knowledgeable. To think and to do that with an awareness of the underlying computability content. And to engage with science within such a perspective at all levels at what is surely an exciting and formative period.

If you would like to read more, there are many recommendable books and articles. Some authors have been brave enough to step outside the safe confines of their own specialisms. Others deal in some special way with technical aspects I have only managed to touch on. You will find a very personal selection in the *Further Reading* section at the end of this book.

One of Turing's few mathematics PhD students, and a close friend of his, was Robin Gandy. I stayed with Robin at his house in Oxford a short while before he died. He was reviewing Roger Penrose's *Shadows of the Mind* at the time, and we got on to Turing's ideas on machines and the mind, and eventually I could not resist asking him his opinion about the mysterious circumstances of Turing's early death — it seems he had been with Turing just days before the end. He said "Sometimes there are things which happen for no reason".

Some years before, Albert Einstein had famously said: "God does not play dice". Only a computability theorist could reconcile such apparently differing views of the Universe.

Further Reading

Computability theory is a very big subject. Here is a small selection of books and articles I have found specially useful or interesting. I have listed most of those mentioned in the text, and concentrated on those of more general interest, listing details of current editions where possible. No doubt you will have particular directions in which you want to develop your interests, but here is something to help you on the way.

Books

Henk P. Barendregt: *The Lambda Calculus: Its Syntax and Semantics*, Revised Edition, Elsevier, Amsterdam, New York, Oxford, Tokyo, 1984.

Gregory J. Chaitin: *Algorithmic Information Theory*, Cambridge University Press, Cambridge, London, New York, Sydney, 1987.

S. Barry Cooper and Sergei S. Goncharov (Editors): *Computability and Models*, Kluwer Academic/Plenum, New York, Boston, Dordrecht, London, Moscow, 2003.

S. Barry Cooper, Theodore A. Slaman, and Stanley S. Wainer (Editors), *Computability, Enumerability, Unsolvability: Directions in Recursion Theory*, Cambridge University Press, Cambridge, London, New York, Sydney, 1996.

Nigel J. Cutland: *Computability: An Introduction to Recursive Function Theory*, Cambridge University Press, Cambridge, London, New York, Sydney, 1980.

Joseph Warren Dauben: *Georg Cantor: His Mathematics and Philosophy of the Infinite*, Princeton University Press, Princeton, N.J., 1990.

Martin Davis: *Computability and Unsolvability*, Dover, New York, 1983.

Martin Davis: *The Universal Computer: The Road from Leibniz to Turing* (hardback) or *Engines of Logic: Mathematicians and the Origin of the Computer* (paperback), W.W. Norton, New York, London, 2000.

Martin D. Davis, Ron Sigal, and Elaine J. Weyuker: *Computability, Complexity, and Languages: Fundamentals of Theoretical Computer Science*, Second Edition, Academic Press/Morgan Kaufmann, San Diego, San Francisco, New York, Boston, London, Sydney, Tokyo, 1994.

John W. Dawson: *Logical Dilemmas: The Life and Work of Kurt Gödel*, AK Peters, Wellesley, Mass.,1997.

Richard L. Epstein and Walter A. Carnielli, *Computability: Computable Functions, Logic, and the Foundations of Mathematics*, Second Edition, Wadsworth, Stamford, CT, London, Singapore, Ontario, 2000.

Yu. L Ershov, S.S. Goncharov, A. Nerode, J.B. Remmel (Editors): *Handbook of Recursive Mathematics*, Volumes 1 and 2, Elsevier, Amsterdam, New York, Oxford, Tokyo, 1998.

Edward R. Griffor (Editor): *Handbook of Computability Theory*, Elsevier, Amsterdam, New York, Oxford, Tokyo, 1999.

Jacques Hadamard: *The Psychology of Invention in the Mathematical Field*, Dover, New York, 1945; or *The Mathematician's Mind*, Princeton University Press, Princeton, N.J., 1996.

Rolf Herken (Editor): *The Universal Turing Machine: a half-century survey*, Second Edition, Springer-Verlag, Berlin, Heidelberg, New York, London, Paris, Tokyo, 1994.

Andrew Hodges: *Alan Turing: The Enigma*, Vintage, London, Melbourne, Johannesburg, 1992 or Walker, Knoxville, TN, 2000.

Stephen C. Kleene: *Introduction to Metamathematics*, 11th reprint, North-Holland, Amsterdam, New York, Oxford, Tokyo, 1996.

Imre Lakatos: *Proofs and Refutations: The Logic of Mathematical Discovery*, Cambridge University Press, Cambridge, London, New York, 1977.

Manuel Lerman: *Degrees of Unsolvability: Local and Global Theory*, Springer-Verlag, Berlin, Heidelberg, New York, London, Paris, Tokyo, 1983.

Yuri Matiyasevich: *Hilbert's Tenth Problem*, MIT Press, Cambridge, Mass., London, 1993.

Elliott Mendelson: *Introduction to Mathematical Logic*, Fourth Edition, CRC/Lewis Publishers, Boca Raton, FL, 1997.

Piergiorgio Odifreddi: *Classical Recursion Theory*, Volumes I and II, North-Holland/Elsevier, Amsterdam, New York, Oxford, Tokyo, 1989 and 1999.

Emil L. Post: *Solvability, Provability, Definability: The Collected Works of Emil L. Post* (Martin Davis, Editor), Birkhäuser, Boston, Basel, Berlin, 1994.

Constance Reid: *Hilbert*, Springer-Verlag, Berlin, Heidelberg, New York, London, Paris, Tokyo, 1986 (paperback edition).

Roger Penrose: *The Emperor's New Mind: Concerning Computers, Minds, and the Laws of Physics*, Oxford University Press, Oxford, New York, Melbourne, 2002.

Roger Penrose: *Shadows of the Mind: A Search for the Missing Science of Consciousness*, Oxford University Press, Oxford, New York, Melbourne, reprint edition, 1996.

Marian B. Pour-El and J. Ian Richards: *Computability in Analysis and Physics*, Springer-Verlag, Berlin, Heidelberg, New York, London, Paris, Tokyo, 1989.

Hartley Rogers, Jr: *Theory of Recursive Functions and Effective Computability*, MIT Press, Cambridge, Mass., London, 1987.

Gerald E. Sacks: *Higher Recursion Theory*, Springer-Verlag, Berlin, Heidelberg, New York, London, Paris, Tokyo, 1991.

Gerald E. Sacks: *Selected Logic Papers*, World Scientific, Singapore, New Jersey, London, Hong Kong, 1999.

Ravi Sethi: *Programming Languages: Concepts and Constructs*, Second Edition, Addison-Wesley, Reading, Mass., London, Amsterdam, Ontario, Sydney, 1996.

Stephen G. Simpson: *Subsystems of Second Order Arithmetic*, Springer-Verlag, Berlin, Heidelberg, New York, London, Paris, Tokyo, 1999.

Robert I. Soare: *Recursively Enumerable Sets and Degrees*, Springer-Verlag, Berlin, Heidelberg, New York, London, Paris, Tokyo, 1987.

Andrea Sorbi (Editor): *Complexity, Logic, and Recursion Theory*, Marcel Dekker, New York, Basel, Hong Kong, 1997.

Doron Swade: *The Cogwheel Brain: Charles Babbage and the Quest to Build the First Computer*, Abacus (paperback), 2001; or *The Difference Engine: Charles Babbage and the Quest to Build the First Computer*, Penguin USA, New York, 2002.

Alan M. Turing: *Collected Works: Mathematical Logic* (R.O. Gandy and C.E.M. Yates, Editors), Elsevier, Amsterdam, New York, Oxford, Tokyo, 2001.

Klaus Weihrauch: *Computable Analysis: An Introduction*, Springer-Verlag, Berlin, Heidelberg, New York, London, Paris, Tokyo, 2000.

Benjamin Woolley: *The Bride of Science: Romance, Reason, and Byron's Daughter*, Macmillan, London, 1999, or McGraw Hill/Contemporary Books, New York, 2000.

Articles

Theodore P. Baker, John Gill and Robert Solovay: Relativisations of the P=? NP question, *SIAM Journal of Computing*, Vol. 4, 1975, pages 431–442.

Christian Calude, Peter Hertling, Bakhadyr Khoussainov and Yongge Wang: Recursively enumerable reals and Chaitin's Ω numbers, *Theoretical Computer Science*, Vol. 255, 2001, pages 125–149.

Doug Cenzer and Jeff Remmel: Π_1^0 classes in mathematics, in Yu. L Ershov, S.S. Goncharov, A. Nerode, J.B. Remmel (Editors), *Handbook of Recursive Mathematics*, Volume 2, pages 623–821.

Adam Cichon: A short proof of two recently discovered independence results using recursion theoretic methods, *Proceedings of the American Mathematical Society*, Vol. 87, 1983, pages 704–706.

S. Barry Cooper and Piergiorgio Odifreddi: Incomputability in Nature, in S.B. Cooper and S.S. Goncharov, *Computability and Models*, pages 137–160.

Rod Downey: On presentations of algebraic structures, in Andrea Sorbi, *Complexity, Logic and Recursion Theory*, pages 157–206.

Robin Gandy, The confluence of ideas in 1936, in R. Herken, *The Universal Turing Machine: A Half-Century Survey*, pages 51–102.

William Gasarch: A survey of recursive combinatorics, in Yu. L. Ershov, S.S. Goncharov, A. Nerode, J.B. Remmel, *Handbook of Recursive Mathematics*, Volume 2, pages 1041–1176.

Valentina Harizanov: Pure computable model theory, in Yu. L. Ershov, S.S. Goncharov, A. Nerode, J.B. Remmel, *Handbook of Recursive Mathematics*, Volume 1, pages 3–113.

Carl Jockusch: Ramsey's theorem and recursion theory, *Journal of Symbolic Logic*, Vol. 37, 1972, pages 268–280.

Richard M. Karp, Reducibility among combinatorial problems, in *Complexity of Computer Computation* (R.E. Miller and J.W. Thatcher, Editors), Plenum, New York, London, 1972, pages 85–103.

Stephen C. Kleene and Emil L. Post: The upper semi-lattice of degrees of recursive unsolvability, *Annals of Mathematics*, Vol. 59, 1954, pages 379–407.

Julia Knight: Degrees of models, in Yu. L. Ershov, S.S. Goncharov, A. Nerode, J.B. Remmel, *Handbook of Recursive Mathematics*, Volume 1, pages 289–309.

Antonín Kučera: Measure, Π_1^0-classes and complete extensions of PA, in *Recursion Theory Week* (H.D. Ebbinghaus, G.H. Müller and G.E. Sacks, Editors), Springer-Verlag, Berlin, Heidelberg, New York, London, Paris, Tokyo, 1985.

Antonín Kučera: On the use of diagonally nonrecursive functions, in *Logic Colloquium '87, Granada* (H.D. Ebbinghaus, Editor), North-Holland, Amsterdam, New York, Oxford, Tokyo, 1989, pages 219–239.

Richard E. Ladner: On the structure of polynomial time reducibility, *Journal of the Association for Computing Machinery*, Vol. 22, 1975, pages 155–171.

Jeff Paris and Reza Tavakol: Goodstein algorithm as a super-transient dynamical system, *Physics Letters A*, Vol. 180, 1993, pages 83–86.

Emil L. Post: Recursively enumerable sets of positive integers and their decision problems, *Bulletin of the American Mathematical Society*, Vol. 50, 1944, pages 284–316; reprinted in E.L. Post, *Solvability, Provability, Definability: The Collected Works of Emil L. Post*, pages 461–494.

Hartley Rogers, Jr: Some problems of definability in recursive function theory, in *Sets, Models, and Recursion Theory* (J.N. Crossley, Editor), North-Holland, Amsterdam, New York, Oxford, Tokyo, 1967, pages 183–201.

Gerald E. Sacks: Forcing with perfect closed sets, in *Axiomatic Set Theory I* (D. Scott, Editor), *Proceedings of Symposia in Pure Mathematics*, Vol. 13, American Mathematical Society, 1971, pages 331–355; reprinted in Gerald E. Sacks, *Selected Logic Papers*, pages 156–180.

Joseph R. Shoenfield: A theorem on minimal degrees, *Journal of Symbolic Logic*, Vol. 31, 1966, pages 539–544.

Andrea Sorbi: The Medvedev lattice of degrees of difficulty, in S.B. Cooper, T.A. Slaman, S.S. Wainer, *Computability, Enumerability, Unsolvability: Directions in Recursion Theory*, pages 289–312.

Viggo Stoltenberg-Hansen and John Tucker: Computable rings and fields, in Edward R. Griffor, *Handbook of Computability Theory*, pages 363–447.

Alan Turing: On computable numbers, with an application to the Entscheidungsproblem, *Proceedings of the London Mathematical Society*, Vol.
42, 1936–37, pages 230–265; reprinted in A.M. Turing, *Collected Works:
Mathematical Logic*, pages 18–53.

Alan Turing: Systems of logic based on ordinals, *Proceedings of the
London Mathematical Society*, Vol. 45, 1939, pages 161–228; reprinted
in A.M. Turing, *Collected Works: Mathematical Logic*, pages 81–148.

Index